T0305601

Encyclopedia of Fluid Mechanics

This book was developed using material from teaching courses on fluid mechanics, high-speed flows, aerodynamics, high-enthalpy flows, experimental methods, aircraft design, heat transfer, introduction to engineering, and wind engineering. It precisely presents the theoretical and application aspects of the terms associated with these courses. It explains concepts such as cyclone, typhoon, hurricane, and tornado, by highlighting the subtle difference between them. The text comprehensively introduces the subject vocabulary of fluid mechanics for use in courses in engineering and the physical sciences.

This book

- Presents the theoretical aspects and applications of high-speed flows, aerodynamics, high-enthalpy flows, and aircraft design.
- Provides a ready reference source for readers to learn essential concepts related to flow physics, rarefied, and stratified flows.
- Comprehensively covers topics such as laser Doppler anemometer, latent heat of fusion, and latent heat of vaporisation.
- Includes schematic sketches and photographic images to equip the reader with a better view of the concepts.

This is ideal study material for senior undergraduate and graduate students in the fields of mechanical engineering, aerospace engineering, flow physics, civil engineering, automotive engineering, and manufacturing engineering.

Encyclopedia of Fluid Mechanics

Ethirajan Rathakrishnan

CRC Press
Taylor & Francis Group
Boca Raton London New York

CRC Press is an imprint of the
Taylor & Francis Group, an **informa** business

First edition published 2023
by CRC Press
6000 Broken Sound Parkway NW, Suite 300, Boca Raton, FL 33487-2742

and by CRC Press
4 Park Square, Milton Park, Abingdon, Oxon, OX14 4RN

CRC Press is an imprint of Taylor & Francis Group, LLC

© 2023 Ethirajan Rathakrishnan

Reasonable efforts have been made to publish reliable data and information, but the author and publisher cannot assume responsibility for the validity of all materials or the consequences of their use. The authors and publishers have attempted to trace the copyright holders of all material reproduced in this publication and apologize to copyright holders if permission to publish in this form has not been obtained. If any copyright material has not been acknowledged please write and let us know so we may rectify in any future reprint.

Except as permitted under U.S. Copyright Law, no part of this book may be reprinted, reproduced, transmitted, or utilized in any form by any electronic, mechanical, or other means, now known or hereafter invented, including photocopying, microfilming, and recording, or in any information storage or retrieval system, without written permission from the publishers.

For permission to photocopy or use material electronically from this work, access www. copyright.com or contact the Copyright Clearance Center, Inc. (CCC), 222 Rosewood Drive, Danvers, MA 01923, 978-750-8400. For works that are not available on CCC please contact mpkbookspermissions@tandf.co.uk

Trademark notice: Product or corporate names may be trademarks or registered trademarks and are used only for identification and explanation without intent to infringe.

ISBN: 978-1-032-39101-4 (hbk)
ISBN: 978-1-032-38238-8 (pbk)
ISBN: 978-1-003-34840-5 (ebk)

DOI: 10.1201/9781003348405

Typeset in Sabon
by codeMantra

This book is dedicated to my parents

Mr. Thammanur Shunmugam Ethirajan

and

Mrs. Aandaal Ethirajan

Contents

2 Backflow to Bypass Ratio

4 d'Alembert's Paradox to Dynamics   107

7 Gas to Gyroscopic Effect   173

9   ICBM to Isotopes   207

10   Jelly to Joule–Thomson Effect   225

13   Mach to Munk's Theorem of Stagger   265

14   Natural Convection to Nusselt Number   289

15 OASPL to Ozone Hole   301

16 Pacific Ocean to Pyrometers   311

19   Sake to System   369

20 Tail-First Aircraft to Typhoon   427

23 Wake to Working Fluid   479

Preface

This book is developed to introduce the subject vocabulary of flow physics to the beginners entering the university courses in engineering and physical sciences, which is a compulsory course for the students of all engineering disciplines, all over the world.

From senior high school students to university students in the engineering and physics disciplines at the Bachelor's, Master's, and doctoral levels, to the common public interested in the field of technology and science would find this book as a useful one catering to their knowledge with many popular terms used in their education as well as day-to-day life.

Some of the terms meaning almost the same, such as cyclone, typhoon, hurricane, and tornado, are explained highlighting the subtle difference between them. Also, some of the popular terms, which will be of interest to the common public like body temperature, heart pulse rate, atmospheric pressure, temperature, tsunami, barometer, thermometer, volcano, rain, fog, dust storm, and so on, are also presented in detail. This book was developed using the course material used in teaching the courses on fluid mechanics, high-speed flows, aerodynamics, high-enthalpy flows, experimental methods, aircraft design, heat transfer, introduction to engineering, and wind engineering. It precisely presents the theoretical and application aspects of the terms associated with these courses.

Since this book is developed based on the class-tested material and general material, it presents a vast quantum of definitions, descriptions, and explanations for the terms commonly prevailing in the field of fluid flow. The chapters in this book are arranged in the alphabetical order, covering the words used in this subject. At all appropriate locations, illustrations in the form of schematic sketches and photographic images are provided to enable the reader to get a better glimpse of those terms.

All these terms are introduced in such a manner that the students encountering these for the first time can comfortably follow and assimilate their meanings.

This Practical Guide for Students and Common Public (Encyclopedia of Fluid Mechanics) can serve as a ready reckoner to the engineering and physical science students of all universities in the world. Fluid Mechanics being a compulsory course for all engineering students – a book on this topic effectively presenting the subject items is long due. Indeed, whenever the author taught this course, he was forced to prepare proper handouts for distribution after almost every lecture. This process resulted in the accumulation of the material for this book. Also, since the material of this book has been class-tested repeatedly, this book containing the material in the present form will prove to be an effective handbook for a complete course on this topic. This book is a kind of encyclopedia/handbook that can serve as a ready reference source for all vital features of the vocabulary associated with all branches of flow physics: incompressible, compressible, Newtonian, non-Newtonian, continuum, rarefied and stratified flows, enthalpy, entropy, and shock and expansion waves. In addition to the technical information, general information which will be of interest to the common public, such as rain, cloud, hurricane, tsunami, cyclone, typhoon, tornado, thermal, volcano, and so on is also presented in detail along with pictures at appropriate locations.

These terms will fascinate the common public since most people are interested in knowing them.

My sincere thanks to my student Dr. S. M. Aravindh Kumar, Research Assistant Professor, Department of Aerospace Engineering, SRM Institute of Science and Technology, Chennai, for his immense help in checking the manuscript thoroughly at every stage, including the proof checking.

Ethirajan Rathakrishnan

Author

Ethirajan Rathakrishnan is a Professor of Aerospace Engineering at the Indian Institute of Technology Kanpur, India. He is well known internationally for his research in the area of high-speed jets. The limit for the passive control of jets, called the *Rathakrishnan Limit*, is his contribution to the field of jet research, and the concept of *breathing blunt nose (BBN)*, which simultaneously reduces the positive pressure at the nose and increases the low pressure at the base, is his contribution to drag reduction at hypersonic speeds. Positioning the twin-vortex Reynolds number at around 5,000, by changing the geometry from the cylinder, for which the maximum limit for the Reynolds number for positioning the twin-vortex was found to be around 160, by von Karman, to a flat plate, is his addition to vortex flow theory. He has published a large number of research articles in many reputed international journals. He is a *Fellow of* many professional societies including the *Royal Aeronautical Society*. Rathakrishnan serves as the *Editor-in-Chief* of the *International Review of Aerospace Engineering (IREASE)* and *International Review of Mechanical Engineering (IREME)*.

He has authored the following books: *Gas Dynamics*, 7th ed. (PHI Learning, New Delhi, 2020); *Fundamentals of Engineering Thermodynamics*, 2nd ed. (PHI Learning, New Delhi, 2005); *Fluid Mechanics: An Introduction*, 4th ed. (PHI Learning, New Delhi, 2022); *Gas Tables*, 3rd ed. (Universities Press, Hyderabad, India, 2012); *Theory of Compressible Flows* (Maruzen Co., Ltd. Tokyo, Japan, 2008); *Gas Dynamics Work Book*, 2nd ed. (Praise Worthy Prize, Napoli, Italy, 2013); *Elements of Heat Transfer* (CRC Press, Taylor & Francis Group, Boca Raton, Florida, USA, 2012); *Theoretical Aerodynamics* (John Wiley, New Jersey, USA, 2013); *High Enthalpy Gas Dynamics* (John Wiley & Sons, Inc., 2015); *Dynamique Des Gaz* (Praise Worthy Prize, Napoli, Italy, 2015); *Instrumentation, Measurements and Experiments in Fluids*, 2nd ed. (CRC Press, Taylor & Francis Group, Boca Raton, Florida, USA, 2017); *Helicopter Aerodynamics* (PHI Learning,

New Delhi, 2019); *Applied Gas Dynamics*, 2nd ed. (John Wiley & Sons, Inc., 2019); *Introduction to Aerospace Engineering: Basic Principles of Flight* (Wiley, New Jersey, 2021); *Japan* (Praise Worthy Prize, Napoli, Italy, 2021); and *Mind Power: The Sixth Sense* (Routlege, Taylor and Francis Group, UK, 2023).

Chapter 1

Ablation to Axisymmetric Flow

1.1 ABLATION

Ablation is removal or destruction of material from an object by vaporization, chipping, or other erosive processes.

1.2 ABLAZE

Ablaze means burning strongly and completely on fire, as illustrated in Figure 1.1.

1.3 ABNORMAL WEATHER

Abnormal weather is one of the serious adverse effects in the course of a flight. The unsteady conditions caused by abnormal weather conditions such as turbulence can be quite significant as they must be taken into account when designing a commercial aircraft. The aircraft designed for performing aerobatics should be strong enough to withstand any loads caused by adverse weather conditions. The conditions that are likely to inflict the most adverse loads consist of strong gusty winds, hot sun, intermittent clouds, especially thunderclouds with considerable turbulence, and uneven ground conditions.

1.4 ABOVE GROUND LEVEL

In aviation, atmospheric sciences, and broadcasting, a height above ground level (AGL) is a height measured with respect to the underlying ground surface. This is as opposed to the altitude/elevation above mean sea level (AMSL), or (in broadcast engineering) height above average terrain (HAAT). In other words, these expressions (AGL, AMSL, HAAT) indicate where the 'zero level' or 'reference altitude' is located.

DOI: 10.1201/9781003348405-1

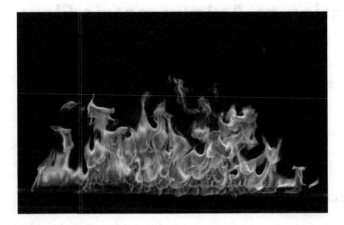

Figure 1.1 Ablaze.

1.5 ABSOLUTE ENTROPY

This statement about the zero entropy state is known as the third law of thermodynamics. It provides an absolute reference point for the determination of entropy. The entropy determined relative to this point is called absolute entropy. Absolute entropy plays a dominant role in the thermodynamic analysis of chemical reactions.

1.6 ABSOLUTE HUMIDITY

Absolute humidity is the measure of water vapour (moisture) in the air regardless of the temperature. It is expressed as grams of moisture per cubic metre of air (g/m^3). The maximum absolute humidity of cold air at 0°C is ~5 g of water vapour per cubic metre (5 g/m^3).

1.7 ABSOLUTE INCIDENCE

Absolute incidence is the angle between the axis of zero lift of the profile and the direction of motion of the aerofoil.

1.8 ABSOLUTE PRESSURE

Pressure measured from zero pressure (vacuum) as the reference is known as absolute pressure. The standard sea level pressure in the international standard atmosphere is 760 mm of mercury (absolute). This pressure in other units is 101,325 pascal in SI units, 101,325 N/m² in MKS units, and 14.7 psi in British units. Pressure measured with reference to some known pressure is termed gauge pressure. Gauge pressure can be converted to absolute pressure by adding the local atmospheric pressure,

$$p_{absolute} = p_{gauge} + p_{atm}$$

1.9 ABSOLUTE TEMPERATURE SCALE

The Kelvin scale is called the absolute temperature scale in SI units. The temperature measured on this scale is called absolute temperature. On the Kelvin temperature scale,

- The temperature ratio depends on the ratio of the heat transfer between a reversible heat engine and the reservoir.
- The temperature ratio is independent of the physical properties of any substance.
- Temperature varies between zero and infinity.
- In the Celsius scale, the temperature is expressed as degree Celsius (°C). A temperature in °C can be converted to Kelvin by adding 273.15. For example, the standard sea level temperature in the international standard atmosphere is 15°C = 15 + 273.15 = 288.15 K.

1.10 ABSOLUTE VELOCITY

The concept of absolute velocity generally refers to a standard uniform velocity of the various objects of a physical system relative to a postulated immobile space that exists independently of the physical objects contained therein (i.e., an absolute space).

1.11 ACCELERATION

Acceleration in general is the *time rate of change in velocity*. But for fluid elements (particles), we can define *local acceleration, convective acceleration*, and *material acceleration*. These three can be related as

$$\frac{DV}{Dt} = \frac{\partial V}{\partial t} + V \cdot (\nabla V)$$

where $V = i\,V_x + j\,V_y + k\,V_z$ is the flow velocity, t is the time, and $\nabla = i\frac{\partial}{\partial x} + j\frac{\partial}{\partial y} + k\frac{\partial}{\partial z}$. This relation is known as Euler's acceleration formula.

1.12 ACCELERATION DUE TO GRAVITY

Acceleration due to gravity is the acceleration gained by an object due to gravitational force. Its SI unit is m/s^2.

1.13 ACCELERATION WORK

Acceleration work may be defined as the work associated with a change in the velocity of a system.

1.14 ACCEPTABLE NOISE LEVEL

Normal conversation is about 60 dB. A lawn mower is about 90 dB, and a loud rock concert is about 120 dB. In general, sounds above 85 dB are harmful, depending on how long and how often you are exposed to them and whether you wear hearing protection, such as earplugs or earmuffs.

1.15 ACCURACY VS. PRECISION

In a set of measurements, accuracy is the closeness of the measurements to a specific value, while precision is the closeness of the measurements to each other.

1.16 ACETONE

Acetone, a colourless liquid also known as propane, is a solvent used in the manufacture of plastics and other industrial products.

1.17 ACID

An acid is a molecule or ion capable of donating a proton or, alternatively, capable of forming a covalent bond with an electron pair.

1.18 ACKERET'S THEORY

The thin aerofoil theory is also called Ackeret's theory. When a thin aerofoil is kept at a small angle of attack, i.e., if the flow inclinations are small, the compression waves over that can be approximated as isentropic compression waves. This approximation will result in simple analytical expressions for lift and drag.

1.19 ACOUSTICS

Acoustics is a branch of physics that deals with the study of mechanical waves in gases, liquids, and solids including topics such as vibration, sound, ultrasound, and infrasound.

1.20 ACOUSTICAL ENGINEERING

Acoustical engineering (also known as acoustic engineering) is the branch of engineering dealing with sound and vibration. It includes the application of acoustics, the science of sound and vibration, in technology. Acoustical engineers are typically concerned with the design, analysis and control of sound.

1.21 ACTIVE CONTROL

Many active jet control methods use energised actuators to dynamically manipulate flow phenomena by employing open- or closed-loop algorithms. Pulsed jets, piezoelectric actuators, microjets, and oscillating jets are among the most effective controls for active mixing enhancement. The design of an active flow control system requires knowledge of flow phenomenon and the selection of appropriate actuators, sensors, and a control algorithm.

1.22 ACTIVE TRANSPORT

Active transport is the movement of molecules across a cell membrane from a region of lower concentration to a region of higher concentration-against the concentration gradient.

1.23 ACTUATOR DISC THEORY

Momentum theory of helicopter is referred to as actuator disc theory.

1.24 ADHESION

Adhesion is the attraction between the particles of one substance and the particles of another substance.

1.25 ADIABATIC EFFICIENCY OF NOZZLES

The adiabatic efficiency of a nozzle is defined as the ratio of the actual kinetic energy of the fluid at the nozzle exit to the kinetic energy value at the exit of an isentropic nozzle for the same inlet state and exit pressure.

1.26 ADIABATIC FLAME TEMPERATURE

The adiabatic flame temperature is the maximum temperature that the combustion products can attain when there is no heat loss from the system to the surroundings, during the combustion process.

1.27 ADIABATIC PROCESS

Adiabatic process – a process in which no heat is added or taken away from the system. In other words, adiabatic process is that with stagnation temperature as invariant, that is, T_0 remains constant. In thermodynamics, an adiabatic process is a type of thermodynamic process, which occurs without transferring heat or mass between the system and its surroundings. Unlike an isothermal process, an adiabatic process transfers energy to the surroundings only as work.

1.28 ADVECTION

Advection is the movement of some material dissolved or suspended in the fluid. An example of advection is the transport of pollutants or silt in a river by bulk water flow downstream. Another commonly advected quantity is energy or enthalpy. Here the fluid may be any material that contains thermal energy, such as water or air.

1.29 ADVECTION PROCESS

During advection, a fluid transports some conserved quantity or material via bulk motion. It does not include transport of substances by molecular diffusion.

1.30 ADVERSE PRESSURE GRADIENT

Increase of pressure in the flow direction is termed adverse pressure gradient.
   A positive pressure gradient is termed the adverse pressure gradient. Fluid might find it difficult to negotiate an adverse pressure gradient.

In fluid dynamics, an adverse pressure gradient occurs when the static pressure increases in the direction of the flow. Mathematically this is expressed as: $dp/dx > 0$ for a flow in the positive x-direction. This is important for boundary layers. Increasing the fluid pressure is akin to increasing the potential energy of the fluid, leading to a reduced kinetic energy and a deceleration of the fluid. Since the fluid in the inner part of the boundary layer is slower, it is more greatly affected by the increasing pressure gradient.

1.31 ADVERSE YAW

Adverse yaw is the natural and undesirable tendency for an aircraft to yaw in the opposite direction of a roll. It is caused by the difference in lift and drag of each wing.

1.32 AEROACOUSTICS

Aeroacoustics is a branch of acoustics that studies noise generation via either turbulent fluid motion or aerodynamic forces interacting with surfaces. Noise generation can also be associated with periodically varying flows. A notable example of this phenomenon is the Aeolian tones produced by wind blowing over fixed objects.

1.33 AEROBALLISTICS

Aeroballistics is the study of the effects of aerodynamic forces upon missiles and projectiles during flight.

1.34 AEROBATICS

Flight modes such as loop, spin, roll, nosedive, upside down, sideslip, inverted spin, and inverted loop are called aerobatics. Aerobatics should be performed only with aircraft suitable for them. These flight modes provide good training for the accuracy and precision in manoeuvre, which are essential for combat flying.

1.35 AEROBRAKING

Aerobraking is a spaceflight manoeuvre that reduces the high point of an elliptical orbit (apoapsis) by flying the vehicle through the atmosphere at the low point of the orbit (periapsis). The resulting drag slows the spacecraft.

Aerobraking is used when a spacecraft requires a low orbit after arriving at a body with an atmosphere, and it requires less fuel than the direct use of a rocket engine.

1.36 AEROCAPTURE

Aerocapture is an orbital transfer manoeuvre used to reduce the velocity of a spacecraft from a hyperbolic trajectory to an elliptical orbit around the targeted celestial body.

1.37 AERODROME

An aerodrome or airdrome is a location from which aircraft flight operations take place, regardless of whether they involve air cargo, passengers, or neither, and regardless of whether it is for public or private use.

1.38 AERODYNAMIC CENTRE

The aerodynamic centre, defined as the point on the aerofoil where the moments are independent of the angle of incidence, has very important implications.

The aerodynamic centre of a low-speed aerofoil is always at or near the quarter-chord point, that is, at a distance of $c/4$ aft of the leading edge, where c is the chord of the aerofoil.

The torques or moments acting on an aerofoil moving through a fluid can be accounted for by the net lift and net drag applied at some point on the aerofoil, and a separate net pitching moment about that point whose magnitude varies with the choice of where the lift is chosen to be applied. The aerodynamic centre is the point at which the pitching moment coefficient for the aerofoil does not vary with lift coefficient (i.e. angle of attack), making analysis simpler.

1.39 AERODYNAMIC EFFICIENCY

The lift-to-drag ratio, called aerodynamic efficiency, is a measurement of the efficiency of the aerofoil.

1.40 AERODYNAMIC FORCES

The forces acting on the bodies moving through air are termed aerodynamic forces.

Aerodynamic force acting on an aircraft is the force due to the pressure distribution around it, caused by the motion of the aircraft. Thus, the gravity does not enter into the specification of aerodynamic force.

1.41 AERODYNAMIC FORCE AND MOMENT COEFFICIENTS

The important aerodynamic forces and moment associated with a flying machine, such as an aircraft, are the lift L, drag D, and pitching moment M. The lift and drag forces can be expressed as dimensionless numbers, popularly known as lift coefficient C_L and drag coefficient C_D by dividing L and D with $(\frac{1}{2}\rho V^2 S)$. Thus,

$$C_L = \frac{L}{\frac{1}{2}\rho V^2 S}$$

$$C_D = \frac{D}{\frac{1}{2}\rho V^2 S}$$

The pitching moment, which is the moment of the aerodynamic force about an axis perpendicular to the plane of symmetry, will depend on the particular axis chosen. Denoting the pitching moment about the chosen axis by M, we define the pitching moment coefficient as

$$C_M = \frac{M}{\frac{1}{2}\rho V^2 S\ c}$$

where c is the chord of the wing.

1.42 AERODYNAMIC HEATING

Aerodynamic heating is the heating of a solid body produced by its high-speed passage through air (or by the passage of air past a static body), whereby its kinetic energy is converted to heat by adiabatic heating, and by skin friction on the surface of the object at a rate that depends on the viscosity and speed of the air.

Aerodynamic heating increases with the speed of the vehicle. Its effects are minimal at subsonic speeds, but are significant enough at supersonic speeds beyond about Mach 2.2 that they affect design and material considerations for the vehicle's structure and internal systems.

1.43 AERODYNAMIC PROBLEMS

Aerodynamic problems can be classified according to the flow environment, point on a vibrating body or wave measured from its equilibrium position. External aerodynamics is the study of flow around solid objects of various shapes. Evaluating the lift and drag on an aeroplane or the shock waves that form in front of the nose of a rocket are examples of external aerodynamics. Internal aerodynamics is the study of flow through passages in solid objects. For instance, internal aerodynamics encompasses the study of the airflow through a jet engine.

Aerodynamic problems can also be classified according to whether the flow speed is below, near or above the speed of sound. A problem is called subsonic if all the speeds in the problem are less than the speed of sound, transonic if speeds both below and above the speed of sound are present, supersonic if the flow speed is greater than the speed of sound, and hypersonic if the flow speed is more than five times the speed of sound.

The influence of viscosity in the flow dictates a third classification. Some problems may encounter only very small viscous effects on the solution; therefore, the viscosity can be considered to be negligible. The approximations made in solving these problems is the viscous effect that can be regarded as negligible. These are called inviscid flows. Flows for which viscosity cannot be neglected are called viscous flows.

1.44 AERODYNAMIC TWIST

An aerofoil is said to have aerodynamic twist when the axes of zero lift of its individual profiles are not parallel. The incidence is then variable across the span of the aerofoil.

1.45 AERODYNAMICS

Aerodynamics is the study of motion of air, particularly when affected by a solid object, such as an aeroplane wing. It is a sub-field of fluid dynamics and gas dynamics, and many aspects of aerodynamics theory are common to these fields. The term aerodynamics is often used synonymously with gas dynamics, the difference being that 'gas dynamics' applies to the study of the motion of all gases, and is not limited to air. The formal study of aerodynamics began in the modern sense in the 18th century, although observations of fundamental concepts such as aerodynamic drag were recorded much earlier. Most of the early efforts in aerodynamics were directed toward achieving heavier-than-air flight, which was first demonstrated by Otto Lilienthal in 1891. Since then, the use of aerodynamics through mathematical analysis, empirical approximations, wind tunnel experimentation, and computer simulations has formed a rational basis for the development of heavier-than-air flight and a number of other technologies. Recent work in aerodynamics has focused on issues related to compressible flow, turbulence, and boundary layers and has become increasingly computational in nature.

1.46 AERODYNAMICS IN OTHER FIELDS

Aerodynamics is important in a number of applications other than aerospace engineering. It is a significant factor in any type of vehicle design, including automobiles. It is important in the prediction of forces and moments in sailing. It is used in the design of mechanical components such as hard drive heads. Structural engineers also use aerodynamics, and particularly aeroelasticity, to calculate wind loads in the design of large buildings and bridges. Urban aerodynamics seeks to help town planners and designers improve comfort in outdoor spaces, create urban microclimates, and reduce the effects of urban pollution. The field of environmental aerodynamics studies the ways atmospheric circulation and flight mechanics affect ecosystems. The aerodynamics of internal passages is important in heating/ventilation, gas piping, and automotive engines where detailed flow patterns strongly affect the performance of the engine.

1.47 AERODYNAMICALLY UNTWISTED AEROFOIL

When the axes of zero lift of all the profiles of the aerofoil are parallel, each profile meets the freestream wind at the same absolute incidence, the incidence is the same at every point on the span of the aerofoil, and the aerofoil is said to be aerodynamically untwisted.

1.48 AEROELASTICITY

Aeroelasticity is the branch of physics and engineering studying the interactions between the inertial, elastic, and aerodynamic forces occurring while an elastic body is exposed to a fluid flow. The study of aeroelasticity may be broadly classified into two fields: static aeroelasticity dealing with the static or steady state response of an elastic body to a fluid flow, and dynamic aeroelasticity dealing with the body's dynamic (typically vibrational) response.

1.49 AEROFOIL

An aerofoil is a streamlined body that would experience the maximum aerodynamic efficiency (that is, maximum lift-to-drag ratio) compared with

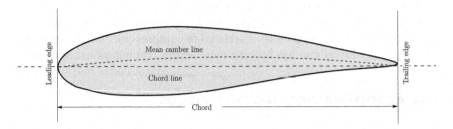

Figure 1.2 Aerofoil.

any other body under identical flow conditions. A typical aerofoil is shown in Figure 1.2.

Aerofoil is a streamlined body that would experience the largest value of lift-to-drag ratio, in a given flow, compared to any other body in the same flow. In other words, in a given flow, the aerodynamic efficiency (L/D) of an aerofoil will be the maximum.

The geometry is so shaped to ensure streamlined flow as far as possible, from the leading edge. Indeed, it is preferable to have streamlined from the leading edge to the trailing edge. The leading edge is rounded to ensure smooth flow. The trailing edge is sharp so that the Kutta condition (a body with a sharp trailing edge that is moving through a fluid will create about itself a circulation of sufficient strength to hold the rear stagnation point at the trailing edge) may be satisfied, the wake is kept thin, and any region of separated flow is restricted to as small as possible. These features help to achieve high lift-to-drag ratio.

1.50 AEROFOIL NOMENCLATURE

The geometry of many aerofoil sections is uniquely defined by the NACA designation for the aerofoil. The NACA aerofoils are aerofoil shapes for aircraft wings developed by the National Advisory Committee for Aeronautics (NACA).

1.51 AEROFOIL SECTION

The geometrical section of a wing obtained by cutting it by a vertical plane parallel to the centreline of the aircraft is called aerofoil section.

An aerofoil is a streamlined body that would experience the maximum aerodynamic efficiency (that is, maximum lift-to-drag ratio) compared with any other body under identical flow conditions. To satisfy these characteristics, the body is shaped to have geometry, as shown in Figure 1.3.

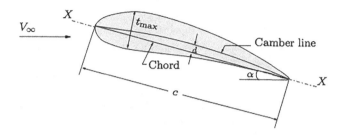

Figure 1.3 A cambered aerofoil at an angle of attack.

The major geometrical parameters of an aerofoil are chord line, chord, maximum thickness, camber line, and camber.

1.52 AEROFOILS OF SMALL ASPECT RATIO

For aerofoils with aspect ratio less than about unity, the agreement between theoretical and experimental lift distribution breaks down. The reason for this breakdown is found to be the consequence of Prandtl's hypothesis that the free vortex lines leave the trailing edge in the same line as the main stream. This assumption leads to a linear integral equation for the circulation.

1.53 AERONAUTICS

Aeronautics is the science involved with the study, design, and manufacturing of air flight-capable machines, and the techniques of operating aircraft and rockets within the atmosphere.

1.54 AERO GRAVITY

Aero gravity is the largest vertical wind tunnel in the world and the only one in Italy. It allows you to experience, every day of the year, the experience of flying in an 8-metre tall crystal cylinder, using six massively powerful turbines and an air flow of up to 370 km/h capable of overcoming the force of gravity and supporting you in flight. Aero gravity is the cutting-edge indoor skydiving facility, designed by a team of world-class engineers to guarantee you the utmost fun with no risks.

1.55 AEROPLANE

Aeroplane is a vehicle that can fly through the air, with wings and one or more engines, as shown in Figure 1.4.

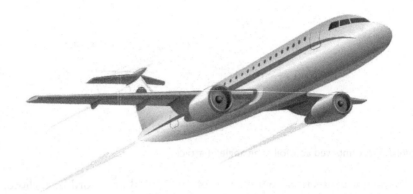

Figure 1.4 An aeroplane.

1.56 AEROSOL

Aerosols are a suspension of fine solid or liquid particles in gas. Aerosols may also be defined as minute particles suspended in the atmosphere.

1.57 AEROSPACE

Aerospace is a term used to collectively refer to the atmosphere and outer space.

1.58 AEROSPACE ENGINEERING

Aerospace engineering is the primary field of engineering concerned with the development of aircraft and spacecraft. It has two major and overlapping branches: aeronautical engineering and astronautical engineering.

1.59 AEROTHERMODYNAMICS

Aerothermodynamics deals with the aerodynamic forces and moments and the heat distribution of a vehicle that flies at hypersonic speeds.

Aerothermodynamics is the study of how high-velocity gases behave, employed as a tool for optimising engine design and also for modelling the process of atmospheric ascent and re-entry.

1.60 AFTERBURNER

The afterburner increases the temperature of the gas ahead of the nozzle. The result of this increase in temperature is an increase of about 40% in thrust at take-off and a much larger percentage at high speeds once the plane is in the air.

An afterburner is an additional combustion component used on some jet engines, mostly those on military supersonic aircraft. Its purpose is to increase thrust, usually for supersonic flight, take-off, and combat. Afterburning injects additional fuel into a combustor in the jet pipe behind the turbine, 'reheating' the exhaust gas. Afterburning significantly increases thrust as an alternative to using a bigger engine with its attendant weight penalty, but at the cost of very high fuel consumption (decreased fuel efficiency), which limits its use to short periods. This aircraft application of reheat contrasts with the meaning and implementation of reheat applicable to gas turbines driving electrical generators and which reduces fuel consumption.

1.61 AILERONS

These are the control surfaces used to vary differentially the lift on the wings and so to control the rolling moment. Ailerons are located at the rear portions of the wing tips.

That is, an aileron (French for 'little wing' or 'fin') is a hinged flight control surface usually forming part of the trailing edge of each wing of a fixed-wing aircraft. Ailerons are used in pairs to control the aircraft in roll (or movement around the aircraft's longitudinal axis), which normally results in a change in flight path due to the tilting of the lift vector. Movement around this axis is called 'rolling' or 'banking'.

1.62 AIR

Air is the general name for the mixture of gases that makes up the Earth's atmosphere. This gas is primarily nitrogen (78%), mixed with oxygen (21%), water vapour (variable), argon (0.9%), carbon dioxide (0.04%), and trace gases. Pure air has no discernible scent and no colour.

1.63 AIR BRAKES

In aeronautics, air brakes or speed brakes are a type of flight control surface used on an aircraft to increase the drag on the aircraft. Air brakes differ from spoilers in that air brakes are designed to increase drag while making little change to lift, whereas spoilers reduce the lift-to-drag ratio and require a higher angle of attack to maintain lift, resulting in a higher stall speed.

1.64 AIR CONDITIONERS

Air conditioners are basically refrigerators whose refrigerated space is very large, like a room or a building, instead of a small volume as in the household refrigerators. A window air conditioner cools a room by absorbing heat from the room air and discharging it to the outside atmosphere. If the operation is reversed, the same air conditioner can be used as a heat pump in winter. In this mode, the unit will be installed backward and it will pick up heat from the cold outside air and deliver it to the room.

1.65 AIR-COOLER

Air-cooler is a device used for cooling and reducing the temperature inside buildings. It works on the principle of air being passed over water.

1.66 AIR FLOW METER

An air flow metre is a device that measures airflow, i.e. how much air is flowing through a tube. It does not measure the volume of the air passing through the tube; it measures the mass of air flowing through the device per unit time.

1.67 AIR-FUEL RATIO

It is the ratio of the mass of air to the mass of fuel for a combustion process.

1.68 AIR INTAKE SYSTEM

The function of the air intake system is to allow air to reach your car engine. Oxygen in the air is one of the necessary ingredients for the engine combustion process. A good air intake system allows for clean and continuous airflow into the engine, thereby achieving more power and better mileage for your car.

1.69 AIR POLLUTION

Solid and liquid particles and certain gases that are suspended in the air cause air pollution. These particles and gases can come from car and truck exhaust, factories, dust, pollen, mould spores, volcanoes and wildfires. The solid and liquid particles suspended in the air are called aerosols.

1.70 AIR TRAFFIC CONTROL (ATC)

Air traffic control (ATC) is a service provided by ground-based air traffic controllers who direct aircraft on the ground and through controlled airspace, and can provide advisory services to aircraft in non-controlled airspace. The primary purpose of ATC worldwide is to prevent collisions, organise and expedite the flow of air traffic, and provide information and other support for pilots. In some countries, ATC plays a security or defensive role, or is operated by the military.

1.71 AIRCRAFT

An aircraft is a machine that is able to fly by gaining support from the air or, in general, the atmosphere of a planet. It counters the force of gravity by using either static lift or dynamic lift of an aerofoil or in a few cases the downward thrust from jet engines.

1.72 AIRCRAFT CLASSIFICATION

Aircraft are classified as propeller aircraft, jet aircraft, and rotorcraft.

1.73 AIRCRAFT CONFIGURATIONS

Aircraft configurations, shown in Figure 1.5, which describe the aerodynamic layout or specific components of an aircraft, vary widely. Aircraft configurations, in general, include fuselage, tail, and power plant configurations. This type of aircraft is referred to as fixed-wing aeroplanes. Another category of flying machine with rotating wing is called rotary-wing aircraft or simply rotorcraft. Helicopter, cyclogyro/cyclocopter, autogyro (or gyrocopter, gyroplane, or rotaplane), gyrodyne, and rotor kite (or gyro glider) are all rotary-wing aircraft.

1.74 AEROPLANE

An aeroplane is a fixed-wing aircraft that is propelled forward by thrust from a jet engine, propeller, or rocket engine.

1.75 AIRPORT

An airport is a place where aeroplanes take-off and land that has all the services and buildings needed to take care of the aeroplanes, passengers, etc.

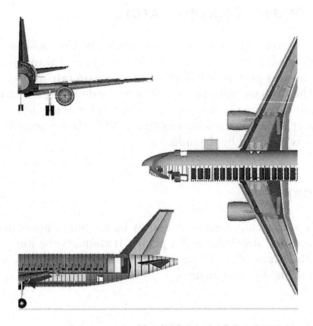

Figure 1.5 Some views of an aircraft configuration.

1.76 AIRSHIP

An airship (shown in Figure 1.6), dirigible balloon or blimp is a type of aerostat or lighter-than-air aircraft that can navigate through the air under its own power.

Aerostats gain their lift from a lifting gas that is less dense than the surrounding air.

1.77 AIRSPEED

Airspeed is the vector difference between the ground speed and the wind speed. On a perfectly still day, the airspeed is equal to the ground speed. But if the wind is blowing in the same direction that the aircraft is moving, the airspeed will be less than the ground speed.

1.78 AIRSPEED INDICATOR (ASI)

The airspeed indicator (ASI) or airspeed gauge is a flight instrument indicating the airspeed of an aircraft in kilometres per hour (km/h), knots

Figure 1.6 Airship.

(kn), miles per hour (MPH) and/or metres per second (m/s). The recommendation by ICAO is to use km/h, however knot is currently the most used unit. The ASI measures the pressure differential between static pressure from the static port, and total pressure from the pitot tube. This difference in pressure is registered with the ASI pointer on the face of the instrument.

1.79 AIRWAY

The airway or breathing passage is the pathway through which air flows into your lungs. This starts from your nose and mouth, it includes your throat, windpipe and lungs.

1.80 ALCOHOL

Alcohol is an organic compound that carries at least one hydroxyl functional group (–OH) bound to a saturated carbon atom. The term alcohol originally referred to the primary alcohol ethanol (ethyl alcohol), which is used as a drug and is the main alcohol present in alcoholic drinks.

1.81 ALCOHOLIC FERMENTATION

Alcoholic fermentation is a biotechnological process accomplished by yeast, some kinds of bacteria, or a few other microorganisms to convert sugars into ethyl alcohol and carbon dioxide.

1.82 ALFA LAVAL NOZZLE

Alfa Laval BRUX, a VLB-validated separator, is specifically designed for continuous in-line removal of surplus yeast as well as microorganisms and cell debris.

1.83 ALFA LAVAL SPRAY NOZZLE

The Alfa Laval rotary spray head tank-cleaning machine is a free-rotating spinner that uses low volumes of cleaning media at low pressure for effective cleaning.

1.84 ALTIMETER

Altimeter is an instrument to indicate the altitude at which an aircraft flies. All aircraft are required to be fitted with one or more altimeters to indicate the altitude at which the aircraft is flying. All altimeters consist of an evacuated aneroid capsule fixed by one side to the instrument casing and the other side of the capsule being free to move as the capsule expands or contracts. The movement of the free side of the capsule is communicated by a gearing system to two or three needles that move over a circular scale calibrated in feet or metres. Atmospheric static pressure is applied to the interior of the instrument case through a static tube mounted on the exterior of the aircraft. The equilibrium distention of the capsule is determined by the pressure difference between the instrument case and the evacuated capsule and thus is directly related to the atmospheric pressure. Therefore, the altitude indicated on the scale by the needle is related to the pressure of the atmosphere in which the aircraft is flying.

1.85 ALTITUDE-COMPENSATING NOZZLE

An altitude-compensating nozzle is a class of rocket engine nozzles that are designed to operate efficiently across a wide range of altitudes.

1.86 AMAGAT'S LAW OF ADDITIVE VOLUMES

The volume of a gas mixture is equal to the sum of the volumes each gas would occupy if it existed alone at the mixture temperature and pressure.

1.87 AMMONIA

Ammonia is a compound of nitrogen and hydrogen with the formula NH_3. A stable binary hydride and the simplest pnictogen hydride, ammonia is a colourless gas with a distinct characteristic of a pungent smell.

1.88 AMMONIA AND HYDROCHLORIC ACID

When ammonia vapour is passed over hydrochloric acid, white fumes of ammonium chloride are formed. The corrosive nature of the fume and the tendency to form white deposits make this method of smoke production undesirable, except on rare occasions.

1.89 AMMONIA SOLUTION

Ammonia solution, also known as ammonia water, ammonium hydroxide, ammoniacal liquor, ammonia liquor, aqua ammonia, aqueous ammonia, or ammonia, is a solution of ammonia in water.

1.90 AMPLITUDE

Amplitude, in physics, the maximum displacement or distance moved by a point on a vibrating body or wave measured from its equilibrium position.

1.91 AMPLITUDE OF SOUND

The amplitude of a sound wave is the measure of the height of the wave. The amplitude of a sound wave can be defined as the loudness or the amount of maximum displacement of vibrating particles of the medium from their mean position when the sound is produced.

1.92 ANALOGUE METHODS

Methods developed to study the fluid flow problems without actually going into the complexities involved are termed analogue methods.

Some of the well-known analogy methods for fluid flow problems are the Hele-Shaw analogy, electrolytic tank, and surface waves in a ripple tank.

Figure 1.7 Anechoic chamber.

1.93 ANECHOIC CHAMBER

An anechoic chamber (anechoic meaning 'non-reflective, non-echoing, echo-free') is a room (as in Figure 1.7) designed to completely absorb reflections of either sound or electromagnetic waves. They are also often isolated from waves entering from their surroundings. This combination means that a person or detector exclusively hears direct sounds (no reverberant sounds), in effect simulating being inside an infinitely large room.

1.94 ANEMOMETER

Anemometer is an instrument meant for the measurement of wind or air velocity.

1.95 ANEMOMETER CUP

An anemometer cup is an instrument that measures wind speed and wind pressure. As the wind blows, the cups rotate, making the rod spin. The stronger the wind blows, the faster the rod spins. The anemometer counts the number of rotations, or turns, which is used to calculate wind speed.

1.96 ANEROID BAROMETER

The basic element of aneroid barometer is the corrugated circular cell made of very thin metallic sheet and is hollow. The cell is partially evacuated, as

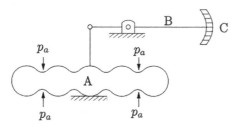

Figure 1.8 Aneroid barometer.

shown in Figure 1.8. At the outer surface of the cell atmospheric pressure p_a is acting. The cell surface will be depressed if the atmospheric pressure increases and the cell surface will move outwards when the atmospheric pressure decreases.

This movement is used as a measure of change of atmospheric pressure from one level to another. Since this movement is very small it is magnified by means of a lever mechanism B. A simple lever mechanism, appropriate for this application is illustrated in Figure 1.8, with a pointer moving over a scale C. In practice lever mechanisms of larger movement magnification would be required and the scale is usually circular with the pointer rotating about the centre of the scale.

1.97 ANGLE OF ATTACK

Angle of attack is the angle between a reference line on a body and the vector representing the relative motion between the body and the fluid through which it is moving, as illustrated in Figure 1.9. Angle of attack is the angle between the body's reference line and the oncoming flow.

1.98 ANGLE OF ATTACK OF HELICOPTER

The angle of attack of helicopter is the angle at which the aerofoil meets the oncoming airflow (or vice versa). In the case of a helicopter, the object is the rotor blade (aerofoi) and the fluid is the air. Lift is produced when a mass of air is deflected, and it always acts perpendicular to the resultant relative wind.

1.99 ANHEDRAL

If the wings are inclined downwards, the angle is negative and termed anhedral, as in Figure 1.10.

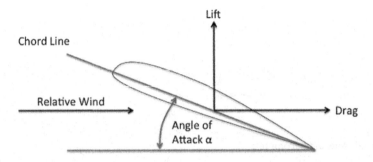

Figure 1.9 Angle of attack.

Figure 1.10 Anhedral.

1.100 ANNULAR FLOW

Annular flow is a type of multi-phase flow in which the lighter fluid generally gas flows in the centre of the pipe while the heavier fluid either oil or water flows as thin film over the walls of the pipe. This kind of flow is also called a channel flow.

1.101 ANNULAR FLOW REYNOLDS NUMBER

The annular flow Reynolds number should be calculated by substituting the difference between the hole diameter and the pipe diameter (i.e., the so-called hydraulic diameter) for the diameter of the pipe in the classic calculation.

1.102 ANTISEPTICS

Antiseptics are antimicrobial substances that are applied to living tissue/ skin to reduce the possibility of infection, sepsis, or putrefaction.

Figure 1.11 Arch dam.

1.103 AQUA REGIA

Aqua regia is a mixture of concentrated nitric and hydrochloric acids. It is a highly corrosive liquid able to attack gold and other resistant substances.

1.104 AQUIFER

An aquifer is a body of rock and/or sediment that holds groundwater. Groundwater is the word used to describe precipitation that has infiltrated the soil beyond the surface and collected in empty spaces underground. There are two general types of aquifers: confined and unconfined.

1.105 ARCH DAM

An arch dam is a concrete dam that is curved upstream in plan. The arch dam is designed so that the force of the water against it, known as hydrostatic pressure, presses against the arch, causing the arch to straighten slightly and strengthening the structure as it pushes into its foundation or abutments. A photographic view of an arch dam is shown in Figure 1.11.

1.106 ARCHIMEDES' PRINCIPLE

Archimedes' principle, physical law of buoyancy, discovered by the ancient Greek mathematician and inventor Archimedes, stating that any body completely or partially submerged in a fluid (gas or liquid) at rest is acted upon by an upward, or buoyant, force, the magnitude of which is equal to the weight of the fluid at least.

1.107 AREA-MACH NUMBER RELATION

For isentropic flow of a perfect gas through a duct, the Area-Mach number relation may be expressed, assuming one-dimensional flow, as

$$\left(\frac{A}{A^*}\right)^2 = \frac{1}{M^2}\left[\frac{2}{\gamma+1}\left(1+\frac{\gamma-1}{2}M^2\right)\right]^{\frac{\gamma+1}{\gamma-1}}$$

where A^* is called the sonic or critical throat area. From this equation, we get the striking result $M = f\left(A/A^*\right)$, that the Mach number at any location in the duct is a function of the ratio of the local area of the duct to the sonic throat area. The local area A of the duct must be larger than or at least equal to A^*, the case in which $A < A^*$ is physically impossible for an isentropic flow. Further, for any given $A/A^* > 1$, two values of Mach number M are obtained, a subsonic value and a supersonic value. Once the variation of Mach number through the nozzle is known, the variation of static temperature, pressure, and density can be determined from isentropic relations. The pressure, temperature, and density decrease continuously throughout the nozzle. Also, the exit pressure, density, and temperature ratios depend only on the exit area ratio A_e/A^*. That is, if the nozzle is part of a supersonic wind tunnel, then the test-section conditions are determined by A_e/A^* (geometry of the nozzle) and the stagnation pressure p_0 and temperature T_0 (properties of the gas in the reservoir).

A variety of flow field can be generated in a convergent-divergent nozzle by independently governing the backpressure downstream of the nozzle exit.

1.108 AREA-VELOCITY RELATION

This is a relation between the stream-tube cross-section area (A) and flow velocity (V), for a steady quasi-one-dimensional flow. The relation is

$$\frac{dA}{dV} = -\frac{A}{V}\left(1 - M^2\right)$$

where M is the Mach number given by the ratio of local flow velocity to local speed of sound.

The following information can be derived from the area-velocity relation:

1. For incompressible flow limit, that is for $M \to 0$, the area-velocity relation shows that $AV=$constant. This is the famous volume conservation equation or continuity equation for incompressible flow.

2. For $0 \le M \le 1$, a decrease in area results in an increase of velocity and vice versa. Therefore, the velocity increases in a convergent duct and decreases in a divergent duct. This result for compressible subsonic flows is the same as that for incompressible flows.

3. For $M > 1$, an increase in area results in an increase of velocity and vice versa, that is the velocity increases in a divergent duct and decreases in a convergent duct. This is directly opposite to the behaviour of subsonic flow in divergent and convergent ducts.

4. For $M = 1$, $dA/A = 0$, which implies that at the location where the Mach number is unity the area of the passage is either minimum or maximum. We can easily show that the minimum in area is the only physically realistic solution.

1.109 ARGAND DIAGRAM

The picture, as shown in Figure 1.12, in which the complex number is represented by a point is called the Argand diagram.

1.110 ARGON

Argon is a chemical element with the symbol Ar and atomic number 18. It is in group-18 of the periodic table and is a noble gas. Argon is the third-most abundant gas in the Earth's atmosphere, at 0.934%.

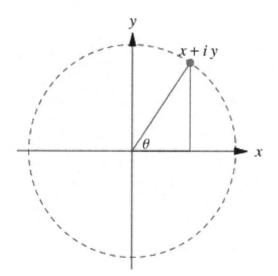

Figure 1.12 Argand diagram.

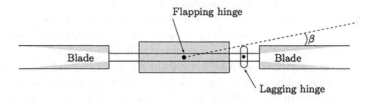

Figure 1.13 Helicopter rotor with flapping and lagging hinges.

1.111 ARROW

A thin piece of wood or metal, with one pointed end and feathers at the other end, that is shot by pulling back the string on a curved piece of wood (a bow) and letting go.
   or
   A linear body having a wedge-shaped end, as one used on a map or architectural drawing, to indicate direction or placement.

1.112 ARTICULATED ROTOR SYSTEM

Articulated rotor system is that in which the blades are hinged with the hub of the rotor so that the blades can have free vertical motion. A typical articulated rotor system with the flapping and lagging hinges is shown in Figure 1.13.

1.113 ASPECT RATIO

Aspect ratio, *AR*, is the ratio of the span and the average chord.
   The ratio of span-to-chord of a body is termed aspect ratio. For example, aspect ratio for an aircraft wing is the ratio of the wingspan to chord. For a wing the shortest distance between the wing tips is the span, and the shortest distance between the leading edge and training edge of the aerofoil is the chord. If all the aerofoils along the wingspan are identical, the chord is uniquely defined. This kind of wings is called rectangular profiles. For wings, which are not rectangular, the mean chord is usually taken as the representative one.
   The *AR* is a fineness ratio of the wing, and it varies from 35 for sailplanes to about two for supersonic fighter planes.

1.114 ATLANTIC HURRICANE

An Atlantic hurricane or tropical storm is a tropical cyclone that forms in the Atlantic Ocean, primarily between the months of June and November.

A hurricane differs from a cyclone or typhoon only on the basis of location. A hurricane is a storm that occurs in the Atlantic Ocean and northeastern Pacific Ocean, a typhoon occurs in the northwestern Pacific Ocean, and a cyclone occurs in the South Pacific Ocean or Indian Ocean.

1.115 ATOM

An atom is the smallest unit of ordinary matter that forms a chemical element. Every solid, liquid, gas, and plasma is composed of neutral or ionised atoms. Atoms are extremely small, typically around 100 picometers across.

1.116 ATOMIC MASS

The atomic mass is a weighted average of all of the isotopes of that element, in which the mass of each isotope is multiplied by the abundance of that particular isotope. (Atomic mass is also referred to as atomic weight, but the term 'mass' is more accurate.)

1.117 ATOMIC NUMBER

The atomic number or proton number of a chemical element is the number of protons found in the nucleus of every atom of that element. The atomic number uniquely identifies a chemical element. It is identical to the charge number of the nucleus.

1.118 ATOMIC PHYSICS

Atomic physics is the field of physics that studies atoms as an isolated system of electrons and an atomic nucleus. It is primarily concerned with the arrangement of electrons around the nucleus and the processes by which these arrangements change. This comprises ions, neutral atoms and, unless otherwise stated, it can be assumed that the term atom includes ions.

1.119 ASTRONOMICAL UNIT

The astronomical unit is a unit of length, roughly the distance from the Earth to the Sun and equal to about 150 million kilometres.

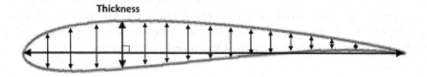

Figure 1.14 Asymmetrical aerofoil.

1.120 ASTRONOMY

Astronomy is the study of everything in the universe beyond the Earth's atmosphere. That includes objects we can see with our naked eyes, like the Sun, the Moon, the planets, and the stars. It also includes objects we can only see with telescopes or other instruments, like faraway galaxies and tiny particles.

1.121 ASTROPHYSICS

Astrophysics is a branch of space science that applies the laws of physics and chemistry to explain the birth, life and death of stars, planets, galaxies, nebulae and other objects in the universe. It has two sibling sciences, astronomy and cosmology, and the lines between them blur.

1.122 ASYMMETRICAL AEROFOIL

An aerofoil whose shape on either side of the chord is not the same, as illustrated in Figure 1.14.

Asymmetric aerofoils can generate lift at zero angle of attack, while a symmetric aerofoil may better suit frequent inverted flight as in an aerobatic aeroplane.

1.123 ATMOSPHERE

The atmosphere may be regarded as an expanse of fluid (air) substantially at rest. Hydrostatic theory may be used to calculate its macroscopic properties. The atmosphere is a mixture of gases of which nitrogen and oxygen are the main constituents. It also contains small amounts of other gases, including hydrogen and helium and the rare inert gases argon, krypton, neon, etc. Over the range of altitudes involved in conventional aerodynamics, the properties of the constituents vary little, and the atmosphere may be regarded as a homogeneous gas of uniform composition.

It is well established that the atmosphere may conveniently be divided into two district continuous regions. The lower of these regions is called the *troposphere*, and it is found that the temperature with in the troposphere decreases approximately linearly with height. The upper region is the *stratosphere* wherein the temperature remains almost constant with height. The supposed boundary between the two regions is termed the *tropopause*. The sharp distinction between the two, implied above, does not exist in reality, but one merges gradually into the other. Nevertheless, this distinction represents a useful convention for the purposes of calculation.

1.124 ATMOSPHERIC BOUNDARY LAYER

The atmospheric boundary layer (ABL) or peplosphere is the lowest part of the atmosphere, and its behaviour is directly influenced by its contact with a planetary surface.

1.125 ATMOSPHERIC BOUNDARY LAYER THICKNESS

The thickness of the boundary layer is quite variable in space and time. Normally 1 or 2 km thick (i.e. occupying the bottom 10%–20% of the troposphere), it can range from tens of metres to 4 km or more.

1.126 ATMOSPHERIC PRESSURE

Atmospheric pressure, also known as barometric pressure (after the barometer), is the pressure within the atmosphere of the Earth.

At the standard sea level, the pressure is 101,325 Pa or 760 mm of mercury.

1.127 ATMOSPHERIC TEMPERATURE

Sea level temperature of the atmosphere at the Earth's surface is 15°C.

1.128 ATMOSPHERIC TURBULENCE

Atmospheric turbulence usually refers to the three-dimensional, chaotic flow of air in the Earth's atmosphere with a time scale of <1 seconds to typically 1 hour.

1.129 AUDIBLE RANGE OF HUMAN EAR

Humans can detect sounds in a frequency range from about 20 Hz to
20 kHz. (Human infants can actually hear frequencies slightly higher than
20 kHz, but lose some high-frequency sensitivity as they mature; the upper
limit in average adults is often closer to 15–17 kHz.)

1.130 AUTOMOTIVE COOLING SYSTEM

A typical automotive cooling system comprises (a) a series of channels cast
into the engine block and cylinder head, surrounding the combustion chambers with circulating water or other coolant to carry away excessive heat,
and (b) a radiator, consisting of many small tubes equipped with a honeycomb of fins to radiate heat.

1.131 AUTOROTATION

A natural tendency for the aircraft to rotate on its own accord is called
autorotation.

1.132 AVAILABILITY

Availability is a property that enables us to determine the useful work
potential of a given amount of energy at some specified state. The work
potential of the energy contained in a system at a specified state may be
viewed as the maximum useful work that can be obtained from the system.

1.133 AVERAGE CHORD

This is the geometric average of the chord distribution over the length of
the wingspan.

1.134 AVIATION

The human activity that surrounds aircraft is called aviation. Crewed aircraft are own by an onboard pilot, but unmanned aerial vehicles may be
remotely controlled or self-controlled by onboard computers.

Aviation is the activities surrounding mechanical flight and the aircraft
industry. Aircraft includes fixed-wing and rotary-wing types, morphable

wings, wing-less lifting bodies, and lighter-than-aircraft such as hot-air balloons and airships.

1.135 AVIATION GASOLINE

Aviation gasoline, often referred to as 'avgas' or 100 - LL (low-lead), is a highly refined form of gasoline for aircraft, with an emphasis on purity, anti-knock.

1.136 AVIONICS

Avionics refers to the control, navigation, and communication systems, usually electrical in nature.

1.137 AVOGADRO'S LAW

Avogadro's law states, 'equal volumes of all gases, at the same temperature and pressure, have the same number of molecules'.

1.138 AVOGADRO'S NUMBER

Avogadro's number, number of units in one mole of any substance (defined as its molecular weight in grams), equals to $6.02214076 \times 10^{23}$.

1.139 AXIAL FLOW FAN

An axial flow fan is a type of industrial fan used to cool machines and equipment, which heat up after use.

1.140 AXIAL FLOW PUMPS

Axial flow pumps, also called propeller pumps, are centrifugal pumps, which move fluid axially through an impeller.

1.141 AXIS-SWITCHING

Axis-switching is a phenomenon in elliptic jets which is manifestation of an alteration between the major and minor axes.

1.142 AXISYMMETRIC FLOW

In hydrodynamics and fluid mechanics, axisymmetric flow is a flow in which the streamlines are symmetrically located around an axis. Every longitudinal plane through the axis would exhibit the same streamline pattern.

Chapter 2

Backflow to Bypass Ratio

2.1 BACKFLOW

Backflow is defined as an undesirable reverse flow of water that returns contaminated water from a work site back to the potable water source.

2.2 BACKPRESSURE

The pressure of the environment to which a flow is discharged.

2.3 BACK WORK RATIO

In gas-turbine power plants, the ratio of the compressor work to the turbine work is termed the back work ratio.

2.4 BAKING

Baking is a method of preparing food that uses dry heat, typically in an oven, but can also be done in hot ashes, or on hot stones. The most common baked item is bread but many other types of foods are baked. Heat is gradually transferred from the surface of cakes, cookies, and breads to their centre.

2.5 BALL VALVE

A ball valve is a shut-off valve that controls the flow of a liquid or gas by means of a rotary ball having a bore.

DOI: 10.1201/9781003348405-2

2.6 BALLISTICS

Ballistics is the field of mechanics concerned with the launching, flight behaviour and impact effects of projectiles, especially ranged weapon munitions such as bullets, unguided bombs, rockets, or the like. This is the science or art of designing and accelerating projectiles so as to achieve a desired performance.

2.7 BALLISTIC MISSILE

Ballistic missile is a rocket-propelled self-guided strategic-weapon system that follows a ballistic trajectory to deliver a payload from its launch site to a predetermined target. Ballistic missiles can carry conventional high explosives as well as chemical, biological, or nuclear munitions. A few typical ballistic missiles are shown in Figure 2.1.

2.8 BALLOON

Balloon is a large airtight bag filled with hot air or a lighter-than-air gas, such as helium or hydrogen, to provide buoyancy so that it will rise and float in the atmosphere. Transport balloons have a basket or container hung below for passengers or cargo. A self-propelled steerable balloon is called an airship or a dirigible.

Figure 2.1 Ballistic missiles.

2.9 BANKED TURN

Banked turn is a steady motion in a horizontal circle with the plane of symmetry inclined to the vertical.

2.10 BAR

The bar is a metric unit of pressure but is not part of the International System of Units (SI). It is defined as exactly equal to 100,000 Pa (100 kPa).

2.11 BAROMETER

The basic principle of pressure measurement with a barometer can easily be seen in the system shown in Figure 2.2. There is a reservoir containing a liquid. At the surface of this liquid, atmospheric pressure p_a is acting. An open-end glass tube is inserted into the liquid, as shown. The atmospheric pressure p_a will act down through the tube onto the surface of the liquid contained in the tube. Since only p_a acts at the liquid surface both outside and inside the glass tube, the liquid surface level throughout will be the same. It must be ensured that the tube bore is large enough so that no capillary action occurs due to surface tension. Examine the system shown in Figure 2.2. The top of the glass tube is sealed, and the tube has been evacuated to vacuum. Therefore, there is no pressure acting inside the tube. Since the atmospheric pressure p_a acts in all directions within the liquid, there is an unbalanced pressure in the upward direction inside the tube. The liquid is thus forced up in the tube till the weight of the liquid column above the surface level of the reservoir balances the atmospheric pressure p_a. If h is the

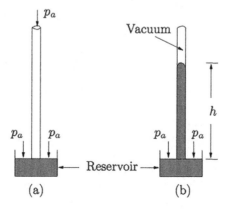

Figure 2.2 Simple barometer.

height of the liquid column and ρ is the density of the liquid, then the pressure is given by $p_a = \rho g h$. This simply implies that $p \propto h$, since ρ and g are constants. The height of the liquid column established in the above manner can therefore be used as a measure of the atmospheric pressure.

This is the measuring principle of a barometer, and was developed by Evangelista Torricelli (1608–47), an Italian physicist. To create necessary vacuum above the surface of the liquid, the tube is firstly inverted and filled with liquid. The open end is now held closed, the tube is turned over so that the open end is at the bottom, and with the end still held closed, it is immersed in the liquid in the reservoir. The end is then opened. The liquid in the tube now adjusts itself to balance the atmospheric pressure by falling down the tube until the required height is attained. A vacuum will thus be formed in the tube above the surface of the liquid. This vacuum is referred to as the Torricelli vacuum. The height to which the liquid column will rise in the tube depends upon the density of the liquid used. Let us assume the liquid to be water and the column height in the tube to be h m. The atmospheric pressure becomes $p_a = \rho g h = 9.81 \times 10^{23}$ Pa, since the density of water is 10^3 kg/m^3. In the international standard atmosphere, the atmospheric pressure is 101,325 Pa (1.01325 bar). Therefore, $101,325 = 9.81 \times 10^3 \times h$. Thus, the water column height of 1 atmosphere (at standard conditions) becomes

$$h = 101325/\left(9.81 \times 10^3\right) = 10.329 \text{ m}$$

A column height of 10.329 m of water is too large for measurement purposes. In actual measurements, this order of column height is unmanageable. To reduce this column height, mercury is used in the barometer. As we know, mercury is 13.6 times heavier than water. Hence, the height of mercury column will be only 1/13.6 times the height of corresponding water column. Thus, the barometric column height, under standard sea level conditions, is

$$h = 10,329/13.6 = 760 \text{ mm (approximately)}$$

This height is convenient to measure. Even though barometric height is useful for measurement purposes, in practice, we express pressure as a force per unit area. Usually, pressure is expressed as N/m^2 or pascal (Pa) in SI units. An important aspect of practical interest that must be noted here is that the barometric height increases with increase of pressure. The mercury necessary to accommodate this increase in height has to come from the reservoir at the bottom of the barometer. Hence, the mercury level in the reservoir falls. Conversely, the mercury level in the reservoir increases when the barometric height falls, as a result of decrease in atmospheric pressure.

In either case, with increase in or fall of level in the reservoir, the barometric height must be measured from the mercury surface level in the reservoir.

2.12 BAROMETRIC PRESSURE

Barometric pressure is the pressure caused by the weight of the air above us. The Earth's atmosphere above us contains air, and although the air is relatively light, having much of air starts to have some weight as gravity pulls the air molecules.

2.13 BAROTROPIC FLUIDS

Barotropic fluids are those for which the density is a function of only pressure, i.e., $\rho = \rho(p)$.

2.14 BARRAGE

A barrage is a type of low-head, diversion dam that consists of several large gates that can be opened or closed to control the amount of water passing through as shown in Figure 2.3. This allows the structure to regulate and stabilise river water elevation upstream for use in irrigation and other systems.

Figure 2.3 A barrage.

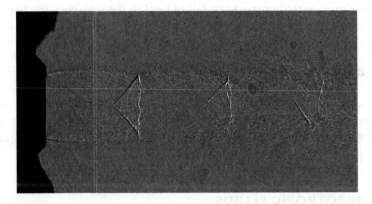

Figure 2.4 Shadowgraph picture of Mach I jet at NPR 3.5.

2.15 BARREL SHOCK

For a two-dimensional supersonic flow, on either side of the axis and around the axis for axisymmetric flow, the shape of the shock assumes the form of a 'barrel' as seen in Figure 2.4, and is thus referred to as a barrel shock. Outside of the barrel shock, the flow is a mixture of supersonic and subsonic Mach numbers.

2.16 BASE DRAG

Base drag is the drag generated in an object moving through a fluid from the shape of its rear end. For example, the negative pressure originating behind the flat base of projectiles presents a drag component that is consequently termed 'base drag'.

2.17 BASE PRESSURE

Base pressure is the pressure in the separated region close behind a body. In general, it is lower than the freestream static pressure and acting on the base surface.

2.18 BASIC DIMENSIONS

In fluid dynamics, mostly the gross, measurable molecular manifestations such as pressure and density as well as other equally important measurable abstract entities, for example, length and time, are dealt with.

Table 2.1 Units

Quantity	Unit	SI	CGS	FPS	MKS
Mass	Kilogram	kg	g	lb	kg
Length	Metre	m	cm	ft	m
Time	Second	s	s	s	s
Force	Newton	N	dyn	pdl	kgf
Temperature	Kelvin	K	°C	°F	°C

These manifestations that are characteristics of the behaviour of a particular fluid, and not of the manner of flow, may be called *fluid properties*. Density and viscosity are examples of fluid properties. To adequately discuss these properties, a consistent set of standard units must be defined. Table 2.1 gives the common system of units and their symbols.

2.19 BASIC LAWS

In the range of engineering interests, four basic laws must be satisfied for any continuous medium. They are

- Conservation of matter (the continuity equation)
- Newton's second law (momentum equation)
- Conservation of energy (first law of thermodynamics)
- Increase of entropy principle (second law of thermodynamics)

In addition to these primary laws, there are numerous subsidiary laws, sometimes called constitutive relations, that apply to specific types of media or flow processes (for example, equation of state for a perfect gas, Newton's viscosity law for certain viscous fluids, isentropic process relation, adiabatic process relation).

2.20 BASIC POTENTIAL EQUATION FOR COMPRESSIBLE FLOW

The basic potential equation for compressible flow is

$$\left(1 - \phi_x^2/a^2\right)\phi_{xx} + \left(1 - \phi_y^2/a^2\right)\phi_{yy} + \left(1 - \phi_z^2/a^2\right)\phi_{zz}$$

$$- 2\left(\frac{\phi_x\phi_y}{a^2}\phi_{xy} + \frac{\phi_y\phi_z}{a^2}\phi_{yz} + \frac{\phi_z\phi_x}{a^2}\phi_{zx}\right) = 0$$

This is a nonlinear equation. The difficulties associated with compressible flow stem from the fact that the basic equation is nonlinear. Hence, the superposition of solutions is not valid. Further, the local speed of sound 'a' in this equation is also a variable.

2.21 BASIC SOLUTIONS OF LAPLACE'S EQUATION

The four basic solutions of Laplace's equation are the following:

For uniform flow (toward positive x-direction), the potential function is $\phi = V_\infty \, x$ for a source of strength Q, the potential function is $\phi = \left(\dfrac{Q}{2\pi} \right) \ln r$, for a doublet of strength μ (issuing in negative x-direction), the potential function is $\phi = \mu \cos\theta / r$, and for a potential (free) vortex (counterclockwise) with circulation Γ, the potential function is $\phi = \left(\dfrac{\Gamma}{2\pi} \right) \theta$.

2.22 BATTERY

A battery is a source of electric power consisting of one or more electro-chemical cells with external connections for powering electrical devices such as flashlights, mobile phones, and electric cars. When a battery is supplying electric power, its positive terminal is the cathode and its negative terminal is the anode. The terminal marked negative is the source of electrons that will flow through an external electric circuit to the positive terminal. When a battery is connected to an external electric load, a redox reaction converts high-energy reactants to lower-energy products, and the free-energy difference is delivered to the external circuit as electrical energy. Historically, the term 'battery' specifically referred to a device composed of multiple cells; however, the usage has evolved to include devices composed of a single cell.

2.23 BECKMANN THERMOMETER

The Beckmann thermometer is used for the accurate determination of small temperature changes, such as those encountered while using the bomb calorimeter. This is meant only for the determination of temperature change and not the temperature itself. The maximum temperature change that can be measured is usually 6°C.

2.24 BEER

Beer is a type of alcoholic drink that is made from grain.

2.25 BENZENE

Benzene is an organic chemical compound with the molecular formula C_6H_6.

2.26 BENZENE RING

Benzene ring: An aromatic functional group characterised by a ring of six carbon atoms bonded by alternating single and double bonds.

2.27 BERNOULLI'S EQUATION

Both incompressible and compressible forms of the Bernoulli equation are simply relations between the total pressure, static pressure, and flow velocity.

Bernoulli's equation is a statement of conservation of *mechanical energy* between any two points along a streamline, that is,

$$\int \frac{dp}{\rho} + \int VdV = \text{constant}$$

The constant in the equation is known as the *Bernoulli constant*. It varies, in general, from one streamline to another but remains constant along a streamline in a steady, *frictionless, incompressible flow*. These four assumptions are needed and must be kept in mind while applying this equation. Each term in the equation has the dimensions $(L/T)^2$ or the units of metre-newton per kilogram

$$\frac{\text{m N}}{\text{kg}} = \frac{\left(\text{m}\frac{\text{kg}}{\text{s}^2}\right)(\text{m})}{\text{kg}} = \frac{\text{m}^2}{\text{s}^2}$$

Therefore, the Bernoulli equation is interpreted as energy per unit mass. When it is divided by g,

$$\frac{p}{\rho g} + \frac{v^2}{2g} + z = \text{constant}$$

it can be interpreted as energy per unit weight, metre-newton per newton. This form is convenient for dealing with liquid problems with a free surface.

Each term in Bernoulli's equation may be interpreted as a form of available energy. This equation is also referred to as a conservation of *mechanical energy* equation. It is essential here to realise that this energy equation was derived from the *conservation of momentum equation*. Thus, the Bernoulli equation is essentially a *momentum equation* and *not an energy equation*.

For incompressible flows, the Bernoulli equation reduces to

$$p + \frac{1}{2}\rho V^2 = p_0$$

where p is static pressure and p_0 is the stagnation pressure.

For compressible flows, since the density ρ is a variable, the integration becomes involved. For isentropic process of compressible flows, the Bernoulli equation takes the form

$$\frac{\gamma}{\gamma-1}\frac{p}{\rho} + \frac{V^2}{2} = \frac{\gamma}{\gamma-1}\frac{p_0}{\rho_0}$$

where ρ and ρ_0 are the static and stagnation densities, respectively.

2.28 BERNOULLI'S PRINCIPLE

In fluid dynamics, Bernoulli's principle states that an increase in the speed of a fluid occurs simultaneously with a decrease in static pressure or a decrease in the fluid's potential energy.

2.29 BETZ MANOMETER

Betz manometer is a U-tube manometer. The fluid column carries a float, and a graduated scale is attached to the float, as shown in Figure 2.5. The scale is transparent and the graduation markings can be projected on a screen. Thus, the manometer readings can be read directly. Even though it is a very convenient instrument to use, it is somewhat sluggish due to the large volume of the reservoir. The measurement range of this instrument is usually about 0–300 mm, and its sensitivity is ±0.1 mm.

2.30 BIHARMONIC EQUATION

$$\nabla^4 \psi = 0$$

This equation governs the stream function for 2-D Stokes flow.

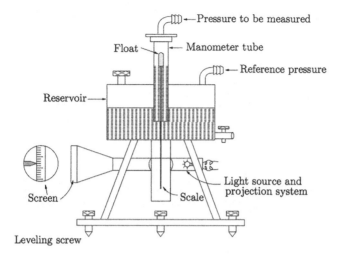

Figure 2.5 Betz manometer.

2.31 BIMETALLIC THERMOMETERS

Bimetallic devices are used for temperature measurements and vary widely as combined sensing and control elements in temperature-control systems, mainly of the on-off type. Also, they are used as overload cut-out switches in electric apparatus by allowing the current to flow through the bimetal, heating and expanding it, and causing a switch to open when excessive current flows. The accuracy of bimetallic elements varies largely, depending on the application requirements. Since a majority of control applications are not extremely critical, requirements can be satisfied with a rather low-cost device. The working temperature range is about −35°C to 550°C. Errors of the order of 0.5%–1% scale range may be expected in bimetal thermometers of high quality.

2.32 BINARY VAPOUR CYCLE

The power cycle that is a combination of cycles, one in the high-temperature region and the other in the low-temperature region, is called the binary vapour cycle.

2.33 BIOGAS

Biogas is the mixture of gases produced by the breakdown of organic matter in the absence of oxygen (anaerobically), primarily consisting of methane and carbon dioxide. Biogas can be produced from raw materials such as agricultural waste, manure, municipal waste, plant material, sewage, green waste, or food waste.

2.34 BIOT NUMBER

Biot number, hL/k, is the ratio of the internal resistance of a body to heat conduction to the resistance to heat convection from the fluid to the body.

It is important to note that the nondimensional group hL/k is also known as the Nusselt number. Although similar in form, the Nusselt and Biot numbers differ in both definition and interpretation. The Nusselt number is defined in terms of the thermal conductivity k of the fluid, whereas the k in the Biot number is the thermal conductivity of the solid.

2.35 BIOT–SAVART LAW

The Biot–Savart law relates the intensity of the magnitude of a magnetic field close to an electric current-carrying conductor to the magnitude of the current. It is mathematically identical to the concept of relating intensity of flow in the fluid close to a vorticity-carrying vortex tube to the strength of the vortex tube.

2.36 BIPLANE

A biplane is a fixed-wing aircraft with two superimposed main wings, as in Figure 2.6.

Figure 2.6 Biplane.

2.37 BLACK HOLE

A black hole is a region of space-time where gravity is so strong that nothing – no particles or even electromagnetic radiation such as light – can escape from it. The theory of general relativity predicts that a sufficiently compact mass can deform space-time to form a black hole.

2.38 BLACK HOLE TYPES

There are four types of black holes: stellar, intermediate, supermassive, and miniature. The most commonly known way a black hole forms is by stellar death.

2.39 BLACK TEA

Black tea is a type of tea that is more oxidised than oolong, yellow, white, and green teas. Black tea is generally stronger in flavour than other teas.

2.40 BLADE ANGLE

Blade angle: The angular difference between the chord of the blade and the plane of rotation is the blade angle. The blade angle is also known as the pitch angle. The blade angle is the angle that the chord of the propeller section at any particular place makes with the horizontal plane where the propeller is laid at on its base in this horizontal plane, its axis being vertical.

2.41 BLADE ELEMENT THEORY

Blade element theory is essentially an application of the airfoil theory to the rotating blade.

The first and foremost assumption in the blade element theory is that the forces acting on each airfoil element of the rotor blade are independent. The blade, which is flexible, is assumed to be rigid.

2.42 BLADE LOADING

Blade loading: The ratio of the gross weight of the helicopter to the combined area of the rotor blades is called blade loading. Since the blade area does not change, blade loading is a constant in flight (ignoring the change in weight and gravitational acceleration).

2.43 BLADE STALL

As the forward speed, V, of the helicopter increases, the velocity of the retreating blade, V_R, decreases and the pitch on the retreating side increases, hence, some blades on the rotor may stall.

In the helicopter, stall begins at the tip of the retreating blade, spreading inboard as forward speed is increased. In the autogyro, the stall begins at the root of the retreating blade, spreading outboard as the speed is increased.

2.44 BLASIUS BOUNDARY LAYER

In physics and fluid mechanics, a Blasius boundary layer (named after Paul Richard Heinrich Blasius) describes the steady two-dimensional laminar boundary layer that forms on a semi-infinite plate that is held parallel to a constant unidirectional flow. Falkner and Skan later generalised Blasius' solution to wedge flow (Falkner–Skan boundary layer), i.e., flows in which the plate is not parallel to the flow.

2.45 BLASIUS EQUATION

The Blasius equation is used to describe the stream function in a steady two-dimensional boundary layer that forms on a semi-infinite plate held parallel to a constant unidirectional flow.

2.46 BLASIUS SOLUTION

The first exact solution to the laminar boundary layer equations, discovered by Blasius (1908), was for a simple constant value of $U(s)$ and pertains to the case of a uniform stream of velocity, U, encountering an infinitely thin flat plate set parallel with that stream.

2.47 BLASIUS THEOREM

The Blasius theorem relates the force and moment acting on the cylinder to the complex potential in the surrounding ideal flow.

2.48 BLAZE

To burn with bright strong flames.

Figure 2.7 Blended winglet.

2.49 BLENDED WINGLETS

Blended winglets are upward-swept extensions to airplane wings, as shown in Figure 2.7. They feature a large radius and a smooth chord variation in the transition section.

2.50 BLENK'S METHOD

This method is meant for wings of finite aspect ratio and is based on the lifting line theory of Prandtl, and hence limited to aerofoils moving in the plane of symmetry and with a trailing edge that could be regarded as approximately straight. This method considers the wing as a lifting surface, that is to say, the wing is replaced by a system of bound vortices distributed over its surface rather than along a straight line coinciding with the span. However, this method has the limitation that the wing is assumed to be thin and practically plane. The following are the two main approaches employed in Blenk's method.

1. Given the load distribution and the plan, find the profiles of the sections.
2. Given the plan and the profiles, find the load distribution (that is, the vorticity distribution).

2.51 BLOCKAGE EFFECT

If the ratio of the flow field area, A_{flow}, to the probe projected area, A_{probe}, normal to the flow direction is more than 64, the blockage of the probe will not cause any appreciable change to the flow properties. Therefore, the limiting minimum size of the probe below which the area blockage due to the probe will not introduce any significant error in the measurement is $A_{flow}/A_{probe} > 64$.

2.52 BLOOD

Blood is a body fluid in humans and other animals that delivers necessary substances such as nutrients and oxygen to the cells and transports metabolic waste products away from those same cells. In vertebrates, it is composed of blood cells suspended in blood plasma.

2.53 BLOOD IN THE HUMAN BODY

The average adult weighing 150–180 pounds should have about 1.2–1.5 gallons of blood in the body. This is about 4,500–5,700 mL. Pregnant women: To support their growing babies, pregnant women usually have anywhere from 30% to 50% more blood volume than women who are not pregnant.

Around 7%–8% of an adult's body weight is blood. The body can easily replace a small amount of lost blood, which makes blood donation possible.

2.54 BLOOD PLASMA

Blood plasma is a yellowish liquid component of blood that holds the blood cells of the whole blood in suspension. It is the liquid part of the blood that carries cells and proteins throughout the body. It makes up about 55% of the body's total blood volume.

2.55 BLOOD PRESSURE

Blood pressure is a measure of the force that your heart uses to pump blood around your body. Blood pressure of <120/80 mm Hg is considered within the normal range.

2.56 BLOWDOWN TUNNEL OPERATION

Constant Reynolds number operation, constant pressure operation, and constant throttle operation are the three methods of operation usually adopted for blowdown tunnel operation.

2.57 BLOWER

Blower is a machine for supplying air at a moderate pressure, as to supply forced drafts or supercharge and scavenge diesel engines.

2.58 BLUFF BODY

A bluff body is that for which the wake drag accounts for the major portion of the total drag and the skin friction drag is insignificant. The opposite of a blunt body is the streamlined body for which the skin friction drag is the dominant portion of the total drag.

2.59 BOAT

Boat is a small vehicle that is used for travelling across water, as shown in Figure 2.8.

2.60 BOATTAIL

Boattail: The progressively narrowing rear end of a bullet or ballistic missile that is designed to reduce drag.

Figure 2.8 Boat.

2.61 BODY FORCES

All external forces acting on any material, which are developed without physical contact, are called body forces. The forces that come across in continuum fluid mechanics may broadly be divided into body forces and surface forces. Gravitational force, the effect of the Earth on a mass manifesting itself as a force distribution throughout the material, directed toward the Earth's centre, is a body force. Body forces are usually expressed as per unit mass of the material acted on. All forces exerted on a boundary by its surroundings through direct contact are termed as surface forces, e.g., pressure.

2.62 BOILING

Boiling is a liquid-to-vapour phase change process. In other words, boiling is a process in which a liquid phase is converted to a vapour phase.

2.63 BOILING TEMPERATURE OF WATER

The boiling point of water is 100°C at a pressure of 1 atm.

2.64 BOLTZMANN CONSTANT

The Boltzmann constant (k_B or k) is the proportionality factor that relates the average relative kinetic energy of particles in a gas with the thermodynamic temperature of the gas. It is defined to be exactly 1.380649×10^{-23} J/K. The Boltzmann constant relates the average kinetic energy for each degree of freedom of a physical system in equilibrium to its temperature.

2.65 BOLTZMANN DISTRIBUTION

The Boltzmann distribution is the limiting case of the most probable distribution at high temperature where the molecules are distributed over many energy levels.

2.66 BOLTZMANN EQUATION FOR ENTROPY

The Boltzmann equation for entropy is

$$S = k_B \ln W$$

where W is the number of different ways or microstates in which the energy of the molecules in a system can be arranged on energy levels.

2.67 BOND ENERGY

The internal energy associated with the bonds in a molecule is called bond energy.

2.68 BOOSTER PUMP

A booster pump is a high-pressure pump, which has an impeller inside which improves its water flow and pressure.

2.69 BOSE–EINSTEIN STATISTICS

In quantum statistics, the Bose–Einstein statistics describes one of the two possible ways in which a collection of non-interacting, indistinguishable particles may occupy a set of available discrete energy states at thermodynamic equilibrium.

2.70 BOSONS

For bosons, the number of molecules that can be in any one degenerate state is unlimited.

Molecules and atoms with an even number of elementary particles obey a certain statistical distribution called Bose–Einstein statistics, and they are usually referred to as bosons.

2.71 BOUND VORTEX

The circulation about the aerofoil with a vortex lying over the aerofoil due to the boundary layer at the surface is called the bound vortex.

It is a vortex that is considered to be tightly associated with the body around which a liquid or gas flows, and equivalent with respect to the magnitude of speed circulation to the real vorticity that forms in the boundary layer owing to viscosity.

A curve which surrounds the aerofoil only has the same circulation as the free vortex, but with opposite sign, and therefore, the aerofoil experiences a lift.

2.72 BOUNDARY CONDITIONS

Boundary conditions (b.c.) are constraints necessary for the solution of a boundary value problem. A boundary value problem is a differential equation (or system of differential equations) to be solved in a domain on whose boundary a set of conditions is known.

2.73 BOUNDARY LAYER

The concept of a boundary layer is important in many aerodynamic problems. The viscosity and fluid friction in the air are approximated as being significant only in this thin layer. This principle makes aerodynamics much more tractable mathematically.

Boundary layer may be defined as that *thin layer adjacent to a solid boundary within which the flow velocity increases from 0 to 99% of its freestream value.* Boundary layer may also be defined as that *fluid layer which has had its velocity affected by the boundary shear.* This definition refers to the momentum boundary layer. In an identical manner, we can define the thermal boundary layer as *the thin layer adjacent to a solid surface within which the temperature varies from the stagnation level to the freestream value.* In other words, the layer at the body surface with the temperature at the surface as the total temperature and the temperature at the edge of the layer as the freestream static temperature is termed the thermal boundary layer.

In an incompressible flow, it is possible to separate the calculation of the velocity boundary layer from that of the thermal boundary layer. But in compressible flow, it is not possible, since the velocity and thermal layers interact intimately, and therefore, they must be considered simultaneously. This is because, for high-speed flows (compressible flows), heating due to friction as well as temperature changes due to compressibility must be taken into account. Further, it is essential to include the effects of viscosity variation with temperature. Usually, large variations of temperature are encountered in high-speed flows.

The boundary layer may also be defined as a thin layer adjacent to a solid surface where the viscous effects are predominant. Thus, inside the boundary layer, the effect of viscosity is predominant. Outside the boundary layer, the effect of viscosity is negligible.

2.74 BOUNDARY LAYER BLOWING

The principle of boundary layer blowing is similar to that of the leading edge slot. High-speed air is blown into the boundary layer through a narrow slit, as illustrated in Figure 2.9, in the upper surface of the wing,

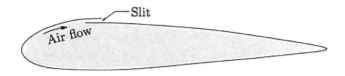

Figure 2.9 Illustration of boundary layer blowing through a slit near the nose of an aerofoil.

where it reenergises the boundary layer and prevents separation. Since the velocity of the air fed in this way is so much higher than the speed of the air passing through the leading edge slot, or a slotted flap, blowing will generally prove to be much more effective. The stall can be delayed almost indefinitely by this means. In addition, the jet of air has the effect of increasing the circulation around the wing, thus giving a direct lift increment at all incidences.

2.75 BOUNDARY LAYER CONTROL

Boundary layer control is a method of increasing the lift generated by a wing without changing its geometry. Thus, boundary layer control is an artificial high-lift device. The aim is to prevent or delay separation by blowing air into the boundary layer to energise it or suck away the boundary layer.

2.76 BOUNDARY LAYER FENCING

Boundary layer fencing: These are small plates employed to prevent the outward drift of the boundary layer that is a factor in causing the tip stall. Notches at the leading edge of the wing also tend to produce a similar effect.

2.77 BOUNDARY LAYER SUCTION

The principle of suction of boundary layer is the removal of slowly moving air in the boundary layer so that there is no layer to separate.

A series of holes, flushed with the surface, are made in the surface of the aerofoil upstream of the separation point, as shown in Figure 2.10, and the air in the boundary layer is sucked into the wing through these holes. However, from this point onward, the boundary layer will re-form and thicken, and separation may still occur at some point downstream. To prevent this, a series of suction holes must be made at various chordwise positions, as shown in the figure. The logical extension of this idea

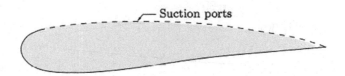

Figure 2.10 An aerofoil with suction ports over its upper surface.

is the use of a porous wing surface, with suction applied everywhere on the surface.

In addition to preventing separation, suction may also be used to prevent transition, and hence, to keep the drag low. Such a device would appear to be of particular interest in conjunction with the use of low-drag wing sections. The principle behind the design of a low-drag section is the maintenance of laminar flow. The disadvantage associated with such a design is that separation occurs easily when the incidence is increased even by a small amount above the design value.

2.78 BOUNDARY LAYER THICKNESS

Boundary layer thickness δ may be defined as the distance from the wall in the direction normal to the wall surface, where the fluid velocity is within 1% of the local mainstream velocity. The boundary layer thickness may be shown schematically as in Figure 2.11.

2.79 BOUNDARY LAYER TRIPPING

Tripping the boundary layer refers to the action of artificially transitioning a laminar boundary layer into a turbulent one.

2.80 BOW SHOCK

A bow shock, also called a detached shock or normal shock, is a curved propagating disturbance (as in Figure 2.12) wave characterised by an abrupt, nearly discontinuous, change in pressure, temperature, and density. It occurs when a supersonic flow encounters a body, around which the necessary deviation angle of the flow is higher than the maximum achievable deviation angle for an attached oblique shock. Then, the oblique shock transforms into a curved detached shock wave. As bow shocks occur for high flow deflection angles, they are often seen forming around blunt bodies because of the high deflection angle that the body imposes on the flow around it.

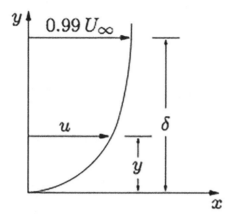

Figure 2.11 Illustration of boundary layer thickness.

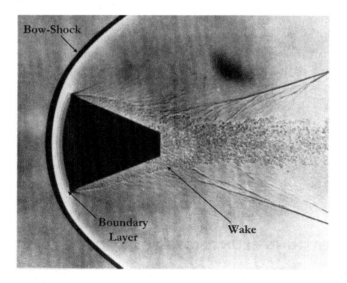

Figure 2.12 Bow shock.

2.81 BOYLE'S LAW

Boyle's law (isothermal law) states that for constant temperature, the density varies directly as the absolute pressure.

2.82 BRAKE HORSEPOWER

The power required to drive the pump is the brake horsepower.

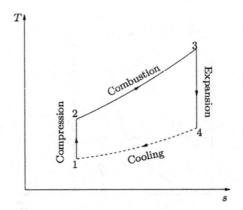

Figure 2.13 T-s diagram of Brayton cycle.

2.83 BRAYTON CYCLE

The Brayton cycle is the fundamental constant pressure gas heating cycle used by gas turbines. It consists of isentropic compression, constant pressure heating, isentropic expansion, and constant pressure cooling (absent in open cycle gas turbines), as illustrated in the *T-s* diagram in Figure 2.13.

2.84 BREATHING

Breathing (or ventilation) is the process of moving air out and into the lungs to facilitate gas exchange with the internal environment, mostly to flush out carbon dioxide and bring in oxygen.

2.85 BREATHING BLUNT NOSE TECHNIQUE

Breathing blunt nose technique is one of the promising methods for reducing the drag of blunt-nosed body at hypersonic speeds.

2.86 BREEZE

A breeze is a gentle, cool wind.

2.87 BROADBAND NOISE

Broadband noise is a noise consisting of a mixture of a wide range of frequencies. Most industrial noise is broadband noise.

2.88 BROWNIAN MOTION

Brownian motion is the stochastic motion of particles induced by random collisions with molecules and becomes relevant only for certain conditions.

2.89 BUBBLE

A bubble is a globule of one substance in another, usually gas in a liquid. Due to the Marangoni effect, bubbles may remain intact when they reach the surface of the immersive substance.

2.90 BUBBLE FLOW

Bubble flow is a two-phase flow where small bubbles are dispersed or suspended as discrete substances in a liquid continuum.

2.91 BUCKET FRICTION LOSS COEFFICIENT

The nozzle loss is usually expressed as $k_n V_j^2 / 2g$, where k_n is the bucket friction loss coefficient. Typical values of k_n vary from 0.2 to about 0.6.

2.92 BUCKINGHAM π-THEOREM

The Buckingham π-theorem states that "the number of dimensionless groups that may be employed to describe a phenomenon known to involve n variables is equal to the number $(n - r)$, where r is usually the number of basic dimensions needed to express the variables dimensionally".

2.93 BUNGEE JUMPING

Bungee jumping is an activity that involves a person jumping from a great height while connected to a large elastic cord. The launching pad is usually erected on a tall structure such as a building or crane, a bridge across a deep ravine, or on a natural geographic feature such as a cliff.

2.94 BUOYANCY

Buoyancy is the net upward force or vertical force acting on it due to the fluid or fluids in contact with the body. A body in flotation is in contact only

Figure 2.14 Butter.

with fluids and the surface force from the fluids is in equilibrium with the force of gravity on the body.

2.95 BUTTER

Butter is a dairy product made from the fat and protein components of milk or cream. It is a semi-solid emulsion (as in Figure 2.14) at room temperature, consisting of ~ 80% butterfat. It is used at room temperature as a spread, melted as a condiment, and used as an ingredient in baking, sauce making, pan-frying, and other cooking procedures.

2.96 BUTTERMILK

Buttermilk is a fermented dairy drink. Traditionally, it was the liquid left behind after churning butter out of cultured cream; however, the most modern butter is made not with cultured cream, but with sweet cream.

2.97 BYPASS RATIO

The ratio of the mass flow rate of air bypassing the combustion chamber to that of air flowing through it is called the bypass ratio. Increasing the bypass ratio of a turbofan engine increases thrust.

Chapter 3

Calcium to Cylindrical Rectangular Aerofoil

3.1 CALCIUM

Calcium is a mineral that is necessary for life. In addition to building bones and keeping them healthy, calcium enables our blood to clot, our muscles to contract, and our heart to beat. About 99% of the calcium in our bodies is in our bones and teeth.

3.2 CALCIUM CARBONATE

Calcium carbonate is a chemical compound with the formula $CaCO_3$. It is a common substance found in rocks as the minerals calcite and aragonite (most notably as limestone, which is a type of sedimentary rock consisting mainly of calcite) and is the main component of eggshells, snail shells, sea-shells and pearls.

3.3 CALIBRATION

The calibration of low-speed tunnels involves the determination of speed setting; calibration of true airspeed in the test-section, flow direction; determining the flow angularity (pitch and yaw) in the test-section, turbulence-level, velocity distribution; determination of flow quality and wake survey; and determination of flow field in the wake of any model.

The calibration of a supersonic wind tunnel includes determining the test-section flow Mach number throughout the range of operating pressure of each nozzle, determining flow angularity and determining an indication of turbulence-level effects.

3.4 CALORIC EQUATIONS OF STATE

$c_p = (\partial h / \partial T)_p$ and $c_v = (\partial u / \partial T)_v$ are caloric equations of state.

DOI: 10.1201/9781003348405-3

3.5 CALORICAL PROPERTIES

The thermodynamic properties, which are constants and independent of temperature, are called calorical properties. For a perfect gas, both c_p and c_v are constants and are independent of temperature, and hence are calorical properties.

The internal energy, u, enthalpy, h, and entropy, s, which do not depend on the process, are also calorical properties.

3.6 CALORIFIC VALUE

Calorific value is defined as the amount of calories generated when a unit amount of substance is completely oxidised and is determined using the bomb calorimeter.

3.7 CALORIFIC VALUES OF FUEL

Lignite has a calorific value of 5,000 kcal/kg. Bituminous coal has a calorific value of 7,600 kcal/kg. Anthracite coal has a calorific value of 8,500 kcal/kg. Heavy oil has a calorific value of 11,000 kcal/kg. Diesel oil has a calorific value of 11,000 kcal/kg. Petrol oil has a calorific value of 11,110 kcal/kg. Natural gas has a calorific value of 560 kcal/m³. Coal gas has a calorific value of 7,600 kcal/m³.

3.8 CAMBER

The camber of an aerofoil is the maximum displacement of the mean camber line from the chord. The mean camber line is the locus of midpoints of lines drawn perpendicular to the chord, as shown in Figure 3.1. In other words, the camber line is the bisector of the aerofoil profile thickness distribution from the leading edge to the trailing edge. Camber is the deviation of the camber line from the chord, namely the shortest line joining the leading and trailing edges of the aerofoil profile.

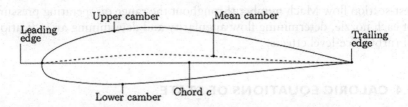

Figure 3.1 Camber.

3.9 CAMBERLINE

Camberline of an aerofoil is essentially the bisector of its thickness.

3.10 CANARD CONFIGURATION

There are some designs in which the tail of the aircraft is located ahead of the wing. This is termed as tail-first or canard configuration.

3.11 CANTED WINGLETS

Canted winglets are short, upward-sloping wedges; they can be found on Airbus $A330$ and $A340$ aircrafts and Boeing $747-400$.

3.12 CAPACITANCE PICKUP

In this transducer, the capacitance between two metallic plates with a dielectric in between is varied. One of the plates is fixed and the other one is connected to a pressure capsule. The change in capacitance is a measure of the pressure. A typical capacitance-type transducer is shown in Figure 3.2.

3.13 CAPILLARY

A capillary is a small blood vessel from 5 to $10\,\mu m$ in diameter and has a wall one-endothelial cell thick. They are the smallest blood vessels in the body. They convey blood between the arterioles and venules.

3.14 CAPILLARY ACTION

A liquid such as water or alcohol, which wets the glass surface, makes an acute angle with the solid, and the level of free surface inside the tube will be higher than that outside. This is termed capillary action.

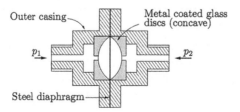

Figure 3.2 Capacitance pressure transducer.

3.15 CAPILLARY WAVES

Waves whose characteristics are governed mainly by surface tension are known as capillary waves.

3.16 CARBONATED WATER

Carbonated water is water containing dissolved carbon dioxide gas, either artificially injected under pressure or occurring due to natural geological processes. Carbonation causes small bubbles to form, giving the water an effervescent quality.

3.17 CARNOT CYCLE

The Carnot cycle is a reversible cycle. Probably, it is the best-known reversible cycle in thermodynamics. Sadi Carnot, a French engineer, proposed it in 1824. We can infer from the second law statements that no actual cycle can be completely reversible. This implies that the Carnot cycle is only a fictitious theoretical cycle.

3.18 CARNOT HEAT ENGINE

The theoretical heat engine that operates on the Carnot cycle is called the Carnot heat engine. The Carnot cycle is composed of four reversible processes – two isothermal and two adiabatic – and it can be executed either in a closed- or a steady-flow (open) system.

3.19 CARNOT HEAT PUMP

A Carnot heat pump is a heat pump that operates on a reversed Carnot cycle.

3.20 CARNOT PRINCIPLES

Carnot principles are two conclusions pertaining to the thermal efficiency of reversible (ideal) and irreversible (actual) heat engines, drawn from the Kelvin–Plank and Clausius statements of the second law of thermodynamics. The Carnot principles are as follows:

1. The efficiency of an irreversible heat engine is always less than the efficiency of a reversible one operating between the same two reservoirs.
2. The efficiencies of all reversible heat engines operating between the same two reservoirs are the same.

3.21 CARNOT REFRIGERATOR

A Carnot refrigerator is a refrigerator that operates on a reversed Carnot cycle.

3.22 CARNOT'S THEOREM

Carnot's theorem, also known as Carnot's rule, or the Carnot principle, can be stated as follows: No heat engine operating between two heat reservoirs can be more efficient than a reversible heat engine operating between the same two reservoirs.

3.23 CASCADE FLOW

With the cascade system, the level sensor provides the feedback to the outer loop controller, which then gives an output that provides the set point input to the second controller, this being used to control the rate of flow of the liquid.

3.24 CAUCHY–RIEMANN EQUATIONS

For irrotational flows (the fluid elements in the flow field are free of angular motion), there exists a function ϕ called *velocity potential or potential function*. For two-dimensional flows, ϕ must be a function of x, y and t. The velocity components are given by

$$V_x = \partial\phi/\partial x \quad V_y = \partial\phi/\partial y$$

From these equations, we can write

$$\partial\psi/\partial y = \partial\phi/\partial x, \quad \partial\psi/\partial x = -\partial\phi/\partial y$$

These relations between stream function and potential function are the famous Cauchy–Riemann equations of complex-variable theory. It can be shown that the lines of constant ϕ or potential lines form a family of curves, which intersect the streamlines in such a manner as to have the tangents of the respective curves always at right angles at the point of intersection. Hence, the two sets of curves given by ψ=constant and ϕ=constant form an orthogonal grid system or flow net.

3.25 CAVITATION

Cavitation is a phenomenon in which the static pressure of the liquid reduces to below the liquid's vapour pressure, leading to the formation of small vapour-filled cavities in the liquid.

A cavitation tunnel is used to investigate propellers. This is a vertical water circuit with large diameter pipes. At the top, it carries the measuring facilities.

3.26 CEILING

Ceiling is the flying altitude at which there is only one possible speed for level flight and the rate of climb is zero. At the ceiling altitude, the engine does not have any extra power, that is, the power available is fully used. Thus, the ceiling is of little use for practical purposes, and therefore the idea of a service ceiling is introduced. The service ceiling is the height at which the rate of climb becomes less than 0.5 m/s or some other specified rate.

With respect to aircraft performance, a ceiling is the maximum density altitude an aircraft can reach under a set of conditions, as determined by its flight envelope.

3.27 CENTRE OF PRESSURE

The point from the leading edge of the aerofoil at which the resultant pressure acts is called the centre of pressure. In other words, centre of pressure is the point where line of action of the lift L meets the chord. Thus, the position of the centre of pressure depends on the particular choice of chord. 'Centre of pressure' is the point at which the pressure distribution can be considered to act analogous to the 'centre of gravity' as the point at which the force of gravity can be considered to act. The aerodynamic force acts along a line whose intersection with the chord line is called the centre of pressure cp of the aerofoil, as shown in Figure 3.3.

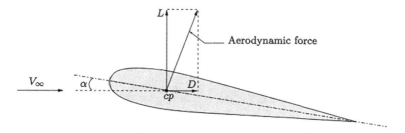

Figure 3.3 Aerodynamic force on an aerofoil.

3.28 CENTRE OF PRESSURE COEFFICIENT

The centre of pressure coefficient is defined as the ratio of the centre of pressure from the leading edge of the chord to the length of chord.

3.29 CENTRELINE DECAY

The jet centreline pitot pressure decay is a measure of jet mixing.

3.30 CENTRIFUGAL FORCE

The body will cause a reaction on whatever makes it travel on a curved path. This reaction is termed centrifugal force.

3.31 CENTRIFUGAL PUMPS

A centrifugal pump is a mechanical device designed to move a fluid by means of the transfer of rotational energy from one or more driven rotors, called impellers. The action of the impeller increases the fluid's velocity and pressure and also directs it towards the pump outlet.

Radial-flow and mixed-flow machines are commonly referred to as centrifugal pumps.

3.32 CENTRIPETAL FORCE

For a body to move on a curved path, it is necessary to supply a force towards the centre of the curved path, this force being directly proportional to the acceleration required. Such a force is called the centripetal force.

3.33 CHANNEL

A passage for water or other liquids to flow along or a part of a river or other area.

3.34 CHAOTIC ADVECTION

Chaotic advection is the complex behaviour a passive scalar –a fluid particle or a passively advected quantity such as temperature or concentration of a second tracer fluid – can attain, driven by the Lagrangian dynamics of the flow.

3.35 CHAOTIC FLOW

Flows with properties that are neither constant in time nor presenting any regular periodicity are normally referred to as chaotic. It is also random, dissipative, and multiple-scaled in time and space. It is a complex system of infinite degrees of freedom.

3.36 CHAOTIC MIXING

In chaos theory and fluid dynamics, chaotic mixing is a process by which flow tracers develop into complex fractals under the action of a fluid flow. The flow is characterised by an exponential growth of fluid filaments.

3.37 CHARACTERISTICS

The Mach lines that introduce an infinitesimal but finite change to flow properties when a flow passes through them are referred to as characteristics, which are not physical unlike the Mach lines and Mach waves. But the mathematical concept of characteristics (taken as identical to the Mach lines), even though not physical, forms the basis for the numerical method termed method of characteristics used to design contoured nozzles to generate uniform and unidirectional supersonic flows.

3.38 CHARACTERISTIC DECAY

For subsonic jets, after the core, the axial velocity decreases continuously, owing to the mixing process. This decay is rapid and is inversely proportional to the axial distance when the jet Mach number is subsonic. This is popularly called characteristic decay.

3.39 CHARACTERISTIC MACH NUMBER

The ratio between local flow speed and critical speed of sound is called the characteristic Mach number.

3.40 CHARLES' LAW

Charles' law states that at constant pressure, the volume of a given mass of gas varies directly as its absolute temperature.

3.41 CHEMICAL COATING

Chemical coating is used to visualise flow with speeds in the range from 40 to 150 m/s. In boundary layer flow, the transition of the flow from laminar to turbulent nature, and so on are usually described by this visualisation technique.

3.42 CHEMICAL ENERGY

The internal energy associated with the bonds in a molecule is called chemical energy.

3.43 CHEMICAL EQUILIBRIUM

A system will be in chemical equilibrium if its chemical composition does not change with time, that is, no net chemical reaction occurs within the system.

3.44 CHOKING

Choking is a limiting flow condition of maximum mass flow rate through a passage. At the choked condition, the mass flow reaches a maximum value and the flow is said to be *choked*, and the flow inside passage, up to the location where the mass flow per unit mass is maximum, is said to be *frozen*.

3.45 CHORD OF AN AEROFOIL

For a cylindrical aerofoil (that is, a wing for which the profiles are the same at every location along the span), the chord of the aerofoil is taken to be the chord of the profile in which the plane of symmetry cuts the aerofoil. In all other cases, the chord of the aerofoil is defined as the mean or average chord located in the plane of symmetry.

3.46 CHORD LINE

The chord line is defined as the shortest (straight) line connecting the leading and trailing edges.

3.47 CHROMATOGRAPHY

Chromatography is a laboratory technique for the separation of a mixture. The mixture is dissolved in a fluid (gas, solvent, water,...) called the mobile phase, which carries it through a system (a column, capillary tube, plate or sheet) on which is fixed a material called the stationary phase.

3.48 CHROMEL

Chromel is an alloy made of ~90% nickel and 10% chromium by weight that is used to make the positive conductors of ANSITypeE (chromel-constantan) and K (chromel-alumel) thermocouples. It can be used at temperatures up to 1,100°C in oxidising atmospheres.

3.49 CHROMEL-ALUMEL THERMOCOUPLE

Chromel-alumel thermocouples are used for temperatures below 1,000°C.

3.50 CHURN

To move, or to make water, mud, etc. move around violently.

3.51 CHURN FLOW

Churn flow, also referred to as froth flow, is a highly disturbed flow of two-phase fluid flow.

3.52 CIDER

Cider is a low alcoholic beverage made from apple juice by alcoholic fermentation.

3.53 CIRCULAR MOTION

Circular motion is a movement of an object along the circumference of a circle or rotation along a circular path. It can be uniform with a constant angular rate of rotation and constant speed or non-uniform with a changing rate of rotation.

3.54 CIRCULAR VORTEX

A circular vortex is that with the shape of its cross section normal to its axis of rotation as circular. For example, a single cylindrical vortex tube (Figure 3.4) whose cross section is a circle of radius 'a', surrounded by unbounded fluid is a circular vortex.

3.55 CIRCULATION

Circulation is an abstract quantity defined as the line integral of the velocity vector between any two points in a flow field. By definition, the circulation is given as

$$\Gamma = \oint_c V \cdot dl$$

where dl is an elemental length, and the loop through the integral sign signifies that the contour is closed. That is, circulation is the line integral of a vector field around a closed plane curve in a flow field.

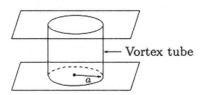

Figure 3.4 Circular vortex.

Circulation implies a component of rotation of flow in the system. The vorticity ζ at a point equals the circulation per unit area, that is,

$$\zeta = \Gamma / A$$

In vector form, ζ becomes

$$\zeta = \nabla \times V = \mathit{curl} \ V$$

For a two-dimensional flow in xy-plane, ζ becomes

$$\zeta_z = \frac{\partial V_y}{\partial x} - \frac{\partial V_x}{\partial y}$$

If the vorticity components are zero, the flow is known as *irrotational flow*. Inviscid flows are essentially irrotational flows.

3.56 CITRIC ACID

Citric acid is an organic compound usually encountered as a white solid. It is a weak organic acid. It occurs naturally in citrus fruits.

If you've ever sunk your teeth into a lemon, you've tasted citric acid. Manufacturers add a manufactured version of it to processed foods.

3.57 CLAPEYRON EQUATION

It is a thermodynamic relation that enables the determination of enthalpy change associated with a phase change, such as the enthalpy of vaporisation h_{fg}, from knowledge of pressure, specific volume and temperature data alone.

3.58 CLASSICAL HYDRODYNAMICS

Through the concept of an 'ideal fluid', mathematical physicists developed the theoretical science known as classical hydrodynamics.

3.59 CLASSICAL THERMODYNAMICS

A macroscopic approach to the study of thermodynamics that does not require knowledge of the behaviour of the individual particles of the substance is called classical thermodynamics.

3.60 CLASSIFICATION OF AUXILIARY LIFT DEVICES

These auxiliary lift devices are broadly classified into the following two categories:

1. Those that alter the geometry of the aerofoil.
2. Those that control the behaviour of the boundary layer over the aerofoil.

3.61 CLASSIFICATION OF HYPERSONIC WIND TUNNELS

Based on run time, hypersonic wind tunnels are classified into the following three categories.

Impulse facilities, which have run times of about 1 s or less.

Intermittent tunnels (blowdown or indraft), which have run times of a few seconds to several minutes.

Continuous tunnels, which can operate for hours (this is only of theoretical interest since continuous hypersonic tunnels are extremely expensive to build and operate).

3.62 CLASSIFICATION OF PRESSURE MEASURING DEVICES

The pressure measuring devices meant for measurements in fluid flow may broadly be grouped into manometers and pressure transducers.

3.63 CLASSIFICATION OF WIND TUNNEL BALANCE

Based on the constructional details, the wind tunnel balances are broadly classified into Wire-type balance, Strut-type balance, Platform-type balance, Yoke-type balance and Strain gauge-type balance.

3.64 CLAUSIUS STATEMENT OF THE SECOND LAW IS THERMODYNAMICS

The Clausius statement of the second law is that "It is impossible to construct a device that operates in a cycle and produces no effect other than the transfer of heat from a lower temperature body to a higher temperature body".

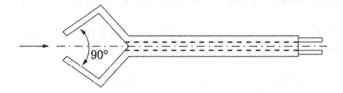

Figure 3.5 Claw yaw probe.

3.65 CLAW YAW PROBE

It is used to measure the rotation and direction of flow near a model at any point because of its slender nature and low blockage. A typical claw yaw probe is shown in Figure 3.5.

3.66 CLAY

Clay is a type of fine-grained natural soil material containing clay minerals. Clay develops plasticity when wet due to a molecular film of water surrounding the clay particles, but becomes hard, brittle and nonplastic upon drying or firing.

3.67 CLOSED-CIRCUIT TUNNEL

Closed-circuit or return-flow tunnel is a tunnel with a continuous path for the air, as illustrated in Figure 3.6.

3.68 CLOSED-THROAT TUNNEL

If rigid walls bind the test section, the tunnel is called a closed-throat tunnel.

3.69 CLOSED SYSTEM

A closed system consists of a fixed amount of mass (say a gas of certain amount of mass) and no mass can cross its boundary. A closed system is also referred to as control mass.

3.70 CLOUD

In meteorology, a cloud is an aerosol consisting of a visible mass of minute liquid droplets, frozen crystals or other particles suspended in the

Fan and motor

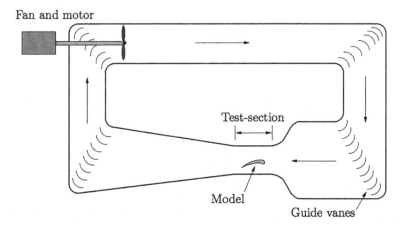

Figure 3.6 Closed-circuit wind tunnel.

atmosphere of a planetary body or similar space. Water or various other chemicals may compose the droplets and crystals.

3.71 CLOUDBURST

Cloudburst is a sudden, very heavy rainfall, usually local in nature and of brief duration.

3.72 COAL

Coal is a combustible black or brownish-black sedimentary rock, formed as rock strata called coal seams. Coal is mostly carbon with variable amounts of other elements, chiefly hydrogen, sulphur, oxygen and nitrogen.

3.73 COANDA EFFECT

The Coanda effect is the tendency of a fluid jet to stay attached to a convex surface.

3.74 COASTAL FLOOD

Coastal flooding normally occurs when dry and low-lying land is submerged by seawater.

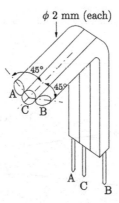

Figure 3.7 Cobra probe.

3.75 COAXIAL

In geometry, coaxial means that several three-dimensional linear or planar forms sharing a common axis. The two-dimensional analogue is concentric.

3.76 COBRA PROBE

This is a two-dimensional probe similar to a three-hole yaw probe, as shown in Figure 3.7.

The three- and five-hole probes cannot be used to get flow direction near any solid surface because of their large size. Therefore, based on the same principle, probes that are small in size can be fabricated for measurements near the wall region of any model. The cobra probe is one such special probe.

3.77 COCONUT OIL

Coconut oil is an edible oil derived from the wick, meat and milk of the coconut palm fruit. Coconut oil is a white solid fat, melting at warmer room temperatures of around 25°C. In warmer climates during the summer months, it is a clear thin liquid oil. Unrefined varieties have a distinct coconut aroma.

3.78 COEFFICIENT OF PERFORMANCE

The efficiency of a refrigerator is expressed in terms of the coefficient of performance (COP). COP can be greater than unity.

3.79 COEFFICIENT OF (CUBIC) THERMAL EXPANSION

Coefficient of (cubic) thermal expansion, α, is defined as

$$\alpha = \frac{1}{V} \left(\frac{\partial V}{\partial T} \right)_p$$

where V is the volume, T is the temperature and p is the pressure.

With this expression, one can, in principle, determine the enthalpy if c_p and V are known as the functions of p and T.

3.80 COEFFICIENT OF VOLUME EXPANSION

Coefficient of volume expansion is an indication of the change in volume that results from a change in temperature while the pressure remains constant.

3.81 COFFEE

Coffee is a brewed drink prepared from roasted coffee beans – the seeds of berries from certain coffee species.

Coffee is the biggest source of antioxidants in the diet. It has many health benefits, such as improved brain function.

3.82 COGENERATION

Cogeneration is the production of more than one useful form of energy from the same energy source.

3.83 COLLOID

A colloid is an intermediate between solution and suspension. It has particles with sizes between 2 and 1,000 nm.

3.84 COLLOIDAL SOLUTIONS

Colloidal solutions, or colloidal suspensions, are nothing but a mixture in which the substances are regularly suspended in a fluid.

3.85 COMBINED GAS-VAPOUR CYCLE

The combined gas-vapour cycle is a modified cycle involving a gas power cycle topping a vapour power cycle. It takes advantage of the gas-turbine cycle at high temperatures and uses the high-temperature exhaust gases as the energy source for bottoming a cycle such as a steam power cycle.

3.86 COMBUSTION

Combustion is a chemical process in which a substance reacts rapidly with oxygen and gives off heat. The original substance is called the fuel and the source of oxygen is called the oxidiser. The fuel can be a solid, liquid or gas, although for airplane propulsion, the fuel is usually a liquid.

3.87 COMBUSTION CHAMBER

A combustion chamber is part of an internal combustion engine in which the fuel/air mix is burned.

3.88 COMBUSTION TEMPERATURE OF CHARCOAL

The maximum combustion temperature of charcoal (forced draft) is $1,390°C$.

3.89 COMBUSTION TEMPERATURE OF COAL

High-volatile coal $670°C$, medium-volatile coal $795°C$, anthracite $930°C$.

3.90 COMBUSTION TEMPERATURE OF GASOLINE

The maximum combustion temperature of gasoline is $1,026°C$.

3.91 COMBUSTION TEMPERATURE OF JET FUEL

The optimum adiabatic combustion temperature for grate combustion is $1,300°C–1,400°C$ to achieve a reasonable trade-off between NO_x and CO, although actual combustion temperatures of course are somewhat lower due to radiation heat loss to the furnace walls.

3.92 COMBUSTION TEMPERATURE OF KEROSENE

The maximum combustion temperature of kerosene is 990°C.

3.93 COMBUSTION TEMPERATURE OF WOOD

The maximum combustion temperature of wood is 1,027°C.

3.94 COMBUSTOR

In the combustor, the air is mixed with fuel and then ignited. There are as many as 20 nozzles to spray fuel into the airstream. The mixture of air and fuel catches fire. This provides a high-temperature, high-energy air-flow. The fuel burns with the oxygen in the compressed air, producing hot expanding gases. The inside of the combustor is often made of ceramic materials to provide a heat-resistant chamber. The heat of the combustion product can reach about 2,700°C.

A combustor is a component or area of a gas-turbine, ramjet or scramjet engine where combustion takes place. It is also known as a burner, combustion chamber or flame holder. In a gas-turbine engine, the combustor or combustion chamber is fed high-pressure air by the compression system. The combustor then heats this air at constant pressure. After heating, air passes from the combustor through the nozzle guide vanes to the turbine. In the case of a ramjet or scramjet engine, the air is directly fed to the nozzle.

3.95 COMPLEX NUMBER

A complex number may be defined as a number consisting of a sum of real and imaginary parts.

3.96 COMPONENTS OF A WIND TUNNEL

All modern wind tunnels have four important components: the effuser, working or test section, diffuser, and driving unit.

3.97 COMPOUND

A compound is a material formed by chemically bonding two or more chemical elements. The type of bond keeping elements in a compound together may vary; covalent bonds and ionic bonds are two common types. The elements are always present in fixed ratios in any compound.

3.98 COMPOUND HELICOPTER

This has the novel features of a conventional helicopter but also has small wings and a separate engine or engines to provide forward speed directly. At high speed, the wings provide most of the lift, allowing the rotor to be rotated relatively slowly or even feathered.

3.99 COMPOUND PENDULUM

A compound pendulum has an extended mass, like a swinging bar, and is free to oscillate about a horizontal axis. A special reversible compound pendulum called Kater's pendulum is designed to measure the value of g, the acceleration due to gravity.

3.100 COMPOUND VORTEX

In the free vortex, $V = c/r$ and thus, theoretically, the velocity becomes infinite at the centre (Figure 3.8). The velocities near the axis would be very high and, skin friction losses vary as the square of the velocity, they will cease to be negligible. Also, the assumption that the total head H remains constant will cease to be true. The portion of fluid around the axis tends to rotate as a solid body. Thus, the central portion essentially forms a forced vortex. The free surface profile of such a compound vortex and the pressure variation with radius on any horizontal plane in the vortex is shown in Figure 3.8.

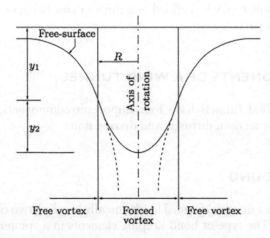

Figure 3.8 Compound vortex.

3.101 COMPRESSED AIR

Compressed air is the air kept under a pressure that is greater than the atmospheric pressure.

3.102 COMPRESSED LIQUID

A liquid at a state where it is not about to vapourise is called a compressed liquid or a supercooled liquid. For example, water, say, at 1 atm and 20°C is a supercooled liquid.

3.103 COMPRESSED NATURAL GAS

Compressed natural gas is a fuel gas made of natural gas, which is mainly composed of methane, compressed to <1% of the volume it occupies at standard atmospheric pressure.

3.104 COMPRESSIBLE AERODYNAMICS

According to the theory of aerodynamics, a flow is considered to be compressible if its change in density with respect to pressure is more than 5%. This means that – unlike incompressible flow – changes in density must be considered. In general, this is the case where the Mach number in part or all of the flow exceeds 0.3. The Mach 0.3 value is rather arbitrary, but it is used because gas flows with a Mach number below 0.3 and demonstrates the changes in density with respect to the change in pressure of <5%. Furthermore, a maximum of 5% density change occurs at the stagnation point of an object immersed in the gas flow, and the density changes around the rest of the object will be significantly lower. Transonic, supersonic and hypersonic flows are all compressible.

3.105 COMPRESSIBLE BERNOULLI'S EQUATION

The compressible form of Bernoulli's equation for inviscid flows is

$$\frac{V^2}{2} + \int \frac{1}{\rho}\frac{\partial p}{\partial s}ds = \text{constant}$$

If ρ is expressible as a function of p only, that is, $\rho = \rho(p)$, the second expression is integrable.

Compressible Bernoulli's equation for isentropic flows is

$$\frac{\gamma}{\gamma-1}\frac{p}{\rho}+\frac{V^2}{2} = \text{constant}$$

This is the form of energy equation commonly used in gas dynamics.

3.106 COMPRESSIBLE FLOW

Compressible flow is defined as a variable density flow; this is in contrast to incompressible flow, where the density is assumed to be invariant. Usually, flows with a Mach number of <0.3 are treated as constant density (incompressible) flows.

Compressible flow is the science of fluid flow in which the density change associated with pressure change is appreciable.

Compressible flow is a flow with significant compressibility.

It is widely accepted that compressibility can be neglected when

$$\frac{\Delta\rho}{\rho_i} \leq 0.05$$

where $\Delta\rho$ is the change in density and ρ_i is the initial density.

3.107 COMPRESSIBLE FLOW REGIMES

The compressible flow can be subdivided into different zones based on the flow velocity and the speed of sound. To do this classification, we can make use of the energy equation as follows.

Consider a streamtube in a steady compressible flow in which the flow does not exchange heat with the fluid in the neighbouring streamtubes. The steady-flow energy equation for the adiabatic flow through such a streamtube is

$$h+\frac{V^2}{2} = h_0$$

For a perfect gas, $h = c_p\,T$. Therefore,

$$c_p\,T+\frac{V^2}{2} = c_p\,T_0$$

$$\frac{\gamma}{\gamma-1}RT+\frac{V^2}{2} = \frac{\gamma}{\gamma-1}\,RT_0$$

$$V^2 + \frac{2}{\gamma - 1} a^2 = \frac{2}{\gamma - 1} a_0^2 = V_{\max}^2$$

since $a = \sqrt{\gamma R T}$ and $c_p = \gamma/(\gamma - 1) R$, where a is the local speed of sound and a_0 is the speed of sound at the stagnation state (where $V=0$) and $V_{\max}$ is the maximum possible flow velocity in the fluid (where the absolute temperature is zero). The above equation represents an ellipse and is called as an adiabatic *steady-flow ellipse*.

Different realms of compressible flow having significantly different physical characteristics are the following.

- **Incompressible flow** is the flow in which the flow velocity, V is small compared to the speed of sound, a in the flow medium. The changes in a are very small compared with the changes in V.
- **Compressible subsonic flow** is that with flow velocity and speed of sound of comparable magnitude, but $V<a$. A change in the flow Mach number M is mainly due to changes in V.
- **Transonic flow** is that in which the difference between the flow velocity and speed of sound is small compared to either V or a. Changes in V and a are of comparable magnitude.
- **Supersonic flow** is that where flow velocity and speed of sound are of comparable magnitude, but $V > a$. Changes in M take place through substantial variation in both V and a.
- **Hypersonic flow** is that where the flow velocity is very large compared with the speed of sound. Changes in flow velocity are very small, and thus, variations in Mach number are almost exclusively due to changes in the speed of sound a.

3.108 COMPRESSIBLE FLOW PRESSURE COEFFICIENT

For compressible flows, the pressure coefficient, which is a dimensionless pressure difference, can be expressed as

$$C_p = \frac{2}{\gamma M_\infty^2} \left(\frac{p}{p_\infty} - 1 \right)$$

where γ is the specific heat ratio, M_∞ and p_∞ are the freestream Mach number and static pressure, respectively, and p is the local static pressure.

3.109 COMPRESSIBLE JETS

Jets with a Mach number of more than 0.3 are termed compressible jets.

3.110 COMPRESSIBLE SUBSONIC FLOW

Compressible subsonic flow is the flow in which the flow velocity and the speed of sound are of comparable magnitude, but $V < a$. The changes in the flow Mach number M are mainly due to the changes in V.

3.111 COMPRESSIBILITY

Compressibility is a phenomenon by virtue of which the flow changes its density with a change in speed. Compressibility may also be defined as the volume modulus of pressure.

The change in volume of a fluid associated with change in pressure is called compressibility. When a fluid is subjected to pressure, it gets compressed and its volume changes. Bulk modulus of elasticity is a measure of how easily the fluid may be compressed and is defined as the ratio of pressure change to the volumetric strain associated with it. The bulk modulus of elasticity K is given by

$$K = \frac{\text{Pressure increament}}{\text{Volume strain}} = -V \frac{dp}{dV}$$

In thermodynamics and fluid mechanics, the compressibility (also known as the coefficient of compressibility or, if the temperature is held constant, the isothermal compressibility) is a measure of the relative volume change of a fluid or solid as a response to a pressure (or mean stress) change.

3.112 COMPRESSIBILITY FACTOR

Compressibility factor Z may be viewed as a measure of deviation from ideal-gas behaviour. It is defined as

$$Z = \frac{pv}{RT}$$

It may also be expressed as

$$Z = \frac{v_{\text{actual}}}{v_{\text{ideal}}}$$

where $v_{\text{ideal}} = RT/p$. That is, $Z = 1$ for ideal gases. For real gases, Z can be greater than or less than unity.

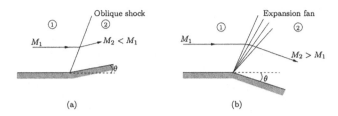

Figure 3.9 Supersonic flow over (a) compression corner and (b) expansion corner.

3.113 COMPRESSIBILITY ISSUES

At low speeds, the compressibility of air is not significant in relation to aircraft design, but as the airflow nears and exceeds the speed of sound, a host of new aerodynamic effects become important in the design of aircraft. These effects, often several of them at a time, made it very difficult for the World War II–era aircrafts to reach speeds beyond 800 km/h.

3.114 COMPRESSION AND EXPANSION CORNERS

The corner (Figure 3.9a) that turns the flow into itself is called the compression or concave corner. In contrast, in an expansion or convex corner, the flow is turned away from itself through an expansion fan, as illustrated in Figure 3.9b. All the streamlines are deflected to the same angle θ after the expansion fan, resulting in a uniform parallel flow downstream of the fan. Across the expansion wave, the Mach number increases, and the pressure, density and temperature decrease.

From Figure 3.9, it is seen that the flow turns suddenly across the shock and the turning is gradual across the expansion fan and hence all flow properties through the expansion fan change smoothly, except for the wall streamline, which changes suddenly.

3.115 COMPRESSION IGNITION ENGINE

If the combustion is by self-ignition as a result of compression of the air-fuel mixture above its self-ignition temperature, the engine is called a compression ignition (CI) engine.

3.116 COMPRESSION RATIO

The ratio of the maximum volume formed in the cylinder to the minimum volume is called the compression ratio of the engine.

3.117 COMPRESSOR

A compressor is a mechanical device that increases the pressure of a gas by reducing its volume. Compressors are similar to pumps – both increase the pressure on a fluid and both can transport the fluid through a pipe. As gases are compressible, the compressor also reduces the volume of a gas.

3.118 COMPUTATIONAL FLUID DYNAMICS

Computational fluid dynamics (CFD) is essentially the numerical solutions of the equations of motion that describe the main governing equations, namely the continuity, momentum and energy equations.

The term CFD implies the integration of two disciplines, namely, fluid dynamics and computation.

In a CFD code developed for solving flow fields, approximations are made both in modelling fluid dynamic phenomena, for example, turbulence, and in the numerical formulation, for example, developing grids to define the configurations and the flow field points where dependent variables are to be computed.

3.119 CONCEPT OF LOST WORK

The irreversibility associated with a thermodynamic process can also be quantified with the concept known as lost work. By definition, lost work W_{lost} is

$$W_{lost} = T_R S_{gen}$$

where T_R is a reference temperature. Since S_{gen} is always positive, lost work is always positive for all real processes.

3.120 CONCEPTUAL

The means related to ideas and concepts formed in the mind is called conceptual.

3.121 CONDENSATION

Condensation is the change of the physical state of matter from the gas phase into the liquid phase and is the reverse of vaporisation.

3.122 CONDENSATION TEMPERATURE

The condensation temperature is that at which a given gas-phase constituent condenses into a liquid.

3.123 CONDENSER MICROPHONES

Condenser microphones are capacitance transducers. They are the most commonly used transducers for dynamic pressure measurements. They measure the change in capacitance in a small air gap between two electrically charged metallic surfaces. One of these surfaces is rigid and the other is a deformable diaphragm or membrane subjected to the source pressure fluctuations.

3.124 CONDUCTION

Conduction is the process by which heat energy is transmitted through collisions between neighbouring atoms or molecules.

3.125 CONFORMAL TRANSFORMATION

The transformation technique that transforms an orthogonal geometric pattern composed of elements of a certain shape into an entirely different pattern, while the elements retain their form and proportion is termed conformal transformation.

3.126 CONING ANGLE

It is the angular difference between the feathering axis and the plane of rotation, as shown in Figure 3.10. It may also be defined as the angular difference between the feathering axis and the tip path plane.

3.127 CONJUGATE

Two complex numbers, which differ from the sign of i, are said to be conjugate.

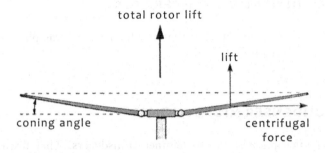

Figure 3.10 Coning angle.

3.128 CONSERVATION OF ENERGY

Although energy can be converted from one form to another, the total energy in a given closed system remains constant:

$$\rho \frac{Dh}{Dt} = \frac{Dp}{Dt} + \nabla \cdot (k \nabla T) + \Phi$$

where h is the enthalpy, k is the thermal conductivity of the fluid, T is the temperature and Φ is the viscous dissipation function. The viscous dissipation function governs the rate at which mechanical energy of the flow is converted to heat. This term is always positive since according to the second law of thermodynamics, viscosity cannot add energy to the control volume. The expression on the left-hand side is a material derivative.

The first law of thermodynamics states that "during an interaction between a system and its surroundings, the amount of energy gained by the system must be exactly equal to the amount of energy lost by the surroundings".

3.129 CONSERVATION OF ENERGY PRINCIPLE

We know that energy can neither be created nor destroyed. It can only change forms. This principle is called the first law of thermodynamics or the conservation of energy principle.

3.130 CONSERVATION OF ENTROPY PRINCIPLE

Entropy is a non-conserved property and there is no such thing as the conservation of entropy principle.

The performance of engineering systems is degraded by the presence of irreversibilities and the entropy generation is a measure of the magnitudes of the irreversibilities present during that process.

3.131 CONSERVATION LAWS

Aerodynamic problems are normally solved using conservation of mass, momentum, and energy, referred to as continuity, momentum, and energy equations. The conservation laws can be written in integral or differential forms.

3.132 CONSERVATION OF MASS

If a certain mass of fluid enters a volume, it must either exit the volume or change the mass inside the volume. In fluid dynamics, the continuity equation is analogous to Kirchhoff's current law (that is, 'the sum of the currents flowing into a point in a circuit is equal to the sum of the currents flowing out of that same point') in electric circuits. The differential form of the continuity equation is

$$\frac{\partial \rho}{\partial t} + \nabla \cdot (\rho u) = 0$$

where ρ is the fluid density, u is a velocity vector and t is the time. Physically, the equation also shows that mass is neither created nor destroyed in the control volume.

3.133 CONSERVATION OF MOMENTUM

The momentum equation applies Newton's second law of motion to a control volume in a flow field, whereby force is equal to the time derivative of momentum. Both surface and body forces are accounted for in this equation.

3.134 CONSERVATIVE SYSTEM

A mechanical system in which the law of conservation of mechanical energy is valid, that is, the sum of the kinetic energy, KE, and the potential energy, PE, of the system is constant: KE + PE = constant.

3.135 CONSTANT CURRENT ANEMOMETER

Constant current anemometer is the one that supplies a constant heating current to the sensor. The variation in the sensor resistance with fluid flow velocity change causes voltage drop variation across the sensor.

3.136 CONSTANT TEMPERATURE ANEMOMETER

Constant temperature anemometer is one that supplies the sensor with a heating current that varies with the fluid velocity to maintain constant sensor resistance, and thus, constant temperature.

3.137 CONSTANT-VOLUME GAS THERMOMETER

This is a device to measure absolute temperature. It consists of a sealed, rigid tank containing gas at a low pressure. The measurement of absolute temperature with constant-volume gas thermometer is based on the principle that the temperature of a gas at low pressure is proportional to its pressure. It can be determined by extrapolation that as the absolute pressure in a constant-volume gas thermometer approaches zero, the absolute temperature reading will also approach zero.

3.138 CONSTANTAN

Constantan is nickel- and copper-based alloy wire that has a high resistivity and is mainly used for thermocouples and electrical resistance heating.

3.139 CONTACT SURFACE

A contact surface may also be idealised as a surface of discontinuity. It can either be stationary or moving. Unlike the shock wave, there is no flow of matter across the contact surface. In the literature, we can find this contact surface being referred to by different names: material boundary, entropy discontinuity, slipstream or slip surface, vortex sheet and tangential discontinuity. It is essential to note that the contact surface is a fluid boundary across which there is no mass transport. Further, the surface can tolerate temperature and density gradients, but not pressure gradients. In other words, the temperature and density on either side of the slipstream can be different but the pressure on both sides must be equal. This may also be stated as follows: The contact surface can tolerate thermal and concentration imbalance but not pressure imbalance.

3.140 CONTINUITY CONCEPT

The foundation of aerodynamic prediction is the continuity assumption. In reality, gases are composed of molecules that collide with one another and solid objects. To derive the equations of aerodynamics, fluid properties such as density and velocity are assumed to be well defined at infinitely small points and to vary continuously from one point to another. That is, the discrete molecular nature of a gas is ignored. The continuity assumption becomes less valid as a gas becomes more rarefied. In these cases, statistical mechanics is a more valid method of solving the problem than continuum aerodynamics. The Knudsen number can be used to guide the choice between statistical mechanics and the continuum formulation of aerodynamics.

3.141 CONTINUITY EQUATION

The continuity equation is essentially a mass balance relation. It represents that the mass within a control volume plus the net outflow of the mass through the surface surrounding the control volume is equal to the rate at which mass is produced in the control volume.

It is statement of mass conservation. In general form, it is

$$\Delta \cdot (\rho V) = 0$$

where $\Delta = i\dfrac{\partial}{\partial x} + j\dfrac{\partial}{\partial y} + k\dfrac{\partial}{\partial z}$, ρ is the flow density, and $V = iu + jv + kw$ is the flow velocity, with u, v and w as the components along the x, y and z directions, respectively. For incompressible flows, the continuity equation becomes

$$\Delta \cdot V = 0$$

in differential form, this becomes

$$\frac{\partial u}{\partial x} + \frac{\partial v}{\partial y} + \frac{\partial w}{\partial z} = 0$$

For compressible flows, the continuity takes the form

$$\frac{\partial(\rho u)}{\partial x} + \frac{\partial(\rho v)}{\partial y} + \frac{\partial(\rho w)}{\partial z} = 0$$

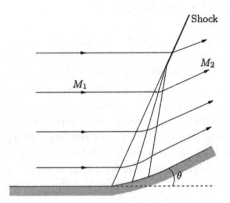

Figure 3.11 Smooth continuous corner.

3.142 CONTINUOUS COMPRESSION

Compression through a large number of weak compression waves, as illustrated in Figure 3.11, is termed continuous compression. These kinds of corners are called continuous compression corners. Thus, the geometry of the corner should have a continuous smooth turning to generate a large number of weak (isentropic) compression waves.

3.143 CONTINUUM

Fluid flows may be modelled on either a macroscopic or microscopic level. The macroscopic model regards the fluid as a *continuum*, and the description is in terms of the variations of macroscopic velocity, density, pressure and temperature with distance and time. On the other hand, the microscopic or molecular model recognises the particulate structure of a fluid as a myriad of discrete molecules and ideally provides information on the position and velocity of every molecule at all times.

The description of a fluid motion essentially involves a study of the behaviour of all the discrete molecules, which constitute the fluid. In liquids, the strong intermolecular cohesive forces make the fluid behave as a continuous mass of the substance and, therefore, these forces need to be analysed by the molecular theory. Under normal conditions of pressure and temperature, even gases have a large number of molecules in unit volume (e.g., under normal conditions, for most gases, the molecular density is 2.7×10^{25} molecules per m^3) and, therefore, they also can be treated as a continuous mass of the substance by considering the average effects of all the molecules within the gas. Such a fluid model is called a continuum.

3.144 CONTROL OF AN AIRCRAFT

The control of an aircraft may broadly be classified as longitudinal control, roll control and directional control. Longitudinal control is provided by the elevators or movement of the whole tailplane. Roll control is provided by the ailerons. Directional control is provided by the rudder.

3.145 CONTROL MASS

A *control mass system* is an identified quantity of matter. It may change its shape, position and thermal condition with time or space or both, but must always entail the same matter.

3.146 CONTROL SURFACE

The boundaries of a control volume are called control surfaces. The control surfaces can be real or imaginary.

3.147 CONTROL SURFACES

The trailing edge flap has another important application. Such flaps are used to vary the lift produced by the various aerofoils of the aircraft (that is, wing, tailplane, and tail fin) in general conditions of flight, so enabling the aircraft to be controlled. When used in this way, they are called flap controls or control surfaces.

3.148 CONTROL VOLUME

Control volume is a definite volume designated in space. The boundary of this volume is known as *control surface*. The amount and identity of the matter in the control volume may change with time, but the shape and size of the control volume are fixed, that is, a control volume may change its position in time or space or both, but its shape and size are always preserved.

3.149 CONVECTION

Convection is mass transfer due to the bulk motion of a fluid. For example, the flow of liquid water transports molecules or ions that are dissolved in the water.
  or

Convection is the mode of heat transfer between a solid surface and the adjacent liquid or gas that is in motion.

or

Convection is a single or multiphase fluid flow that occurs spontaneously due to the combined effects of material property heterogeneity and body forces on a fluid, most commonly density and gravity (see buoyancy). Convection may also take place in soft solids or mixtures where particles can flow.

3.150 CONVECTION EFFECT

When a spatial gradient exists, the fluid motion brings different particles with different values of flow properties to the probe, thereby, modifying the rate of change sensed by the probe. This effect is termed convection effect.

3.151 CONVECTIVE ACCELERATION

The convective acceleration

$$\frac{\partial V}{\partial s}\frac{\partial s}{\partial t} = V\frac{\partial V}{\partial s}$$

is the acceleration between two points in space, that is, change of velocity at a fixed time within space.

3.152 CONVECTIVE HEAT TRANSFER

Convection (or convective heat transfer) is the transfer of heat from one place to another due to the movement of fluid. Although often discussed as a distinct method of heat transfer, convective heat transfer involves the combined processes of conduction (heat diffusion) and advection (heat transfer by bulk fluid flow). Convection is usually the dominant form of heat transfer in liquids and gases.

3.153 CONVECTIVE HEAT TRANSFER COEFFICIENT

The convective heat transfer coefficient, defines, in part, the heat transfer due to convection.

3.154 CONVECTIVE MACH NUMBER

The relative convection speed of the large-scale structures in the shear layer to one of the freestreams normalised by the speed of the sound of that stream is called a convective Mach number.

3.155 CONVECTIVE MASS TRANSFER

The mass transfer associated with significant bulk velocity is termed convective mass transfer.

3.156 CONVERGENT NOZZLE

A convergent nozzle is a nozzle that starts big and gets smaller—a decrease in cross-sectional area, as shown in Figure 3.12. As a fluid enters the smaller cross section, it has to speed up due to the conservation of mass. To maintain a constant amount of fluid moving through the restricted portion of the nozzle, the fluid must move faster.

3.157 CONVERGENT-DIVERGENT NOZZLE

Convergent-divergent (C-D) nozzle is a tube, which is pinched in the middle, making a carefully balanced, asymmetric shape. It is used to accelerate a compressible fluid to supersonic speeds in the axial direction by converting the thermal energy of the flow into kinetic energy.

C-D nozzle can be of two types. One is with a straight wall and the second is with a contoured wall. A straight walled C-D nozzle is capable of delivering only uniform supersonic flow but it will not be unidirectional. But the contoured C-D nozzle, also, referred to as the de Laval nozzle, is capable of delivering uniform as well as unidirectional supersonic flow.

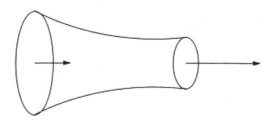

Figure 3.12 Convergent nozzle.

3.158 COOKING OIL

Cooking oil is plant, animal, or synthetic fat used in frying, baking and other types of cooking. It is also used in food preparation and flavouring not involving heat, such as salad dressings and bread dipping like bread dips, and may be called edible oil.

3.159 COOLANT

A coolant is a substance, typically liquid or gas, that is used to reduce or regulate the temperature of a system.

A coolant (also called an antifreeze) is a special fluid that runs through your engine to keep it within its correct operating temperature range. It is made from ethylene glycol or propylene, water and some protection additives and is usually green, blue, or even pink in colour.

3.160 COOLING POND

They are basically large water reservoirs open to the atmosphere to which the waste heat is dumped.

3.161 COPPER-CONSTANTAN THERMOCOUPLE

Copper-constantan thermocouples are suited for measurements in the −200°C to 350°C range.

3.162 CORIOLIS EFFECT

The Coriolis effect describes the pattern of deflection taken by objects not firmly connected to the ground as they travel long distances around the Earth.

3.163 CORIOLIS FORCE

Coriolis force is an inertial or fictitious force that acts on objects that are in motion within a frame of reference that rotates with respect to an inertial frame. In a reference frame with clockwise rotation, the force acts to the left of the motion of the object. In the one with anticlockwise (or counterclockwise) rotation, the force acts to the right. The deflection of an object due to the Coriolis force is called the Coriolis effect.

3.164 CORRECT BANK ANGLE

A bank angle during which the wind will come straight ahead is called a correct bank angle.

3.165 CORRECTLY EXPANDED OPERATION

The backpressure $p_b = p_e$ for which the flow in the nozzle is accelerated throughout the nozzle (both in the convergent and divergent portions), and the nozzle delivers uniform supersonic flow, is termed the design pressure. This is referred to as a correctly expanded operation.

For the correctly expanded condition, the nozzle exit pressure p_e is equal to the backpressure ($p_e = p_b$). Therefore, there is no need for the flow to be compressed or expanded after exiting the nozzle, as in the case of overexpanded or underexpanded states, respectively, to come to equilibrium with the backpressure. This gives the impression that, when the flow is correctly expanded, there are no compression or expansion waves generated at the nozzle exit.

Indeed, we tend to think that the correctly expanded sonic and supersonic flow exiting a nozzle is wave free. But in the actual flow process, it is not true. This is because, even though the flow is correctly expanded, soon after exiting the nozzle, the flow encounters a large space to relax. Therefore, the flow turns away from the nozzle axis in a bid to occupy the space available at the nozzle exit. We know that supersonic flow is essentially wave dominated, and any change in the state (change of p, T, ρ) or mode (direction) of the flow will take place only through these waves. In the case of correctly expanded flow exiting a nozzle, the flow has to turn away from the nozzle axis to occupy the free space available. This leads to the formation of expansion waves at the nozzle exit. It is important to note that the expansion strength in the case of correctly expanded flow is much less than that for an underexpanded flow. Thus, both correctly expanded and underexpanded flows encounter expansion at the nozzle exit.

However, the expansion for a correctly expanded flow is to turn the flow away from the nozzle axis, but for an underexpanded flow, the expansion has to reduce the static pressure and turn the flow away from the nozzle axis. Expansion rays formed at the nozzle exit will travel some distance and be reflected from the jet boundary as compression waves, as illustrated in Figure 3.13.

3.166 CORROSION

Corrosion is defined as the chemical or electrochemical reaction between a material, usually a metal or alloy, and its environment that produces a deterioration of the material and its properties

Figure 3.13 Waves in a correctly expanded Mach 2 jet.

3.167 CORROSIVE GASES

Corrosive gases and vapours are also extremely hazardous. Examples, which can cause severe irritation and bodily injury, include ammonia and hydrogen chloride.

3.168 COSMOLOGY

Cosmology is a branch of astronomy that involves the origin and evolution of the universe from the Big Bang to today and on into the future. According to NASA, the definition of cosmology is 'the scientific study of the large-scale properties of the universe as a whole'.

3.169 COUETTE FLOW

Couette flow is a laminar flow that arises when a viscous material lies between two parallel plates, where one of the plates is in relative motion with the other plate.

or

The laminar flow of a viscous fluid is the space between two cylinders, one of which is rotating relative to the other. The flow is driven by a viscous drag force acting on the fluid because of cylinder rotation and there is no pressure gradient by symmetry. Plane-Couette flow is the limit of small gap width to cylinder radius ratio and is equivalent to the motion produced by the relative tangential motion of two parallel planes.

or

Couette flow is the flow of a viscous fluid in the space between two surfaces, one of which is moving tangentially relative to the other. The relative motion of the surfaces imposes shear stress on the fluid and induces flow.

3.170 CREEPING FLOW

Creeping flow is a fluid flow where advective inertial forces are small compared with viscous forces. The Reynolds number for creeping flow is Re ≪ 1. This is also known as Stokes flow.

Examples of creeping flow include very small objects moving in a fluid, such as the settling of dust particles and the swimming of microorganisms.

3.171 CRITICAL ANGLE OF ATTACK

The critical angle of attack is the angle of attack which produces the maximum lift coefficient. This is also called the 'stall angle of attack'. Below the critical angle of attack, as the angle of attack decreases, the lift coefficient decreases.

3.172 CRITICAL CIRCULATION

The circulation which makes the stagnation points coincide at the surface of the cylinder is called critical circulation.

3.173 CRITICAL MACH NUMBER

It is the freestream Mach number at which flow becomes sonic at some point on the body present in the flow.

3.174 CRITICAL OPERATION

Operation where the shock is just at the inlet tip is called critical operation.

3.175 CRITICAL POINT

Critical point is that point at which the saturated liquid and saturated vapour states are identical.

3.176 CRITICAL PRESSURE

For a pure substance, the critical pressure is defined as the pressure above which liquid and gas cannot coexist at any temperature. The critical temperature for a pure substance is the temperature above which the gas cannot become liquid, regardless of the applied pressure.

3.177 CRITICAL PROPERTIES

The properties of a fluid at a location where Mach number $M=1$ are called critical properties.

3.178 CRITICAL REYNOLDS NUMBER

Critical Reynolds number is that at which the flow field is a mixture of laminar and turbulent flows.

Some of the well-known critical Reynolds numbers are

Pipe flow $Re_d = 2300$: based on mean velocity and diameter d.
Channel flow $Re_h = 1000$ (**two-dimensional**): based on height h and mean velocity.
Boundary layer flow $Re_\theta = 350$: based on freestream velocity and momentum thickness θ.
Circular cylinder $Re_w = 200$ (**turbulent wake**): based on wake width w and wake defect.
Flat plate $Re_x = 5 \times 10^5$: based on length x from the leading edge.
Circular cylinder $Re_d = 1.66 \times 10^5$: based on cylinder diameter d.

3.179 CRITICAL REYNOLDS NUMBER FOR FLAT PLATE

For flow past a flat plate, the transition from laminar to turbulent begins when the critical Reynolds number $(Re_{x,\,cri})$ reaches 5×10^5.

3.180 CRITICAL REYNOLDS NUMBER FOR OPEN CHANNEL FLOW

The critical Reynolds number range for open channel flow is 2,000–3,000.

3.181 CRITICAL REYNOLDS NUMBER FOR PIPE FLOW

The critical Reynolds number for pipe flow is about 2,300.

3.182 CRITICAL STATE

Critical state in a flow field is the state with the local Mach number $M = 1$. It is also called the sonic state.

3.183 CRITICAL TEMPERATURE

The critical temperature of a substance can be defined as the highest temperature at which the substance can exist as a liquid.

3.184 CRITICAL VOLUME

The volume occupied by a unit mass of a gas or vapour in its critical state.

3.185 CROCCO'S THEOREM

$$T \frac{\partial s}{\partial n} = \frac{dh_0}{dn} + V \zeta$$

This is known as Crocco's theorem for two-dimensional flows. From this, it is seen that the rotation depends on the rate of change of entropy and stagnation enthalpy normal to the streamlines.

Crocco's theorem essentially relates entropy gradients to vorticity in steady, frictionless, nonconducting, adiabatic flows. In this form, Crocco's equation shows that if entropy (s) is a constant, the vorticity ζ must be zero. Likewise, if vorticity ζ is zero, the entropy gradient in the direction normal to the streamline (ds/dn) must be zero, implying that the entropy (s) is a constant. That is, isentropic flows are irrotational and irrotational flows are isentropic. This result is true, in general, only for steady flows of inviscid fluids in which no body forces are acting and the stagnation enthalpy is a constant.

3.186 CROSSWIND

A crosswind is any wind that has a perpendicular component to the line or direction of travel.

3.187 CROSSWIND LANDING

A crosswind landing is a landing manoeuvre in which a significant component of the prevailing wind is perpendicular to the runway centre line.

3.188 CRUISE MISSILE

A cruise missile is a guided missile used against terrestrial targets that remains in the atmosphere and flies the major portion of its flight path at an approximately constant speed. Cruise missiles are designed to deliver a large warhead over long distances with high precision.

3.189 CRYOGENICS

In physics, cryogenics is the production and behaviour of materials at very low temperatures.

3.190 CRYOGENIC ENGINE

A cryogenic engine/cryogenic stage is the last stage of space launch vehicles, which makes use of cryogenics. Cryogenics is the study of the production and behaviour of materials at extremely low temperatures (below −150°C) to lift and place heavier objects in space.

3.191 CRYOGENIC FLUIDS

It is generally agreed that cryogenic fluids are those whose boiling points (bp) at atmospheric pressure are about 120 K or lower, although liquid ethylene with its boiling point of 170 K is often included.

3.192 CRYOGENIC FLUID PROPERTIES

All gaseous cryogens are odourless and all liquid cryogens are colourless apart from oxygen, which is pale blue, and fluorine, which is pale yellow. They are all diamagnetic except oxygen, which is quite strongly paramagnetic.

3.193 CRYOGENIC PROPELLANTS

Cryogenic propellants are liquefied gases stored at very low temperatures, most frequently liquid hydrogen (LH_2) as the fuel and liquid oxygen (LO_2 or LOX) as the oxidiser. Hydrogen remains liquid at temperatures of −253°C and oxygen remains in a liquid state at temperatures of −183°C.

3.194 CRYOGENIC TEMPERATURE

The cryogenic temperature range has been defined from −150°C (−238°F) to absolute zero (−273°C or −460°F)— the temperature at which molecular motion comes as close as theoretically possible to ceasing completely.

Figure 3.14 Cup anemometer.

3.195 CUP ANEMOMETER

A cup anemometer is an instrument for measuring wind speed. The four horizontally aligned spherical cups, shown in Figure 3.14, which are fixed crosswise or in a star shape on the vertical axis of the cup anemometer, record the wind speed during a measurement.

3.196 CURD

Curd is a thick soft substance that forms when milk turns sour.

3.197 CURL

It is a cross product of the gradient operator with a vector: $\nabla \times u$ equal to the vorticity when applied to the velocity vector. The vorticity is twice the local angular velocity (rate of rotation) of the fluid.

3.198 CURVILINEAR MOTION

The motion of an object moving in a curved path is called a curvilinear motion. Example: A stone thrown into the air at an angle. Curvilinear motion describes the motion of a moving particle that conforms to a known or fixed curve.

Figure 3.15 Cyclonic storm.

3.199 CUTOFF RATIO

The cutoff ratio is defined as the ratio of the cylinder volumes after and before the combustion process.

3.200 CYCLONE

A cyclone is a general term for a weather system in which winds rotate inwardly to an area of low atmospheric pressure, as shown in Figure 3.15.

Cyclone is any large system of winds that circulates about a centre of low atmospheric pressure in a counterclockwise direction north of the Equator and in a clockwise direction to the south. Cyclonic winds move across nearly all regions of the Earth except the equatorial belt and are generally associated with rain or snow.

3.201 CYCLONE SHELTERS

Cyclone shelters are safe shelters where vulnerable people can seek refuge at the time of cyclone.

3.202 CYCLONIC DEPRESSION

A cyclonic disturbance in which the maximum sustained surface wind speed is between 17 and 33 knots (31 and 61 km/h). If the maximum sustained wind speed lies in the range 28 knots (52 km/h) to 33 knots (61 km/h), the system may be called a 'deep depression'.

3.203 CYLINDRICAL RECTANGULAR AEROFOIL

This is the simplest type, of span $2b$ and chord c, which is constant at all sections. All the sections are similar and similarly situated.

Chapter 4

d'Alembert's Paradox
to Dynamics

4.1 d'ALEMBERT'S PARADOX

The symmetry of the pressure distribution in an irrotational flow implies that 'a steadily moving body experiences no drag'. This result, which is not true for actual (viscous) flows where the body experiences drag, is known as d'Alembert's paradox.

4.2 DALTON'S LAW OF ADDITIVE PRESSURES

The pressure of a gas mixture is equal to the sum of the pressures each gas would exert if it existed alone at the mixture temperature and volume.

4.3 DALTON'S LAW OF PARTIAL PRESSURES

Dalton's law of partial pressures states that the total pressure of a mixture of gases is equal to the sum of the partial pressures of the component gases.

4.4 DARCY FRICTION FACTOR

Darcy friction factor, which is a dimensionless parameter, is defined as

$$f = \frac{8\,\tau_w}{\rho\,V^2}$$

where τ_w is the shear stress at the wall, ρ is the flow density, and V is the flow velocity.

DOI: 10.1201/9781003348405-4

4.5 DARCY NUMBER

In fluid dynamics through porous media, the Darcy number (Da) represents the relative effect of the permeability of the medium versus its cross-sectional area, commonly the diameter squared. The number is named after Henry Darcy and is found by nondimensionalising the differential form of Darcy's Law. This number should not be confused with the Darcy friction factor, which applies to pressure drop in a pipe.

4.6 DARCY RELATION

Darcy's results suggest the formula, termed Darcy relation, is expressed as

$$h_f = \frac{\Delta p}{\rho g} = f\left(\frac{fL}{d}\right)\frac{\bar{u}^2}{2g}$$

where h_f is the head loss due to friction, corresponding (steady flow) to the drop Δp of piezometric pressure over length L of the pipe of diameter d, ρ is the fluid density, $\bar{u}$ is the mean velocity, f is a constant, g is the gravitational acceleration.

4.7 DARCY–WEISBACH EQUATION

It is the expression for head loss, h_f, for flow through ducts of any cross section, expressed as

$$h_f = f\frac{L}{d}\frac{V^2}{2g}$$

where f the friction factor, L is the pipe length, d is the pipe diameter, V is the flow velocity, and g is the gravity acceleration. This equation is valid for both laminar and turbulent flows.

4.8 DEAD STATE

The dead state is the state at which a system is in thermodynamic equilibrium with its surroundings. In this state, the system is at the same T and p as its surroundings. It has no kinetic or potential energy relative to its surroundings, and it does not react with the surroundings.

4.9 DECANTATION

Decantation is a process to separate mixtures by removing a liquid layer that is free of a precipitate, or the solids deposited from a solution.

4.10 DEGREES OF FREEDOM

Degrees of freedom refer to the maximum number of logically independent values, which are values that have the freedom to vary in the data sample.

4.11 DEIONIZED WATER

Deionized (DI) water is the water that has been treated to remove all ions – typically, that means all of the dissolved mineral salts.

4.12 de LAVAL NOZZLE

A nozzle is a passage used to transform pressure energy into kinetic energy. A convergent-divergent nozzle used to generate supersonic flow is sometimes called a de Laval nozzle, after Carl G.P. de Laval, who first used such a configuration in his steam turbines in the late nineteenth century. Therefore, we can say that the de Laval nozzle is the only means to generate supersonic flow. A nozzle that does not have an expanding portion can never produce supersonic flow.

The de Laval nozzle, as shown in Figure 4.1, is a contoured nozzle capable of generating uniform and unidirectional supersonic flow; and the straight convergent-divergent nozzle can generate uniform but not unidirectional supersonic flow.

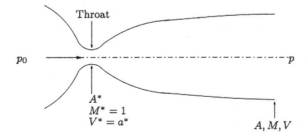

Figure 4.1 de Laval nozzle.

Figure 4.2 An airplane with delta wing.

4.13 DELTA WING

A delta wing is a wing shaped in the form of a triangle, as shown in Figure 4.2. It is named for its similarity in shape to the Greek uppercase letter delta (Δ).

4.14 DENSITY

The total number of molecules in a unit volume is a measure of the density of a substance. The density is a measure of the amount of material contained in a given volume. In other words, the material contained in a unit volume is a measure of the density ρ of the substance. It is expressed as mass per unit volume, say kg/m^3. Mass is defined as weight divided by acceleration due to gravity. At standard atmospheric temperature and pressure (288.15 K and 101,325 Pa), the density of dry air is 1.225 kg/m^3. The density of the fluid at the point on which the sphere is centred can be defined by

$$\rho = \lim_{\delta V \to \infty} \frac{\delta m}{\delta V}$$

4.15 DENSITY OF AIR

At sea level with pressure 101,325 Pa and temperature 15°C, the air density is 1.225 kg/m^3.

4.16 DENSITY OF MERCURY

The mass density of mercury is 13,600 kg/m^3.

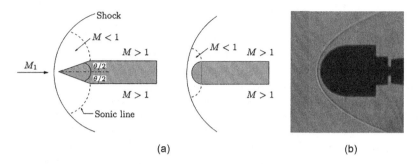

Figure 4.3 Detached shock waves: (a) schematic sketches and (b) detached shock at the nose of a body at Mach 7.

4.17 DENSITY OF PURE WATER

Pure water has its highest density 1,000 kg/m^3 at 4°C.

4.18 DENSITY OF SEAWATER

The density of seawater is about 1,025 kg/m^3.

4.19 DETACHED SHOCK

Detached shock is a compression front made up of approximately a normal shock followed by an infinite number of strong and weak oblique shocks of varying strengths, which finally degenerate, to a Mach wave. Experimentally, it has been observed that a flow with $\theta > \theta_{max}$ will have a configuration as shown in Figure 4.3a, with the shock wave slightly upstream of the nose even for a sharp-nosed body. The shape of the detached shock and its detachment distance depend on the geometry of the object facing the flow and the Mach number M_1.

For an object with a blunt nose, as shown in Figure 4.3b, the shock wave is detached at all supersonic and hypersonic Mach numbers. Therefore, even a streamlined body like a cone is a 'blunt-nosed' body as far as the oncoming flow is concerned, when $\theta > \theta_{max}$.

4.20 DETTOL

Dettol antiseptic disinfectant liquid is a concentrated antiseptic solution. It kills bacteria and protects against germs that can cause infection and illness. It kills germs on the skin and protects against infection from cuts, scratches, and insect bites.

4.21 DEUTERIUM

Deuterium is one of the two stable isotopes of hydrogen. The nucleus of a deuterium atom, called a deuteron, contains one proton and one neutron, whereas the far more common protium has no neutrons in the nucleus. Deuterium has a natural abundance in the Earth's oceans of about one atom in 6,420 of hydrogen.

4.22 DEW

Dew is the moisture that forms as a result of condensation. Condensation is the process a material undergoes as it changes from a gas to a liquid. Dew is the result of water changing from a vapour to a liquid.

4.23 DEW-POINT TEMPERATURE

The dew-point temperature is the temperature at which condensation begins if the air is cooled at constant pressure.

4.24 DIAL-TYPE PRESSURE GAUGE

When large pressures of the order of megapascals have to be measured, the liquid column required for such pressures becomes extremely large and measurement with manometers becomes unmanageable. For such high-pressure measurements, dial-type pressure gauges are appropriate. Dial-type pressure gauges generally operate on the principle of expansion of bellows or Bourdon tube which is usually an elliptical cross-sectional tube having a C-shape configuration. When pressure is applied to the tube, an elastic deformation, proportional to the applied pressure, occurs. The degree of linearity of the deformation depends on the quality of the tube material. The deflection drives the needle on a dial through mechanical linkages. Although pressure gauges of this type may be obtained with accuracies suitable for quantitative pressure measurements, they are not usually used for this purpose. They are primarily used only for visual monitoring of pressure in many pressure circuits. Schematic diagram of a typical Bourdon pressure gauge is shown in Figure 4.4.

4.25 DIAPHRAGM PRESSURE GAUGES

Diaphragm pressure gauges are used to measure gases and liquids. They cover measuring spans from 10 mbar to 40 bar.

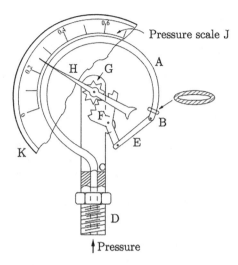

Pressure scale J

Figure 4.4 Bourdon pressure gauge.

4.26 DIESEL CYCLE

It is the ideal cycle for CI reciprocating engines, named after the proposer Rudolph Diesel. In SI or petrol engines, the fuel-air mixture is compressed to a temperature that is below the autoignition temperature of the fuel, and the combustion is initiated by fixing a spark plug. But in CI or diesel engines, the air is compressed to a temperature, which is above the autoignition temperature of the fuel, and the combustion starts on contact as the fuel is injected into the air.

4.27 DIFFERENCES BETWEEN THE JET AND PISTON ENGINE AIRCRAFT

Some of the important differences between jet and piston engine aircraft are the following:

1. On a piston engine aircraft, the engine power is regulated by the throttle control because throttling controls the airflow into the engine.
2. On a jet aircraft, the engine is regulated by adjusting the fuel flow, and it is the thrust that is controlled directly, rather than the power.

Another difference between the piston and jet aircraft is that the higher speed of the jet aircraft means that the speed relative to the speed of sound becomes very important. Therefore, for jet engine performance estimates,

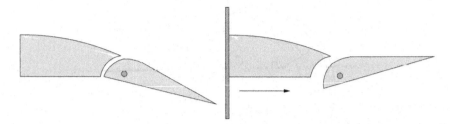

Figure 4.5 Differential aileron.

the data needed to be displayed in a different form than that used for piston aircraft. It is better to work in terms of thrust and drag rather than power required and power available, and we need to know how the thrust and drag vary with Mach number and altitude. That is, the simple performance calculations that are used for low-speed piston aircraft are not appropriate for high-speed jet aircraft.

4.28 DIFFERENTIAL AILERON

Differential aileron is a simple device. Instead of the two ailerons moving equally up or down, a simple mechanical arrangement of the controls causes the aileron that moves upward to move through a larger angle than the aileron that moves downward, as shown in Figure 4.5. The idea here is to increase the drag and decrease the lift on the wing with an up-going aileron, while at the same time the down-going aileron, owing to its smaller movement, will not cause excessive drag.

4.29 DIFFERENTIAL ANALYSIS

Differential analysis is the analysis applied to individual points in the flow field, the resulting equations are differential equations and the method is termed as the differential analysis.

4.30 DIFFERENT TYPES OF WINGLETS

Different types of winglets and wingtip devices: (a) Whitcomb winglet, (b) Tip fence, (c) Canted winglet, (d) Vortex diffuser, (e) Raked winglet, (f) Blended winglet, and (g) Blended split.

4.31 DIFFRACTION

Diffraction refers to various phenomena that occur when a wave encounters an obstacle or opening. It is defined as the bending of waves around the corners of an obstacle or through an aperture into the region of geometrical shadow of the obstacle/aperture.

4.32 DIFFUSER

Diffusers are passages in which flow decelerates.

The diffuser is used to re-convert the kinetic energy of the air stream leaving the working section into pressure energy as efficiently as possible. Essentially it is a passage in which the flow decelerates.

Basically a diffuser is a device to convert the kinetic energy of a flow to pressure energy.

4.33 DIFFUSER STALL

When the diffuser angle is too large and the adverse pressure gradient is excessive, the boundary layer will separate at one or both walls, with the back flow, increased losses, and poor pressure recovery. In diffuser literature, this condition is called diffuser stall. This is usually referred to as the boundary layer separation.

4.34 DIFFUSION

Diffusion is the movement of a substance from an area of high concentration to an area of low concentration. Diffusion happens in liquids and gases because their particles move randomly from place to place. Diffusion is an important process for living things; it is how substances move in and out of cells.

4.35 DIFFUSION FLAME

In combustion, a diffusion flame is a flame in which the oxidiser and fuel are separated before burning. Contrary to its name, a diffusion flame involves both diffusion and convection processes. S.P. Burke and T.E.W. Schumann first suggested the name diffusion flame in 1928, to differentiate it from premixed flame where fuel and oxidiser are premixed before burning. The diffusion flame is also referred to as a non-premixed flame. The burning rate is however still limited by the rate of diffusion. Diffusion flames tend to burn slower and to produce more soot than premixed flames because there may not be sufficient oxidiser for the reaction to go to completion, although there are some exceptions to the rule. The soot typically produced in a diffusion flame becomes incandescent from the heat of the flame and lends the flame its readily identifiable orange-yellow colour. Diffusion flames tend to have a less-localised flame front than premixed flames.

4.36 DIHEDRAL ANGLE

Dihedral angle is the angle between a horizontal plane containing the root chord and a plane midway between the upper and lower surfaces of the wing. If the wing lies below the horizontal plane, it is termed as anhedral angle. The dihedral angle affects the lateral stability of the aircraft.

4.37 DILUTION

Dilution is the addition of solvent, which decreases the concentration of the solute in the solution.

4.38 DIMENSIONAL ANALYSIS

Dimensional analysis is an algebraic operation on the basic dimensions appearing in the dimensional representation of physical quantities.

4.39 DIMENSIONLESS GROUPS

Dimensionless groups are non-dimensional parameters, which make the result of a theoretical or experimental investigation independent of any specific local conditions with which the study was carried out. In other words, any result can be made universal by using dimensionless groups to express them. Some of the popular dimensionless groups in flow physics are the Mach number and Reynolds number.

4.40 DIMENSIONS AND UNITS

Dimensions and units are a consistent set of standards defined to discuss the flow properties adequately. In fluid dynamics, mostly the gross, measurable molecular manifestations such as pressure and density as well as other equally important measurable abstract entities, e.g., length and time, will be dealt with. These manifestations, which are characteristics of the behaviour of a particular fluid, and not of the manner of flow, may be called *fluid properties*. Density and viscosity are examples of fluid properties. To adequately discuss these properties, a consistent set of standard units must be defined.

4.41 DIMENSIONLESS VELOCITY

The dimensionless velocity M^* is one of the most useful parameters in gas dynamics. Generally, it is defined as

$$M^* \equiv \frac{V}{a^*} \equiv \frac{V}{V^*}$$

where V is the local velocity, V^* is the sonic velocity, and a^* is the critical speed of sound.

4.42 DIRECT SIMULATION MONTE CARLO (DSMC)

DSMC is a numerical method for modelling rarefied gas flows, in which the mean free path of a molecule is of the same order (or greater) than a representative physical length scale (i.e., the Knudsen number Kn is >1).

The DSMC method models fluid flows using simulation molecules, which represent a large number of real molecules in a probabilistic simulation to solve the Boltzmann equation. Molecules are moved through a simulation of physical space in a realistic manner that is directly coupled to physical time such that unsteady flow characteristics can be modelled. Intermolecular collisions and molecule-surface collisions are calculated using probabilistic, phenomenological models. Common molecular models include the Hard Sphere model, the Variable Hard Sphere (VHS) model, and the Variable Soft Sphere (VSS) model. The fundamental assumption of the DSMC method is that the molecular movement and collision phases can be decoupled over periods that are smaller than the mean collision time.

4.43 DIRECTIONAL STABILITY

Directional stability of an aircraft is its ability to return to the original mode of flight when it is disturbed from its course.

4.44 DISC LOADING

Disc loading: The ratio of gross weight of the helicopter to the rotating disc area is called disc loading. It is expressed in kg/m^2. Since the disc area is not constant in flight, it follows that disc loading cannot be constant.

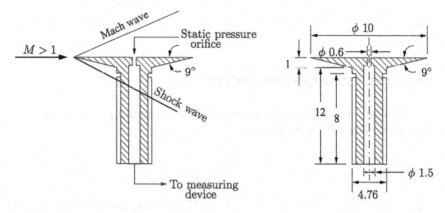

Figure 4.6 Disc-probe (All dimension in mm).

4.45 DISC-PROBE

Disc-probe is a special type of static pressure probe meant for measuring the static pressure in a supersonic free jet. One has to design and fabricate special probes for that purpose. One such special probe is the disc-probe shown in Figure 4.6. The disc-probe shown in the Figure 4.6 has a thin circular disc supported by a stem. The top surface of the disc is smooth and has a static pressure orifice located at its centre. The probe has to be aligned to have its top flat surface perfectly in line with the flow. Such an arrangement will ensure a shock-free supersonic flow over the top surface, and hence, the static pressure measured may be taken as the correct pressure.

4.46 DISCHARGE COEFFICIENT

In a nozzle or other constriction, the discharge coefficient (also known as coefficient of discharge or efflux coefficient) is the ratio of the actual discharge to the theoretical discharge, i.e., the ratio of the mass flow rate at the discharge end of the nozzle to that of an ideal nozzle which expands an identical working fluid from the same initial conditions to the same exit pressures.

4.47 DISPLACEMENT

Displacement is defined as the change in position of an object.

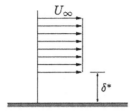

Figure 4.7 Displacement thickness.

4.48 DISPLACEMENT METERS

Displacement meter is a device to measure volume flow rate. Most of the displacement meters are basically positive-displacement meters used for volume flow measurements. Positive-displacement flow meters are generally used in places where consistently high accuracy is desired under steady flow conditions.

4.49 DISPLACEMENT THICKNESS

Displacement thickness, δ^*, may be defined as the distance by which the boundary would have to be displaced if the entire flow field were imagined to be frictionless and the same mass flow is maintained at any section. The displacement thickness is illustrated in Figure 4.7.

4.50 DISSIMILAR METALS

In a vast number of metals like copper, platinum, chromel, and iron, both heat and emf flow in the same direction. But in another group of metals, like constantan, alumel, and rhodium, the direction of heat flow is opposite to that of the emf. These two groups are commonly known as dissimilar metals. Some of the popular dissimilar metals are copper-constantan, iron-constantan, nickel-constantan, chromel-constantan, chromel-alumel, and platinum-rhodium. Copper-constantan is widely used in gas dynamic studies at moderate temperatures since it has a temperature measuring range from –270°C to 400°C. For small temperature changes, the Seebeck voltage is linearly proportional to the temperature.

4.51 DISSOCIATION

Dissociation is the breaking up of a compound into simpler constituents that are usually capable of recombining under other conditions. It is usually reversible.

4.52 DISTILLATION

Distillation refers to the selective boiling and subsequent condensation of a component in a liquid mixture. It is a separation technique that can be used to either increase the concentration of a particular component in the mixture or to obtain (almost) pure components from the mixture.

4.53 DISTILLED WATER

Distilled water is a type of purified water that has both contaminants and minerals removed. Purified water has chemicals and contaminants removed, but it may still contain minerals. Distillation boils the water and then condenses the steam back into a liquid to remove impurities and minerals.

4.54 DIVERGENCE

The inner product of the gradient operator with a vector: $\nabla \cdot u$ is divergence. It is zero for incompressible fluids (or solenoidal vector fields).

4.55 DORSAL FIN

A dorsal fin is a fin located on the back, as in Figure 4.8, of most marine and freshwater vertebrates within various taxa of the animal kingdom.

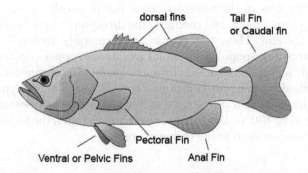

Figure 4.8 Dorsal fin.

4.56 DOUBLET

A doublet or a dipole is a potential flow field due to a source and sink of equal strength, brought together in such a way that the product of their strength and the distance between them remain constant. When the distance between the source and sink is made negligibly small, in the limiting case, the combination results in a doublet.

4.57 DOWNWASH

The tip and trailing vortices cause the flow in the immediate vicinity of the wing, and behind it, to acquire a downward velocity component. This phenomenon is known as induced downwash or simply downwash.

The downwash reduces the effective incidence of the wing. This affects both the lift and drag characteristics of the wing adversely.

The downwash can affect the flow over the tailplane of the aircraft and thus has an important consequence in connection with the stability of the aircraft.

4.58 DOWNWASH CONSEQUENCES

The two important consequences of downwash are as follows:

1. The downwash reduces the effective incidence of the wing. This affects both the lift and drag characteristics of the wing adversely.
2. The downwash can affect the flow over the tailplane of the aircraft and thus has an important consequence in connection with the stability of the aircraft.

4.59 DRAG

The aerodynamic force F_{ad} can be resolved into two component forces, one at right angles to freestream velocity V and the other opposite to V. The force component opposite to V is called drag D.

Drag is the component of the aerodynamic force opposite to the direction of motion.

Drag is a force on a moving body, opposing the motion of the body. When a body moves in a fluid, it experiences forces and moments due to the relative motion of the fluid flow, which is taking place around it. The force on the body along the flow direction is called drag.

The drag is essentially a force opposing the motion of the body. Viscosity is responsible for a part of the drag force, and the body shape generally determines the overall drag. In the design of transport vehicles, shapes experiencing minimum drag are considered to keep the power consumption at a minimum. Low-drag shapes are called *streamlined bodies* and high-drag shapes are termed *bluff bodies*.

Drag arises due to (a) the difference in pressure between the front and back regions; (b) the friction between the body surface and the fluid. Drag force (a) is known as *pressure drag* and (b) drag due to friction in skin is known as *skin friction drag*.

When the flow speed is supersonic, there can be yet another component of drag termed *wave drag*, caused by the compression and expansion waves prevailing in the flow filed around the body in motion.

4.60 DRAG COEFFICIENT

The drag force, D, can be expressed as a dimensionless number popularly known as drag coefficient C_D, expressed as

$$C_D = \frac{D}{\frac{1}{2}\rho V^2 \, S}$$

where ρ, V, and S are the flow density, flight speed, and wing area, respectively.

4.61 DRAG-BODY METERS

Insertion of an appropriately shaped body into a flow stream can serve as a flow meter. The drag experienced by the body becomes a measure of flow rate after suitable calibration. The drag force can be measured by attaching the drag-producing body to a strain gauge force measuring transducer. A typical drag force flow meter using a cantilever beam with bonded strain

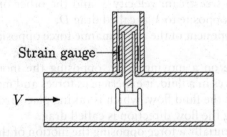

Strain gauge

$V \longrightarrow$

Figure 4.9 Drag-body flow meter.

gauges is shown in Figure 4.9. A hollow tube arrangement with the strain gauges fixed on the outer surface, as shown in Figure 4.9, serves to isolate the strain gauges from the flowing fluid. If the drag force is made symmetric, reversed flows can also be measured. The main advantage of this type of flow meter is the high dynamic response. The type of gauge described above is second order with a natural frequency of 70–200 Hz. However, this type of gauge suffers from small damping, and thus, sharp transients may cause difficulty.

4.62 DRAG-DIVERGENCE MACH NUMBER

The drag-divergence Mach number is the Mach number at which the aerodynamic drag on an airfoil or airframe begins to increase rapidly as the Mach number continues to increase.

The drag-divergence Mach number is not to be confused with the critical Mach number. This increase can cause the drag coefficient to rise to more than 10 times its low-speed value. The value of the drag-divergence Mach number is typically >0.6; therefore, it is a transonic effect.

4.63 DRAINAGE

A system used for making water, etc. flow away from a place.

4.64 DRAINAGE CANAL

An open canal, channel, or ditch, is an open waterway whose purpose is to carry water from one place to another.

4.65 DRIVER SECTION

The shock tube consists of a long duct of constant cross section divided into two chambers by a diaphragm. The high-pressure chamber is called the driver section.

4.66 DROP

A drop or droplet is a small column of liquid bounded completely or almost completely by free surfaces.

4.67 DROPLET

Droplet is a tiny drop (as of a liquid).

4.68 DRUG

A drug is any substance that causes a change in an organism's physiology or psychology when consumed. Drugs are typically distinguished from food and substances that provide nutritional support.

4.69 DRY ICE

Dry ice is the solid form of carbon dioxide (CO_2) – a molecule consisting of a single carbon atom bonded to two oxygen atoms.

4.70 DUAL CYCLE

It is a combination of Otto and Diesel cycles.

4.71 DUCT

Ducts are conduits or passages used in heating, ventilation, and air conditioning (HVAC) to deliver and remove air. The needed airflows include, for example, supply air, return air, and exhaust air. Ducts commonly also deliver ventilation air as part of the supply air. As such, air ducts are one method of ensuring acceptable indoor air quality as well as thermal comfort.

4.72 DUCTILITY

Ductility is a mechanical property commonly described as a material's amenability to drawing. In materials science, ductility is defined by the degree to which a material can sustain plastic deformation under tensile stress before failure.

4.73 DUST STORM

A dust storm, also called a sandstorm, is a meteorological phenomenon common in arid and semi-arid regions. Dust storms arise when a gust front or other strong wind blows loose sand and dirt from a dry surface.

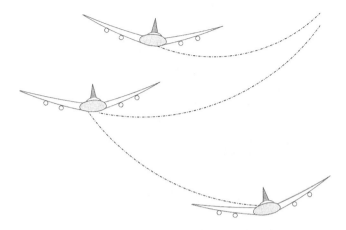

Figure 4.10 Dutch roll.

4.74 DUTCH ROLL

A Dutch roll is a combination of rolling and yawing oscillations that occurs when the dihedral effects of an aircraft are more powerful than the directional stability. It is usually dynamically stable, but it is an objectionable characteristic in an airplane because of its oscillatory nature.

Dutch roll causes cross-control-type movements across two axes, the vertical and longitudinal axes, such that when the airplane rolls right, it yaws to the left, then it swings back the other way and rolls left, while yawing to the right, as in Figure 4.10. This movement is difficult for the pilot to counteract and the inputs from the pilot can amplify these movements.

4.75 DYE

A dye is a coloured substance that chemically bonds to the substrate to which it is being applied. This distinguishes dyes from pigments that do not chemically bind to the material they colour. The dye is generally applied in an aqueous solution and may require a mordant to improve the fastness of the dye on the fibre.

4.76 DYKE

A long thick wall that is built to prevent the sea or a river from flooding low land.

4.77 DYNAMIC ANALYSIS

This study deals with the determination of the effects of the fluid and its surroundings on the motion of the fluid. This involves the consideration of forces acting on the fluid elements in motion with respect to one another. Since there is relative motion between fluid elements, shearing forces must be taken into consideration in the dynamic analysis.

4.78 DYNAMIC INSTABILITY

The instabilities in the form of oscillations or deviations from the desired flight path that vary with time are called dynamic instabilities.

4.79 DYNAMIC PRESSURE

The dynamic pressure is linked to the kinetic energy of the flow and it has the same direction as that of the flow. The dynamic pressure q in compressible flow can be expressed as

$$q = \frac{p_0 - p}{K}$$

where $K = 1 + \dfrac{M^2}{4} + \dfrac{M^4}{40}$ is called the correction coefficient.

4.80 DYNAMIC PRESSURE GAUGES

Dynamic pressure gauges are called microphones or, for liquid application, hydrophones. These are typically elastic transducers having sufficiently fast responses. Obviously, liquid manometers and Bourdon tubes are not suitable for dynamic pressure measurements. A fluctuating pressure causes an elastic membrane to oscillate, and this motion is transduced to an electrical signal preferably linearly proportional to the sensed pressure. Common microphones are classified into five categories, namely piezoelectric transducers, variable capacitance, resistance or reluctance transducers, and linear-variable differential transformers (LVDT).

4.81 DYNAMICS

If the effects of forces on the motion of bodies are accounted for the subject, it is termed dynamics.

Chapter 5

Eckert Number to Extratropical Cyclone

5.1 ECKERT NUMBER

The Eckert number is defined as the ratio of the advective mass transfer to the heat dissipation potential. This number provides a measure of the kinetic energy of the flow relative to the enthalpy difference across the thermal boundary layer. The Eckert number is used to characterise the influence of self-heating of a fluid as a consequence of dissipation effects.

5.2 EDDY

An area of slower-moving fluid that occurs behind an obstacle. In fluid dynamics, an eddy is the swirling of a fluid and the reverse current created when the fluid is in a turbulent flow regime.

5.3 EDDY DIFFUSION

Eddy diffusion, eddy dispersion, multipath or turbulent diffusion is any diffusion process by which substances are mixed in the atmosphere or in any fluid system due to eddy motion. With a spoon, a coffee drinker can create eddies that transport dissolved sugar throughout the cup in <1 s. This is an example of eddy diffusivity.

5.4 EDDY VISCOSITY

Eddy viscosity is the proportionality factor describing the turbulent transfer of energy as a result of moving eddies giving rise to tangential stresses.

DOI: 10.1201/9781003348405-5

5.5 EDDY VISCOSITY TURBULENCE MODELS

Eddy viscosity turbulence models are widely used for industrial, aeronautical, meteorological and oceanographical applications. They are based on a constitutive equation, giving a relation between the Reynolds stress tensor and the mean velocity gradient tensor.

5.6 EFFECTIVE ANGLE OF ATTACK

It is the part of a given angle of attack that lies between the chord of an airfoil and a line representing the resultant velocity of the disturbed airflow.

5.7 EFFECTIVE DOWNWASH

Effective downwash is the downwash velocity that combines with the actual relative wind of speed to produce an effective relative wind, as illustrated in Figure 5.1.

5.8 EFFECTIVE HEAD

Effective head on the turbine (including nozzle) is the static head minus the pipe friction losses.

5.9 EFFECTS OF THE EARTH'S ROTATION

Effects of the Earth's rotation are as follows: (a) The rotation of the Earth causes the day and night. (b) The speed of the Earth's rotation has affected the shape of the Earth. (c) The Earth's rotation affects the movement of water in the oceans. (d) The speed of rotation also affects wind movements.

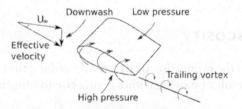

Figure 5.1 Effective downwash.

5.10 EFFECTS OF THE SECOND THROAT

The second throat is used to provide isentropic deceleration and highly efficient pressure recovery after the test section.

5.11 EFFICIENCY OF TURBINES

The efficiency of turbines is the ratio of power delivered to the shaft (brake power) to the power taken from the water.

5.12 EFFUSER

Effuser is a converging passage located upstream of the test section, as shown in Figure 5.2. In this passage, fluid gets accelerated from rest (or from very low speed) at the upstream end to the required conditions at the test section. In general, the effuser contains honeycomb and wire gauze screens to reduce turbulence and produce a uniform air stream at the exit. The effuser is usually referred to as the contraction cone.

5.13 EKMAN BOUNDARY LAYER

In a rotating fluid, the boundary layer formed on the floor of the tank has some special significance. This is termed the Ekman boundary layer and it is spiral in nature. In the Ekman layer, the frictional forces reduce the velocity, and hence, the Coriolis force is no longer in balance with the pressure gradient; hence, the net transport in the Ekman boundary layer is at an angle to the net transport in the interior. The component of velocity in the direction of pressure gradient changes the direction of the fluid flow.

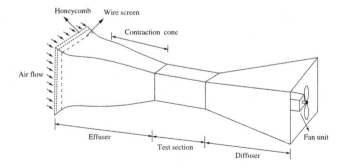

Figure 5.2 Effuser.

5.14 EKMAN SUCTION

Ekman layers are divergent and this creates a suction directed towards the interior whose magnitude is proportional to the vorticity of the flow above this layer. This is known as Ekman suction.

5.15 ELECTROLYSIS

Electrolysis is a process of decomposing ionic compounds into their elements by passing a direct electric current through the compound in a fluid form.

5.16 ELECTROLYTIC TANK

Electrolytic tank is an analogue technique used for solving potential flow problems. In other words, it is used to solve the Laplace equation. Basically it is an analogy method. It makes use of the fact that the equations governing incompressible potential flow and distribution of the electrical potential lines are the same, in their form, to establish an analogy between the two fields.

5.17 ELECTRONIC ENERGY

Electronic energy is due to the motion of electrons about the nucleus of each atom constituting the molecule. The sources of electronic energy are (a) the kinetic energy because of its translational motion throughout its orbit about the nucleus and (b) the potential energy because of its location in the electromagnetic force field established principally by the nucleus. The concepts of geometric and thermal degrees of freedom are usually not useful for describing electronic energy because the overall electron motion is complex.

5.18 ELEVATOR

It is a control surface used to vary the lift on the horizontal tail plane (shown in Figure 5.3) and so to control the pitching moment.

5.19 EMBANKMENT

A wall of stone or earth that is built to stop a river from flooding or to carry a road or railway.

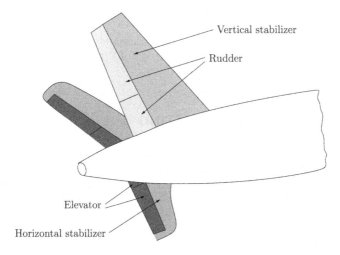

Vertical stabilizer

Rudder

Elevator

Horizontal stabilizer

Figure 5.3 Elevator.

5.20 EMULSIFIER

A substance added to food mixtures to combine them to form a smooth mixture.

5.21 EMULSION

An emulsion is a mixture of two or more liquids that are normally immiscible owing to liquid-liquid phase separation. Emulsions are part of a more general class of two-phase systems of matter called colloids.

5.22 ENDOTHERMIC REACTION

A reaction during which heat is liberated is termed endothermic. Endothermic reactions are chemical reactions in which the reactants absorb heat energy from the surroundings to form products. These reactions lower the temperature of their surrounding area, thereby creating a cooling effect.

5.23 ENDURANCE OF FLIGHT

Endurance is the time duration for which an aircraft can stay in the air for a given quantity of fuel. To get maximum endurance, the least possible fuel should be used in a given time, that is, the power used should be minimum to get maximum endurance.

5.24 ENDURANCE OF JET ENGINE

For jet engines, fuel consumption is approximately proportional to thrust. Therefore, for maximum endurance, it should fly with minimum thrust, that is, with minimum drag.

5.25 ENERGY EFFICIENCY RATING

The performance of refrigerators and air conditioners is expressed in terms of the energy efficiency rating, which is the amount of heat removed from the cooled space in British thermal units for 1 watt-hour of electricity consumed.

5.26 ENERGY EQUATION

This is a statement of conservation of energy along a streamline. The energy conservation principle is popularly known as the *first law of thermodynamics*. For the flow of gases for which the potential energy change is negligible, the energy equation can be expressed as

$$h + \frac{V^2}{2} = h_0$$

where h is the static enthalpy, V is the flow velocity, and h_0 is the total enthalpy.

5.27 ENERGY GRADIENT

Energy gradient is the ratio of head loss over a length h_f to the length l.

5.28 ENERGY RATIO

The ratio of the energy of air stream at the test section to the input energy to the driving unit is a measure of the efficiency of a wind tunnel.

5.29 ENERGY THICKNESS

Energy thickness is the distance by which the boundary, over which the boundary layer prevails, has to be hypothetically shifted so that the kinetic energy of the flow passing through the actual thickness (distance) and the hypothetical thickness will be the same.

5.30 ENGINES

The devices that produce a net power output are called engines.

5.31 ENTHALPY

Enthalpy is a measure of the total energy of a thermodynamic system. It includes the internal energy, which is the energy required to create a system, and the amount of energy required to make room for it by displacing its environment and establishing its volume and pressure.

Enthalpy is a thermodynamic potential. It is a state function and an extensive quantity. The unit for enthalpy is joule.

Enthalpy is the preferred expression of system energy changes in many chemical, biological and physical measurements because it simplifies certain descriptions of energy transfer. This is because a change in enthalpy takes account of the energy transferred to the environment through the expansion of the system under study.

5.32 ENTHALPY OF COMBUSTION

Enthalpy of combustion represents the amount of heat released during a steady-flow combustion process when 1 mole of fuel is burnt completely at a specified temperature and pressure.

or

The enthalpy of combustion of a substance is defined as the heat energy given out when 1 mole of a substance burns completely in oxygen.

5.33 ENTHALPY OF FORMATION

The enthalpy of formation is the enthalpy of a substance at a specified state due to its chemical composition.

5.34 ENTHALPY OF REACTION

It is the difference between the enthalpy of the products at a specified state and the enthalpy of the reactants at the same state for a complete reaction.

5.35 ENTHALPY OF VAPOURISATION

Enthalpy of vaporisation is the amount of energy required to vaporise a unit mass of saturated liquid at a given temperature or pressure.

5.36 ENTRAINMENT

The differential shear at the jet boundary forms vortices and they bring fluid mass from the surrounding environment into the jet field. This transport of mass from the surroundings into the jet is called entrainment. It is the transport of fluid across an interface between two bodies of fluid by a shear-induced turbulent flux.

5.37 ENTRANCE LENGTH

The zone upstream of the boundary layer merging point is called the entrance or flow development length, as shown in Figure 5.4.

5.38 ENTROPY

Entropy may be viewed as a measure of disorder or randomness in a system. Entropy changes only when there are losses in pressure. It does not change with velocity and there is nothing like stagnation or static entropy. A thermodynamic quantity represents the unavailability of a system's thermal energy for conversion into mechanical work often interpreted as the degree of disorder or randomness in the system. The second law of thermodynamics says that entropy always increases with time.

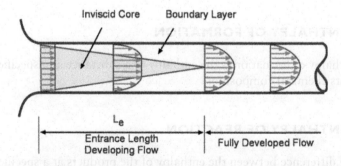

Figure 5.4 Entrance length.

5.39 ENTROPY LAYER

As the gas flows over the body, all of the high-entropy gas initially processed by the bow shock wave ahead of the body is termed the entropy layer.

5.40 ENTROPY TRANSPORT

Mass flow transports both energy and entropy into or out of a control volume. Entropy transfer via mass flow is called entropy transport. For closed systems, the entropy transport is zero since there is no mass flow involved.

5.41 ENZYME

An enzyme is a substance produced by all living things which helps a chemical change to happen more quickly without being changed itself.

5.42 ETHANOL

Ethanol, also called alcohol, ethyl alcohol, or grain alcohol, is a clear, colourless liquid and the principal ingredient in alcoholic beverages like beer, wine, or brandy.

5.43 ETHANOL FUEL

Ethanol fuel is ethyl alcohol, the same type of alcohol found in alcoholic beverages, used as fuel. It is most often used as a motor fuel, mainly as a biofuel additive for gasoline. Ethanol is commonly made from biomass such as corn or sugarcane.

5.44 ETHYL ACETATE

Ethyl acetate is a colourless liquid with an odour similar to glue or nail polish that is used as an industrial solvent.

5.45 EQUATION OF STATE

For air at normal temperature and pressure, the density ρ, pressure p and temperature T are connected by the relation $p = \rho R T$, where R is a constant called gas constant. This is known as the state equation for a perfect gas.

5.46 EQUILIBRIUM

A system is said to be in equilibrium when its properties do not change with time.

5.47 EQUILIBRIUM FLOWS

A flow with an infinitely large reaction rate is called an equilibrium flow. Equilibrium flow, therefore, implies infinite chemical and vibrational rates. At sufficiently high densities, there are a significant number of collisions between particles to allow the equilibrium of energy transfer between various modes. The flow is in thermal equilibrium. For an equilibrium flow, any two thermodynamic properties, for example, p and T, can be used to uniquely define the state. As a result, the remaining thermodynamic properties and the composition of the gas can be determined.

5.48 EQUILIBRIUM MACH NUMBER

The ratio of flow speed, V, to the equilibrium speed of sound, a_e, is called the equilibrium Mach number, M_e.

5.49 EQUILIBRIUM SPEED OF SOUND

If the gas remains in local chemical equilibrium through the internal structure of the sound wave, the gas composition is changed locally within the wave according to the local variations of pressure and temperature. For this situation, the speed of sound wave is called equilibrium speed of sound, denoted by a_e.

5.50 EQUIPARTITION OF ENERGY

The theorem of equipartition of energy of kinetic theory of gases states 'each thermal degree of freedom of the molecule contributes $\frac{1}{2}kT$ to the energy of each molecule, or $\frac{1}{2}RT$ to the energy per unit mass of gas'.

When a large number of nondistinguisable, quasi-independent particles whose energy is expressed as the sum of f squared terms come to equilibrium, the average energy per particle is $f\frac{1}{2}kT$, where k is the Boltzmann constant and T is the temperature. This is the famous principle of the equipartition of energy.

5.51 EQUIVALENCE RATIO

The ratio of the actual amount of air used to the stoichiometric amount of air is called the equivalence ratio.

5.52 EQUIVALENT AIRSPEED (EAS)

Equivalent airspeed (EAS) is calibrated airspeed (CAS) corrected for the compressibility of air at a non-trivial Mach number. It is also the airspeed at sea level in the International Standard Atmosphere at which the dynamic pressure is the same as the dynamic pressure at the true airspeed (TAS) and altitude at which the aircraft is flying. In a low-speed flight, it is the speed which would be shown by an airspeed indicator with zero error.

5.53 EQUIVALENT CHORD

Equivalent chord, c_e, is defined as the chord of the rectangular blade, which would yield the same thrust and torque as the actual blade planform.

5.54 ESCAPE VELOCITY

Escape velocity is the minimum speed needed for a free, non-propelled object to escape from the gravitational influence of a massive body, that is, to eventually reach an infinite distance from it.

5.55 ESCAPE VELOCITY OF EARTH

A spacecraft leaving the surface of Earth, for example, needs to be going about 11 km (7 miles) per s, or over 40,000 km/h (25,000 miles/h), to enter orbit.

5.56 EULER'S ACCELERATION FORMULA

Euler's acceleration formula is the relation between the local and material rates of change. The rate of change of a property measured by a probe at a fixed location is referred to as the *local rate of change* and the rate of change of a property experienced by a material particle is termed as the *material* or *substantive rate of change*. For a fluid flowing with a uniform velocity V_∞, it is possible to write the relation between the local and material rates of change of property η as

$$\frac{\partial \eta}{\partial t} = \frac{D\eta}{Dt} - \frac{\partial \eta}{\partial x}$$

Thus, the local rate of change of η is due to the following two effects.

1. Due to the change of property of each particle with time.
2. Due to the combined effect of the spatial gradient of that property and the motion of the fluid.

5.57 EULER NUMBER

The Euler number is the ratio of pressure force to the inertia force. The Euler number is the ratio of local static pressure to local kinetic pressure. Thus,

$$\text{Eu} = \frac{p}{\rho V^2}$$

5.58 EULERIAN DESCRIPTION

Eulerian description is a field description. It considers the properties of the flow at each position in the flow field by employing spatial coordinates to help identify particles in a flow. The velocity of all particles in a flow, therefore, can be expressed in the following manner:

$$V_x = f(x,y,z,t)$$

$$V_y = g(x,y,z,t)$$

$$V_z = h(x,y,z,t)$$

This is called the Eulerian or field approach. If properties and flow characteristics at each position in space remain invariant with time, the flow is called *steady flow*. A time-dependent flow is referred to as an *unsteady flow*. The steady-flow velocity field would then be given as

$$V_x = f(x,y,z)$$

$$V_y = g(x,y,z)$$

$$V_z = h(x,y,z)$$

5.59 EULER'S EQUATIONS

Euler's equations are momentum equations for potential flow. For potential flows, the Navier–Stokes equations reduce to the form

$$V_x \frac{\partial V_x}{\partial x} + V_y \frac{\partial V_x}{\partial y} + V_z \frac{\partial V_x}{\partial z} = -\frac{1}{\rho}\frac{\partial p}{\partial x}$$

$$V_x \frac{\partial V_y}{\partial x} + V_y \frac{\partial V_y}{\partial y} + V_z \frac{\partial V_y}{\partial z} = -\frac{1}{\rho}\frac{\partial p}{\partial y}$$

$$V_x \frac{\partial V_z}{\partial x} + V_y \frac{\partial V_z}{\partial y} + V_z \frac{\partial V_z}{\partial z} = -\frac{1}{\rho}\frac{\partial p}{\partial z}$$

These are known as Euler's equations.

5.60 EVAPORATION

Evaporation is a type of vaporisation that occurs on the surface of a liquid as it changes into the gas phase.

5.61 EVAPORATIVE COOLING

When water evaporates, the latent heat of vaporisation is absorbed from the water container body and the surrounding air. Evaporative cooling is based on this principle. Both the water and the air are cooled during the evaporative cooling process.

5.62 EXCITED STATE

In quantum mechanics, an excited state of a system (such as an atom, molecule or nucleus) is any quantum state of the system that has higher energy than the ground state (that is, more energy than the absolute minimum). Excitation is an elevation in energy level above an arbitrary baseline energy state. In physics, there is a specific technical definition for energy level that is often associated with an atom being raised to an excited state. The temperature of a group of particles is indicative of the level of excitation (with the notable exception of systems that exhibit negative temperatures).

5.63 EXOSPHERE

The exosphere is the outermost layer of the Earth's atmosphere. It starts at an altitude of about 500 km and goes out to about 10,000 km. Within this region, the particles of the atmosphere can travel for hundreds of kilometres in a ballistic trajectory before bumping into any other particles of the atmosphere.

5.64 EXOTHERMIC REACTION

Exothermic reactions are reactions or processes that release energy, usually in the form of heat or light. In an exothermic reaction, energy is released because the total energy of the products is less than the total energy of the reactants. A reaction during which heat is absorbed is termed endothermic.

5.65 EXPANSION

Expansion is a process in which the flow relaxes. A subsonic flow relaxing in a divergent passage gets decelerated. This means a subsonic flow through a divergent passage diffuses resulting in an increase of pressure and the associated decrease of velocity as per the Bernoulli equation, i.e., $M_2 < M_1$, $p_2 > p_1$. But when a subsonic flow issuing out of a convergent duct relaxing to a free environment at a pressure p_b the flow gets decelerated. The Mach number decreases in the direction downstream of the duct, i.e., $M_2 < M_1$. But unlike in the case of confined diffusion of subsonic flow, in free expansion, the static pressure is the same as the backpressure at locations in the flow field. That is, the subsonic free jet is always correctly expanded and the pressure is invariant.

But when the flow is supersonic, it can encounter two kinds of expansion, namely centred expansion and continuous expansion. As we know, the supersonic flow is essentially wave-dominated. Therefore, an expansion process of a supersonic flow is also wave-dominated. The waves associated with an expansion process are termed expansion waves. Even though like Mach waves, the expansion waves are also isentropic, they distinctly differ from Mach waves in the sense that the change of flow properties across an expansion wave is small but *finite*, whereas across a Mach wave, the change in flow properties is negligibly small.

5.66 EXPANSION LEVELS OF JETS

The NPR – the ratio of the stagnation pressure in the settling chamber – which runs the jet-delivering nozzle to the pressure of the environment to which the jet is discharged, dictates the level of expansion at the nozzle exit.

If the static pressure at the nozzle exit is lower than the pressure of the environment, the jet is termed overexpanded; if the exit pressure is equal to the environmental pressure, the jet is called correctly expanded, and if the exit pressure is greater than the environmental pressure the jet is termed underexpanded.

5.67 EXPANSION SECTION

The shock tube consists of a long duct of constant cross-section divided into two chambers by a diaphragm. The high-pressure chamber is called the driver section and the low-pressure chamber is called the expansion section.

5.68 EXPANSION TUNNEL

Expansion and shock tunnels are aerodynamic testing facilities with a specific interest in high speeds and high-temperature testing. Shock tunnels use steady-flow nozzle expansion whereas expansion tunnels use unsteady expansion with higher enthalpy or thermal energy. In both cases, the gases are compressed and heated until the gases are released, expanding rapidly down the expansion chamber. The tunnels reach speeds from Mach 3 to Mach 30 to create testing conditions that simulate hypersonic to re-entry flight. These tunnels are used by military and government agencies to test hypersonic vehicles that undergo a variety of natural phenomena that occur during a hypersonic flight.

5.69 EXPERIMENT

An experiment is a procedure carried out to support or refute a hypothesis.

5.70 EXPERIMENTAL ACCURACY

Experimental error is the difference between a measured and the true value or between two measured values. Experimental error, itself, is measured by its accuracy and precision. Accuracy measures how close a measured value is to the true value or accepted value.

5.71 EXPERIMENTAL ERROR

Experimental error is the difference between a measurement and the true value or between two measured values. Since a true or accepted value for a physical quantity may be unknown, it is sometimes not possible to determine the accuracy of a measurement.

5.72 EXPERIMENTAL MEAN PITCH

The experimental mean pitch is defined as the forward distance travelled by the propeller in one revolution when it is giving no thrust.

5.73 EXPLOSION

An explosion is a rapid expansion in volume associated with an extremely vigorous outward release of energy, usually with the generation of high temperatures and the release of high-pressure gases. Supersonic explosions created by high explosives are known as detonations and travel through shock waves.

5.74 EXTENSIVE PROPERTIES

Extensive properties are those whose values depend on the size or the extent of the system. Mass, volume and total energy are extensive properties.

5.75 EXTERNAL AERODYNAMICS

External aerodynamics is the study of flow around solid objects of various shapes. Evaluating the lift and drag on an aeroplane or the shock waves that form in front of the nose of a rocket are examples of external aerodynamics.

5.76 EXTERNAL BALANCE

The balance located external to the model and the test section is referred to as external balance.

5.77 EXTERNAL FLOW

In fluid mechanics, external flow is a flow that boundary layers develop freely without constraints imposed by adjacent surfaces. It can be defined as the flow of a fluid around a body that is completely submerged in it.

Figure 5.5 Extratropical cyclone.

5.78 EXTERNALLY REVERSIBLE PROCESS

An externally reversible process is that for which no irreversibilities occur outside the system boundaries during the process. For example, heat transfer between a reservoir and a system is an externally reversible process if the surface of contact between the system and the reservoir is at the temperature of the reservoir.

5.79 EXTRATROPICAL CYCLONE

Extratropical cyclones, sometimes called mid-latitude cyclones or wave cyclones, are low-pressure areas, which along with the anticyclones of high-pressure areas, drive the weather over much of the Earth. Extratropical cyclones are capable of producing anything from cloudiness and mild showers to heavy gales, thunderstorms, blizzards and tornadoes. These types of cyclones are defined as large-scale (synoptic) low-pressure weather systems that occur in the middle latitudes of the Earth. In contrast with tropical cyclones, extratropical cyclones produce rapid changes in temperature and dew point along broad lines called weather fronts about the centre of the cyclone, as shown in Figure 5.5.

Chapter 6

Factors Influencing Pitot-Static Tube Performance to Fusion Pyrometers

6.1 FACTORS INFLUENCING PITOT-STATIC TUBE PERFORMANCE

The major factors influencing pitot-static tube performance are turbulence, velocity gradient, viscosity, wall proximity, misalignment and blockage effect.

6.2 FAN

A fan is the first component in a turbofan engine. The large spinning fan sucks in a large quantity of air. Most blades of the fan are made of titanium. It then speeds up this air and splits it into two parts. One part continues to flow through the 'core' or centre of the engine, where it is acted upon by the other engine components.

6.3 FANNO FLOW

Fanno flow is an adiabatic flow with no external work. For Fanno flow, the wall friction (due to viscosity) is the chief factor bringing about changes in flow properties.

6.4 FAVOURABLE PRESSURE GRADIENT

A negative pressure gradient is termed a favourable pressure gradient. Such a gradient enables the flow.

6.5 FEATHERED

If the blade can be turned beyond the normal fully coarse position until the chord lies along the direction of flight, thus offering the maximum resistance, the propeller is said to be feathered. This condition is very useful on a

DOI: 10.1201/9781003348405-6

multi-engined aircraft for reducing the drag of the propeller as a convenient method of stopping the propeller and so preventing it from 'windmilling'. This reduces the risk of further damage to an engine that is already damaged.

6.6 FEATHERING

Feathering is the movement of the blade about its feathering axis, resulting in pitch angle changes.

6.7 FEATHERING AXIS

This is the straight-line axis between the root of the blade and its tip about which the blade can alter its blade angle.

6.8 FENCE TECHNIQUE

The fence technique is an indirect method for measuring wall shear stress. A fence is just a small narrow strip of material placed on the wall. The pressure difference across the two faces of the fence is a function of wall shear.

6.9 FERMENTATION

Fermentation is a metabolic process that produces chemical changes in organic substrates through the action of enzymes. In biochemistry, it is narrowly defined as the extraction of energy from carbohydrates in the absence of oxygen.

6.10 FERMI-DIRAC STATISTICS

Fermi-Dirac statistics, in quantum mechanics, is one of the two possible ways in which a system of indistinguishable particles can be distributed among a set of energy states: each of the available discrete states can be occupied by only one particle.

6.11 FERMIONS

For fermions, only one molecule may be in any given degenerate state at any instant.

Molecules and atoms with an odd number of elementary particles obey a certain other statistical distribution termed Fermi-Dirac statistics, and they are called fermions.

6.12 FICK'S LAW

$$J = -D \frac{\partial \phi}{\partial x}$$

Fick's first law states that the mass flux via diffusion goes from regions of high concentration to regions of low concentration with a magnitude proportional to the concentration gradient.

6.13 FIELD

A field is a continuous distribution of a scalar, vector, or tensor quantity described by continuous functions of space coordinates and time.

6.14 FIELD APPROACH

Field approach is a treatment followed for fluid flow analysis. It is also referred to as Eulerian description. It considers the properties of the flow at each position in the flow field by employing spatial coordinates to help identify particles in a flow. The velocity of all particles in a flow, therefore, can be expressed in the following manner.

$$V_x = f(x, y, z, t)$$

$$V_y = g(x, y, z, t)$$

$$V_z = h(x, y, z, t)$$

This is called the *Eulerian or field approach*. If properties and flow characteristics at each position in space remain invariant with time, the flow is called *steady flow*. A time-dependent flow is referred to as an *unsteady* flow. The steady flow velocity field would then be given as

$$V_x = f(x, y, z)$$

$$V_y = g(x, y, z)$$

$$V_z = h(x, y, z)$$

6.15 FIGURE OF MERIT

The performance of the helicopter rotor is viewed as its efficiency to use the portion of the total power supplied to generate lift. This ratio of induced power to total power is called the figure of merit.

6.16 FILM TEMPERATURE

In heat transfer and fluid dynamics, the film temperature (T_f) is an approximation to the temperature of a fluid inside a convective boundary layer. It is calculated as the arithmetic mean of the temperature at the surface of the solid boundary wall (T_w) and the freestream temperature (T_∞)
$T_f = (T_w + T_\infty)/2$. The film temperature is often used as the temperature at which fluid properties are calculated when using the Prandtl number, Nusselt number, Reynolds number or Grashof number to calculate a heat transfer coefficient, because it is a reasonable first approximation to the temperature within the convection boundary layer.

6.17 FIN

A fin is a flat, thin part that sticks out of an aircraft (Figure 6.1), a vehicle, etc. to improve its balance and movement through the air or water.

In the study of heat transfer, fins are surfaces that extend from an object to increase the rate of heat transfer to or from the environment by increasing convection. The amount of conduction, convection or radiation of an object determines the amount of heat it transfers. Increasing the temperature gradient between the object and the environment, increasing the

Figure 6.1 Aircraft fin.

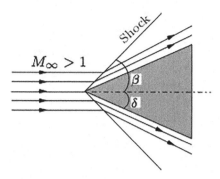

Figure 6.2 Shock wave.

convective heat transfer coefficient, or increasing the surface area of the object increases the heat transfer. Sometimes, it is not feasible or economical to change the first two options. Thus, adding a fin to an object increases the surface area and can sometimes be an economical solution to heat transfer problems.

6.18 FINITE DISTURBANCE

When the wedge angle δ is finite, the disturbances introduced are finite. Then the wave is not called a Mach wave but a shock or shock wave (see Figure 6.2). The angle of shock β is always smaller than the Mach angle. The deviation of the streamlines is finite and the pressure increase across a shock wave is finite.

6.19 FINITE OR THREE-DIMENSIONAL WING

For a finite wing, the span is finite, and the flow at the wing tips can easily establish a cross-flow, moving from the higher pressure to lower pressure. For a wing experiencing positive lift, the pressure over the lower surface is higher than the pressure over the upper surface. This would cause a flow communication from the bottom to the top at the wing tips. This tip communication would establish span-wise variations in the flow.

6.20 FIRE

Fire is the result of applying enough heat to a fuel source when you've got a whole lot of oxygen around.

6.21 FIRST LAW OF THERMODYNAMICS

The first law of thermodynamics states that 'the heat added minus work done by the system is equal to the change in the internal energy of the system', that is, $\delta Q - \delta W = dU$. This is an empirical result confirmed by laboratory experiments and practical experience. The internal energy U is a state variable (thermodynamic property). Hence, the change in internal energy dU is an exact differential and its value depends only on the initial and final states of the system. In contrast (the nonthermodynamic properties), δQ and δW depend on the process by which the system attains its final state from the initial state.

6.22 FISSION ENERGY

Fission occurs when a neutron slams into a larger atom, forcing it to excite and spilt into two smaller atoms also known as fission products. Additional neutrons are also released that can initiate a chain reaction. When each atom splits, a tremendous amount of energy is released.

6.23 FIVE-HOLE YAW PROBES

Five-hole yaw probes are used for measuring the flow direction in three-dimensional flows.

6.24 FIXED ERROR

Fixed (or systematic) error, which makes repeated measurements to be in error by the same amount for each trial. This error is the same for each reading and can be removed by proper calibration or correction.

6.25 FLAMMABLE

Flammable means capable of being easily ignited and of burning quickly.

6.26 FLAMMABLE LIQUID

A flammable liquid is a combustible liquid that can be easily ignited in the air at ambient temperatures, i.e., it has a flash point at or below nominal threshold temperatures defined by several national and international standards organisations.

Figure 6.3 Fire flame.

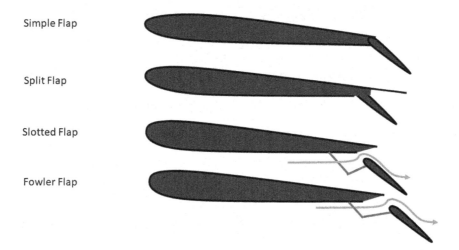

Figure 6.4 Some flaps.

6.27 FLAME

A stream of hot, burning gas from something on fire, as in Figure 6.3.

6.28 FLAP

It is a control surface used to increase the lift on the wing. Some typical flaps are shown in Figure 6.4.

The flap at the trailing edge of an aerofoil is essentially a high-lift device, which when deflected down causes an increase of lift, essentially by increasing the camber of the profile.

6.29 FLAP EFFECT ON AERODYNAMIC CENTRE

Although flap deflection provides a nose-down change in the pitching moment, this is associated simply with a negative change in the zero-lift pitching moment coefficient. There is generally no significant effect on the derivative dC_M/dC_L and therefore no change in the position of the aerodynamic centre.

6.30 FLAPERONS

Flaperons are a specialised type of aircraft flight control surface that combine aspects of both flaps and ailerons.

This has already been used in some types of aircraft, where the ailerons also act as flaps, therefore, known as 'flaperons'.

6.31 FLAPPED AEROFOIL

The flap at the trailing edge of an aerofoil is essentially a high-lift device, which when deflected down causes increase of lift, essentially by increasing the camber of the profile.

6.32 FLAPPING

Flapping is the movement of the blade up or down with respect to the plane of rotation.

6.33 FLASH FLOOD

A flash flood is caused by heavy or excessive rainfall in a short period of time, generally <6 h. Flash floods are usually characterised by raging torrents after heavy rains that rip through riverbeds, urban streets or mountain canyons sweeping everything before them.

Flash floods occur within a few minutes or hours of excessive rainfall, a dam or levee failure or a sudden release of water held by an ice jam. Most flash flooding is caused by slow-moving thunderstorms, thunderstorms repeatedly moving over the same area or heavy rains from hurricanes and tropical storms.

6.34 FLAT TURN

A turn involving pure yaw is called a flat turn.

6.35 FLATTENING OUT FLIGHT

The flattening out flight mode involves a change of direction, acceleration and force towards the centre of the curved path. This force must be provided by the wings. Therefore, the wings should have more speed and more angle of attack. These increased speed and angle lead to higher stalling speed. The steeper the original glide, the greater the change in the flight path involved; the aircraft should have more speed for flattening out. This means that the steeper the glide, the faster must the glide speed be. However, faster gliding speed is not desirable.

6.36 FLEXIBLE

Flexible means able to bend or move easily without breaking.

6.37 FLIGHT

Flight or flying is the process by which an object moves through space without contacting any planetary surface, either within an atmosphere (i.e., air flight or aviation) or through the vacuum of outer space (i.e., spaceflight). This can be achieved by generating aerodynamic lift associated with gliding or propulsive thrust, aerostatically using buoyancy or by ballistic movement.

6.38 FLOAT

A float is a small object attached to a fishing line that floats on the water and moves when a fish has been caught.

6.39 FLOATING ELEMENT METHOD

The floating element method is a direct method for measuring the wall shear stress. This method is used for measuring skin friction directly on a flat plate. A small segment of the flat plate surface is separated from the remaining portion of the plate by a very small gap on either side but kept flush by suspending the element from a set of leaf springs forming a parallelogram linkage, as shown in the Figure 6.5.

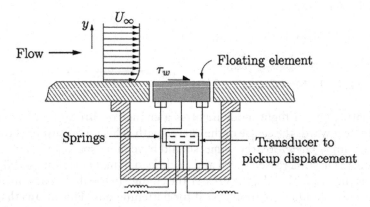

Figure 6.5 Shear stress balance.

When there is a flow over the entire plate, the element is subjected to a shear flow that causes a displacement. The vertical faces of the element being subjected to the same normal pressure all around, the forces acting on these faces get cancelled. The movement of the element is an indication of the shear stress and by pre-calibration of the complete floating system, the actual shear stress can be determined.

6.40 FLOOD

A flood is an overflow of water on normally dry ground. This is most commonly due to an overflowing river, a dam break, snowmelt or heavy rainfall.

6.41 FLOW ANALYSIS

Basically, two treatments are followed for fluid flow analysis. They are *Lagrangian* and *Eulerian* descriptions. The Lagrangian method describes the motion of each particle of the flow field separately and discretely, whereas the Eulerian method describes the variations at all fixed stations as a function of time.

6.42 FLOW CONTROL VALVE

A flow control valve regulates the flow or pressure of a fluid. Flow control valves are used to reduce the rate of flow in a section of a pneumatic circuit, resulting in a slower actuator speed.

6.43 FLOW DEFLECTION ANGLE

All the streamlines passing through an oblique shock are deflected to the same angle θ at the shock, resulting in uniform parallel flow downstream of the shock. The angle θ is referred to as the flow deflection angle.

6.44 FLOW DEVELOPMENT LENGTH

In a pipe flow, the zone upstream of the boundary layer merging point is called the flow development length.

6.45 FLOW-INDUCED VIBRATION

Flow-induced vibration is the result of turbulence in the process fluid, which occurs due to major flow discontinuities such as bends, tees, partially closed valves and small bore connections.

Flow-induced vibration, or vortex shedding, is due to high-flow velocities such as in a piping dead leg of a centrifugal compressor system.

6.46 FLOW MEASUREMENT

Flow measurement is the quantification of bulk fluid movement. Flow can be measured in a variety of ways. Flow measurement methods other than positive-displacement flow meters rely on forces produced by the flowing stream as it overcomes a known constriction to indirectly calculate flow. Flow may be measured by measuring the velocity of fluid over a known area. For very large flows, tracer methods may be used to deduce the flow rate from the change in the concentration of a dye or radioisotope.

6.47 FLOW METER

A flow meter is an instrument used to measure linear, nonlinear, and volumetric or mass flow rate of a liquid or a gas.

6.48 FLOW OF MULTICOMPONENT MIXTURES

This field is simply an extension of basic fluid mechanics. The analysis of the flow of homogeneous fluid consisting of single species, termed basic fluid mechanics, is extended to study the flow of chemically reacting component mixtures made of more than one species. All three means of transport,

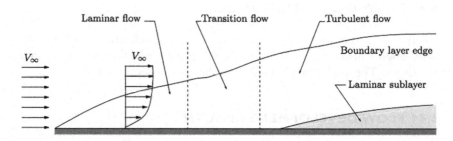

Figure 6.6 Flow past a flat plate.

namely momentum transport, energy transport and mass transport, are considered in this study, unlike the basic fluid mechanics where only transport of momentum and energy are considered.

6.49 FLOW PAST A FLAT PLATE

Different zones of the boundary layer over a flat plate are shown in Figure 6.6. The laminar sublayer is that zone adjacent to the boundary, where the turbulence is suppressed to such a degree that only the laminar effects prevail. The various regions shown in Figure 6.6 are not sharp demarcations of different zones. There is actually a gradual transition from one region, where a certain effect predominates, to another region, where some other effect is predominant.

6.50 FLOW THROUGH PIPES

Fluid flow through pipes with circular and noncircular cross-sections is one of the commonly encountered problems in many practical systems. Flow through pipes is driven mostly by pressure or gravity or both.

6.51 FLOW SEPARATION

Whenever there is relative movement between a fluid and a solid surface, whether externally around a body or internally in an enclosed passage, a boundary layer exists with viscous forces present in the layer of fluid close to the surface. Boundary layers can be either laminar or turbulent. A reasonable assessment of whether the boundary layer will be laminar or turbulent can be made by calculating the Reynolds number of the local flow conditions.

6.52 FLOW VISUALISATION

Visualisation of fluid flow motion proved to be an excellent tool for describing and calculating flow properties in many problems of practical interest, in both subsonic and supersonic flow regimes. Flow visualisation in fluid dynamics is used to make the flow patterns visible to get qualitative or quantitative information on them.

6.53 FLOW WORK

The work done in pushing the fluid element across the boundary is called the flow work. In other words,, the energy required to push fluid mass into and out of a control volume is called the flow work or flow energy.

6.54 FLUID

Fluid is defined as the substance which deforms continuously under the action of tangential (shear) force however small it may be. It may also be defined as a substance that will continue to change shape as long as there is a shear stress present however small it may be. So, the basic feature of a fluid is that it can flow, and this is the essence of any definition of it.

6.55 FLUID DROP

A fluid drop or droplet is a small column of fluid bounded completely or almost completely by free surfaces.

6.56 FLUID ELEMENT

A fluid element can be defined as an infinitesimal region of the fluid continuum in isolation from its surroundings.

6.57 FLUID MECHANICS

Fluid mechanics may be defined as the science of fluid flow in which the temperature change associated with flow speed is negligibly small. Treating any change <5% as negligible, in accordance with engineering practice, any fluid flow with a speed <650 km/h (about 180 m/s), at standard sea-level conditions, can be treated as a fluid mechanic stream.

6.58 FLUID MECHANICS OF PERFECT FLUIDS

Fluid mechanics of perfect fluids – fluids without viscosity and heat (transfer) conductivity – is an extension of equilibrium thermodynamics to moving fluids.

6.59 FLUID MECHANICS OF REAL FLUIDS

Fluid mechanics of real fluids – goes beyond the scope of classical thermodynamics. The transport processes of momentum and heat (energy) are of primary interest here. But, even though, thermodynamics is not fully and directly applicable to all phases of real fluid flow, it is often extremely helpful in relating the initial and final conditions.

6.60 FLUID PARTICLE

The concept of a fluid particle is fundamental to studying the physics of the fluid flow. Three types of fluid particles are currently used: the finite, the infinitesimal and the point particle. The finite particle renders the atomic-molecular structure of matter and is particularly relevant in flow measurements.

6.61 FLUID STATICS

Fluid statics deal with fluid elements at rest with respect to one another and thus free of shearing stress. The static pressure distribution in a fluid and on bodies immersed in a fluid can be determined from a static analysis.

6.62 FLUIDS AND THE CONTINUUM

Fluid flows may be modelled on either microscopic or macroscopic levels. The macroscopic model regards the fluid as a continuum, and the description is in terms of the variations of macroscopic velocity, density, pressure and temperature with distance and time. On the other hand, the microscopic or molecular model recognises the particulate structure of a fluid as a myriad of discrete molecules and ideally provides information on the position and velocity of every molecule at all times.

6.63 FLUID-JET ANEMOMETER

The difficulty in the measurements of small differential pressure with sufficient accuracy imposes a limitation on pitot-static probes for measurement

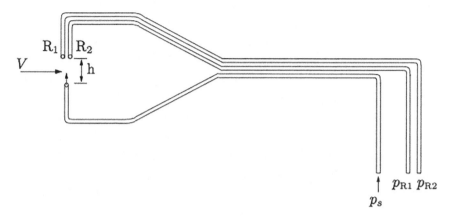

Figure 6.7 Fluid-jet anemometer.

of flow velocities of the order of 2 m/s. This problem may be sorted out by using the device referred to as a fluid-jet anemometer.

As seen in Figure 6.7, a jet is directed across the flow and is deflected to an extent. The deflection of the jet strongly depends on the ratio of the flow velocity V to jet velocity. The pressure difference Δp measured with the tubes R_1 and R_2 is made use of for determining V. The design of appropriate geometry of the probe, together with a suitable choice of jet supply pressure p_s, creates pressure differences Δp which are many times the dynamic pressure $\frac{1}{2} \rho V^2$. This kind of instrument has been used successfully in various industrial-processing plants.

6.64 FLUID THERMOMETERS

A thermometer is a general name given to temperature measuring devices. Some of the commonly used thermometers are mercury-in-glass thermometers (Figure 6.8), Beckmann thermometers and gas thermometers.

6.65 FLUORINATED GASES

Fluorinated gases belong to the category of greenhouse gases. Fluorinated gases are gases that nearly exclusively result from human activities (e.g., industrial/manufacturing processes), i.e., there are no natural sources of fluorinated gases. These gases can be found in different products such as a fridge or aerosol can. While they are not ozone depleting, their Global Warming Potential is very high.

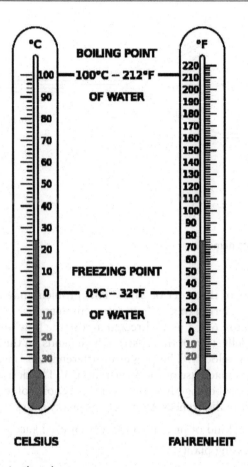

°C   BOILING POINT   °F

100 ——— 100°C -- 212°F ——— 220 210
                                                      200
90   OF WATER   190
                                                      180
80   170
70   160
                                                      150
60   140
                                                      130
50   120
                                                      110
40   100
                                                      90
30   80
20   70
                                                      60
10   FREEZING POINT   50
                                                      40
0 ———————— 0°C -- 32°F ————   30
                                                      20
10   OF WATER   10
                                                      0
20   10
30   20

CELSIUS   FAHRENHEIT

Figure 6.8 Mercury-in-glass thermometer.

6.66 FLUORINATED GREENHOUSE GASES

Fluorinated greenhouse gases are mostly used as cooling agents in refrigeration and air conditioning equipment. They are also used as blowing agents in insulating foam or as components in fire extinguishers.

6.67 FLUORINE

Fluorine is a chemical element with the symbol F and atomic number 9. It is the lightest halogen and exists under standard conditions as a highly toxic, pale yellow diatomic gas. As the most electronegative element, it is extremely reactive, as it reacts with all other elements, except for argon, neon and helium.

6.68 FLUSHING FLOW

Flushing flows are deliberate high-flow releases designed to mimic the effects of floods in removing fine sediments from downstream aquatic habitats.

6.69 FLUSH TOILET

A flush toilet (also known as a flushing toilet, water closet (WC)) is a toilet that disposes of human waste (principally urine and faeces) by using the force of water to flush it through a drainpipe to another location for treatment, either nearby or at a communal facility, thus maintaining a separation between humans and their waste. Flush toilets can be designed for sitting (in which case they are also called 'western' toilets) or for squatting, in the case of squat toilets. Most modern sewage treatment systems are also designed to process specially designed toilet paper. The opposite of a flush toilet is a dry toilet, which uses no water for flushing.

6.70 FLYING SAUCER

A flying saucer is a descriptive term for a type of flying craft having a disc or saucer-shaped body, commonly used generically to refer to an anomalous flying object.

6.71 FOG

Fog is a visible aerosol consisting of tiny water droplets or ice crystals suspended in the air at or near the Earth's surface (see Figure 6.9). Fog can be considered a type of low-lying cloud usually resembling stratus, and is heavily influenced by nearby bodies of water, topography and wind conditions.

6.72 FORCE AND MOMENT OF A FLYING MACHINE

The important aerodynamic forces and moments associated with a flying machine, such as an aircraft, are the lift L, drag D, and pitching moment M.

6.73 FORCED VORTEX

A forced vortex is a rotational flow field in which the fluid rotates as a solid body with a constant angular velocity, and the streamlines form a set of concentric circles. Because the fluid in a forced vortex rotates like a rigid body,

Figure 6.9 Dense fog.

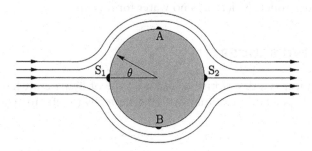

Figure 6.10 Potential flow past a cylinder.

the forced vortex is also called a flywheel vortex. In addition to moving along concentric circular paths, if the fluid elements spin about their axes, the flow field is termed a forced vortex.

6.74 FORM DRAG

Form drag is the drag force due to the difference in pressure between the front and back regions of a body. Drag arises due to (a) the difference in pressure between the front and back regions; (b) the friction between the body surface and the fluid. Drag force (a) is known as *form* or *pressure drag* and (b) is also known as *skin friction drag or shear drag*. If there is no viscosity, and hence, no boundary layer, the flow would have gone around the cylinder like a true potential flow, as shown in Figure 6.10. For this flow, the pressure distribution will be the same on the front and back sides. The net force along the freestream direction is then zero. So, there is no drag acting on the cylinder.

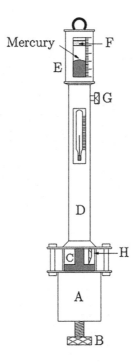

Figure 6.11 Fortin barometer.

6.75 FORTIN BAROMETER

The typical Fortin barometer is illustrated schematically in Figure 6.11. This is a reservoir- or cistern-type barometer, as shown in Figure 6.11. The reservoir of mercury is contained in metal tube A. The bottom of the reservoir, inside tube A, is made of a leather bag and this leather bag can be pushed up or down by rotating the adjusting screw B. The surface level of mercury in the reservoir can be observed through the observation tube C.

The barometer tube passes up through a metal tube D and the surface level of the mercury at barometric height can be observed through a window in the scale tube E. A sliding vernier F is fitted across this window and the height of the vernier can be set by means of the vernier screw G. The zero of the scale is located at the bottom of the barometer. A pointer H is fitted to the top of the observation tube C and the tip of the pointer is zero on the scale. The measuring technique of the barometer is as follows: The surface level of the mercury in reservoir A is adjusted by means of the adjusting screw B until its surface just touches pointer H. Vernier screw G is now rotated until the bottom of the vernier F is in line with the surface meniscus of the mercury at the top of the barometer. To avoid error in this

Figure 6.12 Aircraft with forward-swept wing.

adjustment, scale tube E is usually provided with another window cut in the rear. The vernier is also made in the form of a tube. If the vernier bottom surface and the mercury meniscus are brought in line, then the vernier adjustment can be taken as accurate.

6.76 FORWARD-SWEPT WING

A forward-swept wing is an aircraft wing configuration in which the quarter-chord line of the wing has a forward sweep, as in Figure 6.12. Typically, the leading edge also sweeps forward.

6.77 FOSSIL FUELS

Fossil fuels are hydrocarbons, primarily coal, fuel oil or natural gas, formed from the remains of dead plants and animals. In common dialogue, the term fossil fuel also includes hydrocarbon-containing natural resources that are not derived from animal or plant sources.

6.78 FOUNTAIN

A fountain is a structure that squirts water into a basin to supply drinking water. It is also a structure that jets water into the air for a decorative or dramatic effect, as in Figure 6.13.

Figure 6.13 A fountain.

6.79 FOURIER'S LAW OF HEAT CONDUCTION

Fourier's law of heat conduction $q = -k\nabla T$, where q is the local heat flux [energy/(area×time)], k is the material's thermal conductivity and T is the temperature. The law states that the heat flux via conduction through a material is proportional to the temperature gradient.

6.80 FOWLER FLAP

The Fowler flap is similar to the slotted flap, but a further effect is added. In addition to being deflected downward and opening up of a slot, the flap slides backwards, as shown in Figure 6.14.

Thus, the Fowler flap provides an additional increase in the effective wing area. Therefore, the lift increment caused by the Fowler flap is larger than

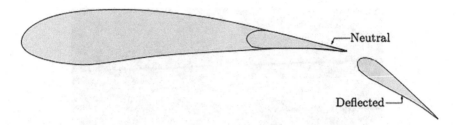

Figure 6.14 A Fowler flap in neutral and deflected positions.

that provided by the other flaps discussed. There is also a reduction in the effective thickness-to-chord ratio, which tends to make the wing stall a little later. The moment effect is large because the flap, which carries much of the increment, is situated so much to the rear. The drag increment, however, is smaller because of the slot effect and reduced thickness-to-chord ratio.

6.81 FRANCIS TURBINE

The Francis turbine is a reaction turbine, which means that the working fluid changes pressure as it moves through the turbine, giving up its energy. A Francis turbine is a large rotary machine that works to convert kinetic and potential energy into hydroelectricity. Francis Turbine is a combination of both impulse and reaction turbine, where the blades rotate using both reaction and impulse force of water flowing through them producing electricity more efficiently.

6.82 FREE-MOLECULAR FLOW

Free-molecular flow is a regime of fluid flow in which the fluid molecules are so widely dispersed that the intermolecular forces can be neglected. For free-molecular flows, the Knudsen number (the ratio of the mean free path to characteristic dimension) is >5.

6.83 FREE SPIRAL VORTEX

A free spiral vortex is essentially the combination of a free cylindrical vortex and a radial flow. The fluid rotates and flows radially forming a free spiral vortex in which a fluid element will follow a spiral path, as shown in Figure 6.15.

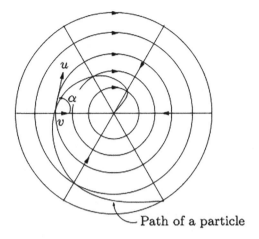

Figure 6.15 A free spiral vortex.

6.84 FREE SURFACE

A surface that is not constrained by a solid boundary. Usually the interface between two fluids.

6.85 FREE SURFACE FLOW

A free surface flow, also called open channel flow, is the gravity-driven flow of a fluid under a free surface, typically water flowing under air in the atmosphere.

6.86 FREE VORTEX

A free vortex is an irrotational flow field in which the streamlines are concentric circles, but the variation of velocity with radius is such that there is no change of total energy per unit weight with radius. Since the flow field is potential, the free vortex is also called the potential vortex.

6.87 FREON

Freon is a non-combustible gas that is used as a refrigerant in air conditioning applications. This freon undergoes an evaporation process over and over again to help produce cool air that can be circulated throughout your AC system.

6.88 FREQUENCY OF SOUND

The frequency of sound is defined as the number of pressure variations per second. The frequency is measured in hertz (Hz).

6.89 FRICTION COEFFICIENT

Friction coefficient is the non-dimensional representation of the viscous friction in fluid flows.

The coefficient of drag, or the coefficient of friction, as it is generally referred to for flow in ducts, is defined as

$$f \equiv \frac{\text{Wall shear stress}}{\text{Dynamic pressure head of the stream}}$$

6.90 FRICTIONAL DRAG

Frictional drag is a tangential traction or drag force acting in the direction of flow velocity.

6.91 FRISE

Frise or other specially shaped ailerons. This aileron when moved downwards the complete top surface of the wing and the aileron will have a smooth, uninterrupted contours causing very little drag. However, when the aileron is moved upwards, it will project below the bottom surface of the wing, and cause excessive drag.

6.92 FRONTAL AREA

It is the frontal area of the body as seen from the flow stream. It is suitable for thick stubby bodies, such as spheres, cylinders, cars, missiles, projectiles and torpedoes.

6.93 FROST

Frost is a covering of ice crystals on the surface produced by the depositing of water vapour to a surface cooler than $0°C$. The deposition occurs when the temperature of the surface falls below the frost point. Similarly, dew forms when the air or surface temperature falls below the dew point temperature.

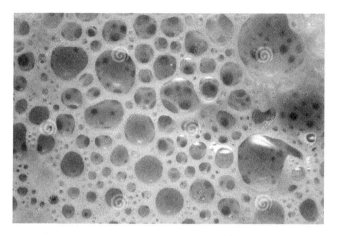

Figure 6.16 Coffee froth.

6.94 FROTH

A mass of small white bubbles on the top of a liquid, etc., as in Figure 6.16.

6.95 FROTH FLOTATION

Froth flotation is a process for selectively separating hydrophobic materials from hydrophilic. This is used in mineral processing, paper recycling and wastewater treatment industries.

6.96 FROUDE NUMBER

Froude number is the ratio of inertia force to gravity force. Usually, it is expressed as

$$Fr = \frac{V}{\sqrt{Lg}}$$

Essentially, it is a similarity parameter.

6.97 FROZEN FLOW

A flow in which the reaction rates are practically zero is termed frozen flow.

6.98 FUEL

Fuel is a material that is burned to produce heat or power.

6.99 FUGACITY

Fugacity is a thermodynamic property that is useful in studying mixtures. It is essentially a pseudo-pressure.

6.100 FULLY DEVELOPED FLOW

In a pipe flow, the zone downstream of the boundary layer merging point is termed a fully developed region. In the fully developed region, the velocity profile remains unchanged. Fully developed flow is that with velocity profile self-similar.

6.101 FULLY DEVELOPED PIPE FLOW

A fully developed pipe flow is the one in which the effects of viscosity are fully present and the pipe entrance effects are not taken into account.

6.102 FULLY DEVELOPED REGION

Beyond the transition region, the jet becomes similar in appearance to a flow of fluid from a source of infinitely small thickness (in an axially symmetric case, the source is a point, and in a plane parallel case, it is a straight line perpendicular to the plane of flow of the jet). In reality, the jet velocity becomes insignificant after about $30D_e$.

6.103 FUMES

Any smoke-like or vaporous exhalation from matter or substances, especially of an odorous or harmful nature (Figure 6.17).

6.104 FUSELAGE

In aeronautics, the fuselage is an aircraft's main body section. It holds crew, passengers or cargo. In single-engine aircraft, it will usually contain an

Figure 6.17 Smoke fume.

engine as well, although, in some amphibious aircraft, the single engine is mounted on a pylon attached to the fuselage, which in turn is used as a floating hull. The fuselage also serves to position the control and stabilisation surfaces in specific relationships to lifting surfaces, which is required for aircraft stability and manoeuvrability.

6.105 FUSION POWER

Fusion power is an experimental form of power generation that generates electricity by using nuclear fusion reactions. In a fusion process, two atomic nuclei combine to form a heavier nucleus while releasing energy. Devices that produce energy in this way are known as fusion reactors.

6.106 FUSION PYROMETERS

The variation of melting point of certain chemical substances is utilised for temperature measurements in fusion pyrometers. Combining certain clays, metals and salts, it is possible to obtain mixtures of these substances which soften and melt at definite temperatures. These mixtures are made generally in the form of slender cones and pyramids, as shown in the Figure 6.18.

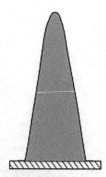
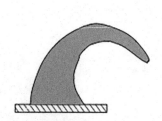

Figure 6.18 Fusion pyrometer cones.

They are used for measuring the temperature of a furnace, kilns, etc. When the required temperature is reached, the cone signifies this fact by beginning to become soft and curling over, as shown in the figure. The disadvantage of this technique is that these cones can be used only once.

Chapter 7

Gas to Gyroscopic Effect

7.1 GAS

The state of matter is distinguished from the solid and liquid states by relatively low density and viscosity, relatively great expansion and contraction with changes in pressure and temperature, the ability to diffuse readily, and the spontaneous tendency to become distributed uniformly throughout any container. So, gas is a fluid (such as air) that has neither independent shape nor volume but tends to expand indefinitely. In the gas phase, the molecules are far apart from each other, and a molecular order is non-existent. Compared to solid and liquid phases, in the gas phase, the molecules are at a considerably higher energy level. Therefore, the gas must release a large amount of heat before it can condense or freeze. Gas is a state of matter that has no fixed shape and no fixed volume. Gases have a lower density than other states of matter, such as solids and liquids. There is a great deal of space between particles, which have a lot of kinetic energy.

7.2 GAS CONSTANT

Gas constant R is the ratio of the universal gas constant, R_u to the molecular weight, M_m of the gas.

$$R = \frac{R_u}{M_m}$$

where R_u is the universal gas constant and is equal to 8,314 J/(kg K).

7.3 GAS DYNAMICS

Gas dynamics may be defined as the science of flow field where a change in pressure is accompanied by density and temperature changes. The theory of gas dynamics deals with the dynamics and thermodynamics of the flow of compressible fluids.

DOI: 10.1201/9781003348405-7

Gas dynamics is a science that primarily deals with the behaviour of gas flows in which compressibility and temperature change become significant. Compressibility is a phenomenon by virtue of which the flow changes its density with a change in speed.

In the theory of gas dynamics, the change of state or flow properties is achieved by the following three means. (a) With area change, treating the fluid to be inviscid and passage to be frictionless. (b) With friction, treating the heat transfer between the surroundings and the system to be negligible. (c) With heat transfer, assuming the fluid to be inviscid.

These three types of flows are called the *isentropic flow*, in which the area change is the primary parameter causing the change of state; the frictional or *Fanno flow*, in which the friction is the primary parameter causing the change of state; and the *Rayleigh flow*, in which the change in the stagnation (total) temperature (that is, heat addition or heat removal) is the primary parameter causing the change of state.

7.4 GAS DYNAMICS DIVISIONS

Regardless of speed ranges, the theory of gas dynamics can be divided into two parts, inviscid gas dynamics and viscous gas dynamics. The inviscid theory is important in the calculation of nozzle characteristics, shock waves, lift, and wave drag of a body, while the viscous theory is applicable to the calculation of skin friction and heat transfer characteristics of a body moving through a gas, such as atmospheric air.

7.5 GAS MIXTURE

One of the properties of gases is that they mix with each other. In gas mixtures, each component in the gas phase can be treated separately. Each component of the mixture shares the same temperature and volume.

7.6 GAS MOLECULES

Gas molecules are made up of several atoms bonded to one another. These interatomic bonds are similar to springs connecting atoms of various masses. This bonding vibrates with a fixed frequency called the natural frequency.

7.7 GAS REFRIGERATION CYCLE

The gas refrigeration cycle is nothing but the reversed Brayton cycle.

7.8 GAS THERMOMETER

It was originally used as the fundamental temperature-measuring device to establish the various temperatures given as the International Temperature Scale in relation to the originally conceived fixed points of the freezing point and boiling point of pure water.

7.9 GAS TURBINE

A gas turbine, also called a combustion turbine, is a type of continuous and internal combustion engine.

7.10 GASOLINE

Gasoline or petrol is a transparent, petroleum-derived flammable liquid that is used primarily as a fuel in most spark-ignited internal combustion engines.

7.11 GASOLINE BOILING POINT

The hydrocarbons of gasoline contain typically 4–12 carbon atoms with a boiling range between 30°C and 210°C, whereas diesel fuel contains hydrocarbons with approximately 12–20 carbon atoms and the boiling range is between 170°C and 360°C.

7.12 GATE VALVE

A gate valve, also known as a sluice valve, is a valve that opens by lifting a barrier (gate) out of the path of the fluid. Gate valves require very little space along the pipe axis and hardly restrict the flow of fluid when the gate is fully opened.

7.13 GEOGRAPHY

Geography is a field of science devoted to the study of lands, features, inhabitants, and phenomena of the Earth and planets. Geography is an all-encompassing discipline that seeks an understanding of Earth and its human and natural complexities not merely where objects are, but also how they have changed and come to be.

7.14 GEOMETRIC TWIST

Geometric twist defines the situation where the chord lines for the spanwise distribution of all the aerofoil sections do not lie in the same plane.

7.15 GEOMETRICAL ANGLE OF ATTACK

The geometrical angle of attack α is the angle between the chord line and the direction of the undisturbed freestream.

7.16 GEOMETRICAL INCIDENCE

For an airplane as a whole, the geometrical incidence will be defined as the angle between the direction of motion and the chord of the aerofoil.

7.17 GEOPHYSICS

Geophysics is a subject of natural science concerned with the physical processes and physical properties of the Earth and its surrounding space environment, and the use of quantitative methods for their analysis. The term geophysics sometimes refers only to solid earth applications: Earth's shape; its gravitational and magnetic fields; its internal structure and composition; its dynamics and surface expression in plate tectonics, the generation of magmas, volcanism, and rock formation. However, modern geophysics organisations and pure scientists use a broader definition that includes the water cycle including snow and ice; fluid dynamics of the oceans and the atmosphere; electricity and magnetism in the ionosphere and magnetosphere and solar-terrestrial physics; and analogous problems associated with the Moon and other planets.

7.18 GEOSTATIONARY ORBIT

A geostationary orbit is a circular geosynchronous orbit 35,786 km in altitude above Earth's equator and following the direction of Earth's rotation.

7.19 GEOSTATIONARY SATELLITE

A geostationary satellite is an Earth-orbiting satellite, placed at an altitude of approximately 35,800 km directly over the equator that revolves in the same direction the Earth rotates (west to east).

7.20 GEOSTROPHIC MOTION

In a slowly rotating fluid, it can be shown theoretically that the Coriolis force is completely balanced by the pressure gradient induced by the centrifugal force. This is known as geostrophic motion.

7.21 GEOTHERMAL ENERGY

Geothermal energy is heat within the earth. The word geothermal comes from the Greek words *geo* (earth) and *therme* (heat). Geothermal energy is a renewable energy source because heat is continuously produced inside the earth. People use geothermal heat for bathing, to heat buildings, and to generate electricity.

7.22 GEOTHERMAL GRADIENT

Geothermal gradient is the rate of temperature change with respect to increasing depth in the Earth's interior.

7.23 GENERAL MOTION

In general motion, translation and rotation are compounded. Such a motion is found, for example, in the wake of a bluff body.

7.24 'g' FACTOR

Pilots often talk about pulling a certain number of g. This quantity is just a number and not acceleration. It has no units and it simply represents a factor, which when multiplied by the weight gives the total force that must be applied to a body to balance the combined effects of gravity and centripetal acceleration. It is really a load factor because it reflects the loads and stresses in the airframe increase during a manoeuvre.

7.25 GIBBS EQUATION

The conservation of energy equation for an internally reversible process of a closed system consisting of a simple compressible substance can be expressed as

$$TdS = dU + p \, dV$$

where T is the temperature, S is the entropy, U is the internal energy, p is the pressure and V is the volume. On a unit mass basis

$$Tds = du + pdv$$

The equation is known as the first $T\ ds$, or Gibbs equation.

7.26 GIBBS FREE ENERGY

In thermodynamics, the Gibbs free energy is a thermodynamic potential that can be used to calculate the maximum reversible work that may be performed by a thermodynamic system at a constant temperature and pressure.

7.27 GLACIER

A glacier is a huge mass of ice that moves slowly over land. The term 'glacier' comes from the French word *glace* (glah-SAY), which means ice. Glaciers are often called 'rivers of ice'. They fall into two groups: alpine glaciers and ice sheets, as in Figure 7.1.

Figure 7.1 Glacier.

7.28 GLADSTONE–DALE EMPIRICAL EQUATION

Gladstone–Dale empirical equation relates the refractive index n with the density of the medium as

$$\frac{n-1}{\rho} = K$$

where K is the Gladstone–Dale constant, and is constant for a given gas and ρ is the gas density.

7.29 GLIDER

A glider is an aircraft without an engine which flies by floating on air currents.

7.30 GLIDING

Gliding is a flight under the action of gravity and without the use of the engine, as in Figure 7.2.

When an aircraft is in a steady gliding, it must be kept in a state of equilibrium by the lift, drag, and weight only. That is, in a gliding flight, the resultant of the lift and drag forces must be equal and opposite of the weight.

Figure 7.2 A glider in flight.

7.31 GLIDING TURN

On a gliding turn, the whole aircraft will move the same distance downward during one complete turn, but the inner wing, because it is turning on a smaller radius, will have descended on a steeper spiral than the outer wing. Therefore, the air will have to come up to meet it at a steeper angle; in other words, the inner wing will have a larger angle of attack and so generate more lift than the outer wing. The extra lift obtained in this way may compensate for the lift generated by the outer wing due to the increase in velocity. Thus, in a gliding turn, there may be no need to hold on to the bank.

7.32 GLOBAL WARMING

Global warming is the phenomenon of a gradual increase in the temperature near the Earth's surface.

7.33 GLOBE VALVE

A globe valve is a type of valve used for regulating flow in a pipeline, consisting of a movable plug or disc element and a stationary ring seat in a generally spherical body.

7.34 GLUCOSE

Glucose is neither acidic nor basic. It is considered to be neutral and its pH value is also 7. It does not donate hydrogen ions on dissolving as most acids do. Neither does it donate hydroxyl ions like the base.

7.35 GLYCERIN

Glycerin, also known as glycerol, is a natural compound derived from vegetable oils or animal fats.

7.36 GLYCOLIC ACID

Glycolic acid is a water-soluble alpha hydroxy acid (AHA) made from sugar cane. It is one of the most widely used AHAs in skincare products. AHAs are natural acids that come from plants. They consist of tiny molecules that are very easy for your skin to absorb.

7.37 GLYCOLS

Glycol is a class of alcohol that possesses two hydroxyl functional groups attached to two adjacent carbon atoms.

7.38 GOTHERT'S RULE

The streamline analogy, which states that the slope of a profile in a compressible flow pattern is larger by the factor

$$1/\sqrt{1-M_\infty^2}$$

than the slope of the corresponding profile in the related incompressible flow pattern is known as Gothert's rule.

7.39 GRADUALLY VARIED FLOW

Gradually varied flow is a non-uniform flow in which changes in depth and velocity take place over a long distance.

7.40 GRASHOF NUMBER

Grashof number, Gr, provides a measure of the ratio of buoyancy force to viscous force in the boundary layer. Its role in free convection is much the same as that of the Reynolds number in forced convection. The Grashof number (Gr) is dimensionless in fluid dynamics and heat transfer, which approximates the ratio of the buoyancy to the viscous force acting on a fluid. It frequently arises in the study of situations involving natural convection and is analogous to the Reynolds number.

7.41 GRAVITATIONAL WORK

Gravitational work can be defined as the work done by or against the gravitational force field.

7.42 GRAVITY

Gravity, or gravitation, is a natural phenomenon by which all things with mass or energy, including planets, stars, galaxies, and even light are attracted to one another. On Earth, gravity gives weight to physical objects, and the Moon's gravity causes ocean tides.

7.43 GRAVITY DAM

A gravity dam is a dam constructed from concrete or stone masonry and designed to hold backwater by using only the weight of the material and its resistance against the foundation to oppose the horizontal pressure of water pushing against it.

7.44 GRAVITY WAVES

Waves whose properties are determined by gravity effects are referred to as gravity waves.

7.45 GREASE

Grease is a solid or semisolid lubricant, as in Figure 7.3, formed as a dispersion of thickening agents in a liquid lubricant.

Figure 7.3 Grease.

7.46 GREEN TEA

Green tea is made from the *Camellia sinensis* plant. The dried leaves and leaf buds of *Camellia sinensis* are used to produce various types of teas.

7.47 GREENHOUSE EFFECT

The greenhouse effect is a natural process that warms the Earth's surface. Greenhouse gases include water vapour, carbon dioxide, methane, nitrous oxide, ozone, and some artificial chemicals such as chlorofluorocarbons (CFCs). The absorbed energy warms the atmosphere and the surface of the Earth.

7.48 GREENHOUSE GAS

A gas that contributes to the greenhouse effect by absorbing infrared radiation. Carbon dioxide and chlorofluorocarbons are examples of greenhouse gases.

A greenhouse gas is a gas that absorbs and emits radiant energy within the thermal infrared range, causing the greenhouse effect. The primary greenhouse gases in Earth's atmosphere are water vapour (H_2O), carbon dioxide (CO_2), methane (CH_4), nitrous oxide (N_2O), and ozone (O_3). Without greenhouse gases, the average temperature of Earth's surface would be about $-18°C$, rather than the present average of $15°C$.

7.49 GROSS HEAD

The gross head for a power plant is the difference in elevation between headwater and tailwater.

7.50 GROSS WING AREA

The wing area that includes the portion of the wing that is effectively cut out to make room for the fuselage is called the gross wing area.

7.51 GROUND EFFECT

When the helicopter is close to the ground, the air column between the ground and the helicopter acts as a cushion and adds to the lift. Ground effect prevails up to a distance, which is approximately equal to the diameter of the rotor.

7.52 GROUND SPEED

Ground speed is the horizontal speed of an aircraft relative to the Earth's surface. For accurate navigation to a destination, it is vital for pilots to estimate the ground speed they will reach in flight. An aircraft diving vertically would have a ground speed of zero.

7.53 GROUND STATE

The ground state of a quantum-mechanical system is its lowest-energy state; the energy of the ground state is known as the zero-point energy of the system. An excited state is any state with energy greater than the ground state. In quantum field theory, the ground state is usually called the vacuum state or the vacuum.

7.54 GROUNDWATER

Groundwater is the water that occurs below the surface of Earth, where it occupies all or part of the void spaces in soils or geologic strata. It is also called subsurface water to distinguish it from surface water, which is found in large bodies like the oceans or lakes or which flows overland in streams.

7.55 GUN TUNNEL

A gun tunnel is quite similar to a shock tunnel in operation. It has a high-pressure-driver section and a low-pressure-driven section with a diaphragm separating the two. A piston is placed in the driven section, adjacent to the diaphragm, so that when the diaphragm ruptures, the piston is propelled through the driven tube, compressing the gas ahead of it. The piston used is so light that it can be accelerated to velocities significantly above the speed of sound in the driven gas. This causes a shock wave to precede the piston through the driven tube and heat the gas. The shock wave will be reflected from the end of the driven tube to the piston, causing further heating of the gas.

7.56 GUSHING FLOW

To flow out or issue suddenly, as in Figure 7.4, copiously, or forcibly, as a fluid from confinement: Water gushed from the broken pipe.

Figure 7.4 Gushing water.

7.57 GYROPLANE

A gyroplane is a flying machine capable of short take-off and landing, similar to a helicopter, as in Figure 7.5. The gyroplane is also referred to as an autogyro. However, a gyroplane differs from a helicopter in the fact that in a helicopter the wings or blades are rotated by the power of the main engine, while in a gyroplane, the rotating wings are driven only by the action of the air upon them and not by any power supply. Thus, forward speed is necessary for a gyroplane, as in a conventional aircraft. The forward speed for a gyroplane is provided by the thrust of an engine and propeller.

7.58 GYROSCOPIC EFFECT

The gyroscopic effect is the ability (tendency) of the rotating body to maintain a steady direction of its axis of rotation. The rotation of the propeller

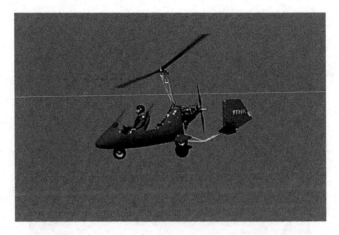

Figure 7.5 Gyroplane.

in a piston engine or the compressor in the case of a jet engine may cause a slight gyroscopic effect. A rotating body tends to resist any change in its plane of rotation, and if such change does take place, there is superimposed a tendency of rotation. Thus if the propeller rotates clockwise when viewed from the cockpit, the nose will tend to drop on a right-hand turn, and the tail to drop on a left-hand turn.

Chapter 8

Hail to Hypothesis

8.1 HAIL

Precipitation in the form of small balls or lumps usually consisting of concentric layers of clear ice and compact snow.

8.2 HAILSTORM

The thunderstorm, which produces hail that reaches the ground, is known as a hailstorm.

8.3 HANG GLIDING

Hang gliding is an air sport or recreational activity in which a pilot flies a light, non-motorised foot-launched heavier-than-air aircraft called a hang glider (see Figure 8.1).

8.4 HARD WATER

Hard water is the water that has high mineral content (in contrast with 'soft water'). Hard water is formed when water percolates through deposits of limestone, chalk, or gypsum, which are largely made up of calcium and magnesium carbonates, bicarbonates, and sulfates.

8.5 HARTMANN NUMBER

A dimensionless number, which gives a measure of the relative importance of drag forces resulting from magnetic induction and viscous forces in Hartmann flow, and determines the velocity profile for such a flow.

DOI: 10.1201/9781003348405-8

Figure 8.1 Hang gliding.

8.6 HARTMANN–SPRENGER TUBE

Hartmann–Sprenger tube is a device in which a jet enters a closed-end tube that is placed at a specific distance from the nozzle. Intense heating is generated in the Hartmann–Sprenger tube by shock waves and friction of the oscillating gas against the tube walls.

8.7 HEADWIND

A headwind is a wind that blows against the direction of travel of an object.

8.8 HEART

The heart is a muscular organ about the size of a fist, located just behind and slightly left of the breastbone. The heart pumps blood through the network of arteries and veins called the cardiovascular system.

8.9 HEAT

Heat is defined as the form of energy that is transformed between two systems (or a system and its surroundings) by virtue of a temperature difference.

The meaning of the term heat in thermodynamics is different from that in our day-to-day usage. In day-to-day usage, heat is often used to mean internal energy, heat content of a fuel, heat rise, the body preserves its heat, etc. In thermodynamics, heat and internal energy are two different entities. Energy is a property, but heat is not a property. A body contains energy, but not heat. Energy is associated with a state; heat is associated with a process.

8.10 HEAT ENGINES

Heat engines are devices to convert heat to work.

8.11 HEAT EXCHANGERS

Heat exchangers are devices where two moving fluid streams exchange heat without mixing.

8.12 HEAT PUMP

Heat pump is a device that transforms heat from a low-temperature medium to a high-temperature medium. Even though heat pumps and refrigerators operate on the same cycle, they differ in their objectives. The refrigerator is meant for maintaining the refrigerated space at a low temperature by removing heat from it. But the objective of a heat pump is to maintain a heated space at a high temperature.

8.13 HEAT TRANSFER

Heat transfer is the process of energy transfer due to temperature differences. Basically, there are three modes of heat transfer: conduction, convection, and radiation.

8.14 HEAT TRANSFER GAUGE

The heat transfer gauge can be used to measure wall shear over a very small area in laminar or turbulent flows with or without pressure gradient. This instrument operates on the principle that the heat transfer from the wall to the flow is a function of wall shear stress.

8.15 HEAT TRANSFER FROM ROCKETS

Heat transfer from rockets involves radiative cooling, radiating heat to space or conducting it to the atmosphere; regenerative cooling, running cold propellant through the engine before exhausting it; boundary-layer cooling, aiming some cool propellant at the combustion chamber walls; and transpiration cooling, diffusing coolant through porous walls.

8.16 HEATING VALUE

Heating value, defined as 'the amount of energy released when a fuel is burned completely in a steady-flow process and the products are returned to the state of the reactants' is another term commonly used in the combustion study of fuels.

8.17 HEAVY WATER

Heavy water is a form of water with a unique atomic structure and properties coveted to produce nuclear power and weapons. Like ordinary water (H_2O), each molecule of heavy water contains two hydrogen atoms and one oxygen atom. The difference, though, lies in the hydrogen atoms. Heavy water is a form of water that contains only deuterium rather than the common hydrogen- 1 isotope that makes up most of the hydrogen in normal water.

8.18 HEISLER CHARTS

Heisler charts are a graphical analysis tool for the evaluation of one-dimensional transient conductive heat transfer in thermal engineering. They are tools used in unsteady-state heat transfer problems, and they relate the temperature to the time of an object under transient heat conduction.

8.19 HELE–SHAW APPARATUS

The Hele–Shaw apparatus, shown in Figure 8.2, produces a flow pattern, which is similar to that of potential flow. It is an analogy experiment known as the Hele–Shaw analogy. The flow in the apparatus is a highly viscous flow between two parallel plates with a very small gap between them. In this flow, the inertia force is negligible compared to the viscous force. Under this condition, the flow equation has the same form as that of Euler's potential flow, however, it does not satisfy the no-slip wall boundary condition. There is a slip on the wall.

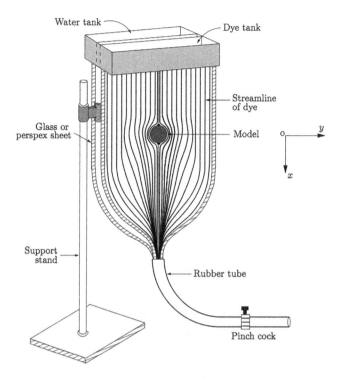

Figure 8.2 Hele–Shaw apparatus.

8.20 HELE–SHAW FLOW

Hele–Shaw flow is a low Reynolds number flow, which has wide application in flow visualisation studies because of its surprising characteristics of resembling the streamlines of potential flows (i.e., infinite Reynolds number flows). Hele–Shaw flow is defined as Stokes flow between two parallel flat plates separated by an infinitesimally small gap, named after Henry Selby Hele–Shaw, who studied the problem in 1898. Various problems in fluid mechanics can be approximated to Hele–Shaw flows and thus the research of these flows is of importance. Approximation to Hele–Shaw flow is specifically important to micro-flows. This is due to manufacturing techniques, which create shallow planar configurations, and the typically low Reynolds numbers of micro-flows.

8.21 HELICOPTER

Helicopter is essentially a rotary wing aircraft (see Figure 8.3). In normal flight, the upward thrust of the revolving blades must be equal to the weight;

Figure 8.3 Helicopter.

forward motion is produced by inclining the effective axis of the rotor forward that normally entails the nose of the helicopter down.

8.22 HELICOPTER CONFIGURATIONS

The configurations of a helicopter can be grouped into three categories. They are single rotor helicopters, twin, or two-rotor helicopters, and three or multiple rotor helicopters.

8.23 HELIUM

Helium is a chemical element with the symbol He and atomic number 2. It is a colourless, odourless, tasteless, non-toxic, inert, monatomic gas, and the first in the noble gas group in the periodic table. Its boiling point is the lowest among all the elements.

8.24 HELIX ANGLE

The angle between the resultant direction of the airflow and the plane of rotation (of the propeller) is called the angle of advance or helix angle, and it is different at each section of the blade. The sections near the tip move on a helix of much greater diameter, and they also move at a much greater velocity than those near the boss.

8.25 HELMHOLTZ'S THEOREMS

The four fundamental theorems governing vortex motion in an inviscid flow are called Helmholtz's theorems. Helmholtz's first theorem states that 'the circulation of a vortex tube is constant at all cross-sections along the tube'. The second theorem demonstrates that 'the strength of a vortex tube (that is, the circulation) is constant along its length'. This is sometimes referred to as the equation of vortex continuity. The third theorem demonstrates that a vortex tube consists of the same particles of fluid, that is, 'there is no fluid interchange between the vortex tube and surrounding fluid'. The fourth theorem states that 'the strength of a vortex remains constant in time'.

8.26 HETEROGENEOUS SYSTEM

A multiphase system is known as a heterogeneous system.

8.27 HEXANE

Hexane is a chemical commonly extracted from petroleum and crude oil. It is a colourless liquid that gives off a subtle, gasoline-like odour. Hexane is highly flammable, yet it can be found in many household products such as stain removers for arts and crafts projects.

8.28 HIGH-ENTHALPY FACILITIES

High-enthalpy facilities are devices to provide hypersonic airflows at high-enthalpy and high-pressure conditions. In such a device, real-gas effects are large causing experimental difficulties to assess the test section freestream characteristics. Some of the popular high-enthalpy facilities are free-piston tunnels, shock tubes, shock tunnels, hotshot tunnels, arc tunnels, and gun tunnels.

8.29 HIGH-ENTHALPY FLOWS

High-enthalpy flows are those that have a specific heats ratio as a function of temperature. The word enthalpy is based on the Greek word *enthalpein*, which means to put heat into.

8.30 HIGH-LIFT AEROFOILS

High-lift aerofoils are essentially aerofoils with some device by which the shape of the aerofoil can be altered during flight. Some of the commonly used high-lift devices are flaps and slots.

8.31 HIGH-LIFT DEVICES

These are auxiliary devices, which can be used to increase the maximum lift coefficient when required for low-speed operation and can be rendered ineffective at higher speeds. These auxiliary lift devices are broadly classified into the following two categories: (a) Those that alter the geometry of the aerofoil. (b) Those that control the behaviour of the boundary layer over the aerofoil.

8.32 HIGH SPEED

A speed of more than 650 kmph is called high speed. A lower limit of high speed might be considered to be the flow with Mach number ~0.5 (about 650 kmph) at standard sea level conditions.

8.33 HIGH-SPEED BIRDS

Some of the well-known high-speed birds and their flying speeds are (a) Peregrine falcon 389 km/h, (b) Golden eagle 240–320 km/h, (c) White-throated needletail 169 km/h, (d) Eurasian hobby 160 km/h, (e) Frigatebird 153 km/h, (f) Spur-winged goose 142 km/h, (g) Red-breasted merganser 129 km/h, (h) Rock dove (pigeon) 148.9 km/h, (i) Grey-headed albatross 127 km/h, and (j) Anna's hummingbird 98.27 km/h.

8.34 HIGH-SPEED WIND TUNNEL

Tunnels with test section speed of more than 650 kmph are called high-speed tunnels. A lower limit of high speed might be considered to be the flow with a Mach number ~0.5 (about 650 kmph) at standard sea level conditions.

8.35 HIGH WING

High wing aircraft are aircraft whose wings are mounted above the fuselage, as in Figure 8.4. The wings on high wing aircraft tend to be relatively flat with little dihedral or anhedral.

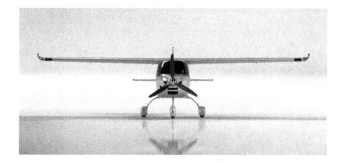

Figure 8.4 High-wing aircraft.

8.36 HIGHLY UNDEREXPANDED JET

A sonic or supersonic jet with a Mach disc in the core is termed a highly underexpanded jet. An underexpansion level of about $2(p_e/p_b \approx 2)$ can be taken as the lower limit of an underexpansion level for the formation of a Mach disc.

8.37 HOHMANN MINIMUM-ENERGY TRAJECTORY

The minimum-energy transfer between circular orbits is an elliptical trajectory called the Hohmann trajectory.

8.38 HOMOGENEOUS MIXTURE

A homogeneous mixture is a mixture in which the composition is uniform throughout the mixture.

8.39 HONEY

Honey is a sweet, viscous food substance made by honeybees and some related insects. Bees produce honey from the sugary secretions of plants or from secretions of other insects, by regurgitation, enzymatic activity, and water evaporation. Bees store honey in wax structures called honeycombs.

8.40 HONEYCOMBS

Wind tunnels have honeycombs in the settling chamber to improve the flow quality in the test section. Usually, the honeycombs are made of octagonal

Figure 8.5 Honeycombs.

or hexagonal (see Figure 8.5) or square or circular cells with length five to ten times their width (diameter).

8.41 HORIZONTAL BUOYANCY

The variation of static pressure along the test section produces a drag force known as horizontal buoyancy. It is usually small in closed test section and is negligible in open jets.

8.42 HORSESHOE VORTEX

Free vortices Γ_t, which are carried away by the flow must be attached at the wing tips. Together with the bound vortex Γ_b and the starting vortex Γ_s, they (the tip vortices) form a closed vortex ring frame in the fluid region cut by the wing, as shown in Figure 8.6. If a long time has passed since start-up, the starting vortex is at infinity (far downstream of the wing), and the bound vortex and the tip vortices together form a horseshoe vortex.

8.43 HOT SPRING

A hot spring is a spring produced by the emergence of geothermally heated groundwater that rises from the Earth's crust. While some of these springs contain water that is safe for bathing, others are so hot that immersion can result in injury or death.

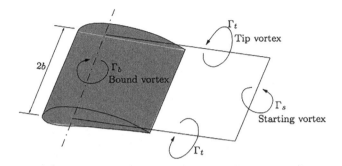

Figure 8.6 Simplified vortex system of a finite wing.

8.44 HOTSHOT TUNNELS

Hotshot tunnels are devices meant for the generation of high-speed flows with high temperatures and pressures for a short duration.

8.45 HOT-WIRE ANEMOMETER

Hot-wire anemometer is a thermal transducer. The principle of operation of a hot-wire anemometer is 'the heat transfer from a fine filament which is exposed to a cross-flow varies with variation in the flow rate'. The modes of operation generally followed are the following: (a) Constant current mode. Here, the current flow through the hot wire is kept constant and variations in the wire resistance caused by the fluid flow are measured by monitoring the voltage drop variations across the filament. (b) Constant temperature mode. Here, the hot-wire filament is placed in a feedback circuit which tends to maintain the hot wire at a constant resistance, and hence, at a constant temperature. Fluctuations in the cooling of the filament are seen as variations in the current flow through the hot wire. (c) Pulsed wire ane-mometer, which measures velocity by momentarily heating a wire to heat the fluid around it. This spot of heated fluid is convected downstream to a second wire that acts as a temperature sensor; the time of flight of the hot spot is inversely proportional to the fluid velocity.

8.46 HOT-WIRE ANEMOMETER WORKING PRINCIPLE

The hot-wire anemometer works on the principle that 'the rate of heat loss from a wire heated electrically and placed in an air stream is proportional to the stream velocity'.

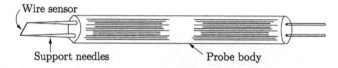

Figure 8.7 How-wire probe.

8.47 HOT-WIRE FILAMENT

The hot-wire filament may be regarded as an infinitely long, straight cylinder in cross-flow.

8.48 HOT-WIRE PROBES

The most important component in a hot-wire system is the hot-wire probe or hot-wire filament. The hot-wire element is usually made of a platinum or tungsten wire of about 1 mm long and about 5 µm in diameter. The wire is soldered to the tips of two needles, as shown in Figure 8.7, which forms the probe.

8.49 HOVERCRAFT

Hovercraft is a vehicle that lifts above the ground or water surface by means of a fan, which produces a raised pressure under the vehicle. A peripheral 'curtain' to high-speed air, normally supplemented by a flexible 'skirt' restricts leakage of the air from the underside.

8.50 HOVERING FLIGHT

During hovering flight, a helicopter maintains a constant position over a selected point, usually a few feet above the ground.

8.51 HUGONIOT EQUATION

This is an equation that expresses that the changes across a normal shock wave can be expressed purely in terms of thermodynamic variables without explicit reference to a velocity or Mach number as

$$e_2 - e_1 = \frac{(p_1 + p_2)}{2}(v_1 - v_2)$$

where e_1, e_2 are the internal energy, p_1, p_2 are the pressure and v_1, v_2 are the specific volume ahead of and behind the shock, respectively.

8.52 HUMAN BODY TEMPERATURE

Normal human body temperature is the typical temperature range found in humans. The normal human body temperature is around 37°C.

8.53 HUMAN BREATHING RATE

Normal respiration rates for an adult person at rest range from 12 to 16 breaths per minute.

8.54 HUMAN HEARING LIMITS

Sound is defined as any pressure variation which can be heard by a human ear. This implies a range of frequencies from 20 Hz to 20 kHz for a healthy human ear. In terms of sound pressure level, audible sound ranges from the threshold of hearing at 0 dB to the threshold of pain which can be over 130 dB.

8.55 HUMIDITY OF AIR

The mass of water vapour present in a unit mass of dry air is absolute or specific humidity.

8.56 HURRICANE

Hurricane or tropical cyclone is a rotating low-pressure weather system that has organised thunderstorms but no fronts (a boundary separating two air masses of different densities).

8.57 HYDEL POWER

Hydroelectric energy, also called hydroelectric power or hydroelectricity, is a form of energy that harnesses the power of water in motion, such as water flowing over a waterfall, to generate electricity.

8.58 HYDRAULICS

Realising that idealised theories were of no practical application without empirical correction factors, engineers developed from experimental studies applied science known as hydraulics for the specific fields of irrigation, water supply, river flow control, hydraulic power, and so on. Hydraulics is a mechanical function that operates through the force of liquid pressure. In hydraulics-based systems, mechanical movement is produced by contained, pumped liquid, typically through cylinders moving pistons.

8.59 HYDRAULIC ANALOGY

The analogy between shallow water flow with a free surface and two-dimensional gas flow is useful for qualitative as well as quantitative study of high-speed flows. In principle, any regulated stream of shallow water can be used for this analogy. In laboratories, it is usually done in a water flow table (channel). This technique is valued highly because many practical problems in supersonic flows, involving shocks and expansion waves, which require a sophisticated and expensive wind tunnel and instrumentation for their analysis may be studied inexpensively with simple water flow channel facility.

The flow pattern around the semi-wedge at 3° angle of attack and Froude number 2.13 ($M = 2.13$) shown in Figure 8.8 exhibits all the waves over the wedge.

The essential feature of this analogy is that the *Froude number of shallow water flow with a free surface is equivalent to a gas stream with a Mach number equal to that Froude number.*

Figure 8.8 A semi-wedge at 3° angle of attack is a water stream at Froude number 2.13.

8.60 HYDRAULIC ANALOGY THEORY

The basic governing equations of flow of an incompressible fluid with a free surface, in which the depth of the flow is small compared to its surface wavelength, form the shallow water theory. The similarity between the governing equations of motion of a two-dimensional isentropic flow of perfect gas and the two-dimensional shallow water flow forms the analogy.

8.61 HYDRAULIC DIAMETER

Hydraulic diameter D for a passage is defined as

$$D \cong \frac{4\left(\text{Cross-sectional area}\right)}{\text{Wetted area}}$$

8.62 HYDRAULIC EFFICIENCY

Hydraulic efficiency of an impulse turbine is the ratio of the power transferred directly to the turbine buckets to the power in the flow at the base of the nozzles.

8.63 HYDRAULIC GRADE LINE (HGL)

The surface or profile of water flowing in an open channel or a pipe flowing partially full. If a pipe is under pressure, the hydraulic grade line is that level water would rise to in a small, vertical tube connected to the pipe. It is a theoretical line showing the sum of the pressure elevation heads of a flow relative to a datum.

8.64 HYDRAULIC HOSE

A hydraulic hose is a specifically designed passage to convey hydraulic fluid to or among hydraulic components, valves, actuators, and tools. It is typically flexible, often reinforced, and usually constructed with several layers of reinforcement since hydraulic systems frequently operate at high or very high pressures.

8.65 HYDRAULIC JUMP

Hydraulic jump is an abrupt change from rapid to tranquil flow, as in Figure 8.9. Hydraulic jumps occur only in shooting water, that is, in the water streams with flow velocity greater than the wave propagation velocity.

Figure 8.9 Hydraulic jump.

8.66 HYDRAULIC LIFT

A hydraulic lift is a type of machine that uses a hydraulic apparatus to lift or move objects using the force created when pressure is exerted on liquid in a piston. The force then produces 'lift' and 'work'.

8.67 HYDRAULIC MACHINERY

Hydraulic machinery is essentially a device to convert one form of fluid energy to another.

8.68 HYDRAULIC MEAN DEPTH

Hydraulic mean depth is the ratio of the flow cross-sectional area to the perimeter.

8.69 HYDRAULIC TURBINES

Hydraulic turbines are machines used primarily for the development of hydroelectric energy.

These are machines used primarily for the development of hydroelectric energy. These turbines extract energy from flowing water and convert it to mechanical energy to drive electric generators.

8.70 HYDROCHLORIC ACID

Hydrochloric acid, also known as muriatic acid, is an aqueous solution of hydrogen chloride. It is a colourless solution with a distinctive pungent smell. It is classified as a strong acid. It is a component of the gastric acid in the digestive systems of most animal species, including humans.

8.71 HYDRODYNAMICA

Hydrodynamica is the book describing the fundamental relationship among pressure, density, and velocity, in particular Bernoulli's principle, which is one method to calculate aerodynamic lift, published by the Dutch–Swiss mathematician Daniel Bernoulli in 1738.

8.72 HYDRODYNAMICS

A branch of physics that deals with the motion of fluids and the forces acting on solid bodies immersed in fluids and in motion relative to them.

8.73 HYDROELECTRIC DAM

A hydroelectric dam is one of the major components of a hydroelectric facility. A dam is a large, human-made structure built to contain some body of water. In addition to producing hydroelectric power, dams are created to control river flow and regulate flooding.

8.74 HYDROFOIL

A hydrofoil is a lifting surface, or foil, that operates in water. They are similar in appearance and purpose to aerofoils used by airplanes. The boats that use the hydrofoil technology are also termed hydrofoils.

8.75 HYDROGEN

Hydrogen is a chemical element with the symbol H and atomic number 1. With a standard atomic weight of 1.008, hydrogen is the lightest element in the periodic table. Hydrogen is the most abundant chemical substance in the universe, constituting roughly 75% of all baryonic mass.

8.76 HYDROGEN FUEL

Hydrogen is a clean fuel that, when consumed in a fuel cell, produces only water. Hydrogen can be produced from a variety of domestic resources, such as natural gas, nuclear power, biomass, and renewable power like solar and wind. These qualities make it an attractive fuel option for transportation and electricity generation applications. It can be used in cars, in houses, for portable power, and in many more applications.

8.77 HYDROMETER

A hydrometer is an instrument used for measuring the relative density of liquids based on the concept of buoyancy.

8.78 HYDROPLANING

An inertially driven phenomenon which occurs when a wheel encounters a layer of water at sufficient speed.

8.79 HYDROPOWER

Hydropower, also known as waterpower, is the use of falling or fast-running water to produce electricity or to power machines.

8.80 HYDROSTATIC PRESSURE DISTRIBUTION

In stationary fluids, the pressure increases linearly with depth. This linear pressure distribution is called hydrostatic pressure distribution.

The hydrostatic pressure distribution is valid for moving fluids, provided there is no acceleration in the vertical direction.

8.81 HYGROSCOPIC

Hygroscopic is the tendency of a substance to absorb water from the air.

8.82 HYPERSONIC EXPERIMENTAL FACILITIES

It is essential to note that there is no single facility capable of simulating a hypersonic flight environment; therefore, different facilities are used to address various aspects of the design problems associated with the hypersonic flight, such as the aerodynamic forces and moments, the heat transfer distribution, and the surface pressure distribution. Some of the primary experimental facilities meant for this are shock tubes, arc tunnels, hypersonic wind tunnels, and ballistic free-flight ranges.

8.83 HYPERSONIC FLOW

A flow with $M > 5$ is called a hypersonic flow. In this flow, the flow velocity is very large compared with the speed of sound. The changes in flow velocity are very small, and thus, variations in the Mach number M are almost exclusively due to the changes in the speed of sound a.

8.84 HYPERSONIC NOZZLE

For hypersonic tunnel and other operations, axisymmetric nozzles are better suited than two-dimensional nozzles. For high Mach numbers of the order 10 the throat size becomes extremely narrow and forming the shape itself becomes very difficult.

8.85 HYPERSONIC SIMILARITY PARAMETER

The hypersonic similarity parameter, K, for slender bodies in the flow is usually expressed as $K = M\theta$, where M is the flow Mach number and θ is the half-angle of the nose.

8.86 HYPERSONIC TUNNELS

Hypersonic tunnels operate with test section Mach numbers above 5. Generally, they operate with stagnation pressures in the range from 10 to 100 atm and stagnation temperatures in the range from 50°C to 2,000°C.

8.87 HYPERVELOCITY FACILITIES

These are experimental aerodynamic facilities that allow testing and research at velocities considerably above those achieved in the wind tunnels. The high velocities in these facilities are achieved at the expense of other parameters, such as the Mach number, pressure, and/or run time.

8.88 HYPERVELOCITY FLOWS

High-enthalpy flows at high speeds are referred to as hypervelocity flows.

8.89 HYPOTHESIS

An idea that is suggested as the possible explanation for something but has not yet been found to be true or correct.

Chapter 9

ICBM to Isotopes

9.1 ICBM

An intercontinental ballistic missile (ICBM) is a missile with a minimum range of 5,500 km (3,400 miles) primarily designed for nuclear weapons delivery (delivering one or more thermonuclear warheads). Short- and medium-range ballistic missiles are known collectively as theatre ballistic missiles.

9.2 ICE TANK

An ice tank is used to develop ice-breaking vessels; this tank fulfils similar purposes as the towing tank does for open water vessels.

9.3 IDEAL FLUID FLOW

Ideal fluid flow is only an imaginary situation where the fluid is assumed to be inviscid or non-viscous and incompressible. Therefore, there is no tangential force between adjacent fluid layers. An extensive mathematical theory is available for the ideal fluid. Although the theory of ideal fluids fails to account for viscous and compressibility effects in the actual fluid flow process, it gives reasonably reliable results in the calculation of lift, induced drag and wave motion for gas at low velocity and for water. This branch of fluid dynamics is called *classical hydrodynamics*.

9.4 IDEAL GAS

Ideal gas is that which obeys the relation, $p \, v = RT$, where p, v, T are the pressure, specific volume and temperature of the gas and R is the gas constant. This is known as the *thermal equation of state*. This equation is also called an ideal gas equation of state or simply the *ideal gas relation*, and a gas that obeys this relation is called an ideal gas.

DOI: 10.1201/9781003348405-9

9.5 IDEAL RAMJET

It is essentially a hypothetical engine with simplified cycles involving isentropic processes. The complicated thermodynamic processes involved in the ramjet flow and combustion can be greatly simplified by making the following assumptions: (a) The compression and expansion processes in the engine are reversible and adiabatic. (b) The combustion process takes place at constant pressure.

9.6 IGNITION SYSTEM

An ignition system generates a spark or heats an electrode to a high temperature to ignite a fuel-air mixture in spark ignition internal combustion engines, oil-fired and gas-fired boilers, rocket engines, etc. Other engines may use a flame, or a heated tube, for ignition.

9.7 IMPACT PRESSURE

Impact pressure is the pressure which a fluid flow will experience if it is brought to rest isentropically. It is also called *total pressure*. Since pressure is the intensity of force, it has the dimensions

$$\frac{[\text{Force}]}{[\text{Area}]} = \left[\text{MLT}^{-2} \right] / \text{L}^2 = \left[\text{ML}^{-1}\text{T}^{-2} \right]$$

and is expressed in the units of Newton per square metre (N/m²) or simply pascal (Pa). At standard sea level conditions, the atmospheric pressure is 101,325 Pa, which corresponds to 760 mm of mercury column height.

9.8 IMPELLER

An impeller (shown in Figure 9.1) is a rotating component of a centrifugal pump that accelerates fluid outward from the centre of rotation, thus transferring energy from the motor that drives the pump to the fluid being pumped.

9.9 IMPULSE FUNCTION

Impulse function F is a parameter, defined as

$$F = pA + \rho A V^2 = pA\left(1 + \gamma M^2\right)$$

Figure 9.1 Centrifugal pump impeller.

where p, V, A, M are the pressure, velocity, flow area, and the Mach number, respectively, and γ is the ratio of specific heats.

9.10 IMPULSE SOUND

An impulse or impulsive sound is a short duration sound of <1 s duration. For example, typewriter and hammering noises are impulse sounds.

9.11 IMPULSE TURBINE

In the impulse turbine, the energy of the fluid supplied to the machine is converted by one or more nozzles into kinetic energy.

An impulse turbine has one or more fixed nozzles through which pressure is converted to kinetic energy as a liquid jet(s)—the liquid is typically water. The jet(s) impinges on the moving plates of the turbine runner that absorbs virtually all of the moving water's kinetic energy.

9.12 IMPURE MATERIALS

Impure materials may be mixtures of elements, mixtures of compounds, or mixtures of elements and compounds.

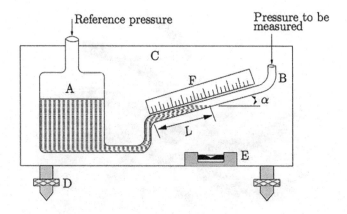

Figure 9.2 Inclined manometer.

9.13 INCLINED MANOMETER

When small pressure differences are involved in an experiment, it becomes difficult to measure accurately the height of the liquid column in the vertical manometer. Inclined manometers are generally used for measuring very small pressure differences. The essential features of an inclined manometer are shown in Figure 9.2.

9.14 INCOMPRESSIBLE AERODYNAMICS

An incompressible flow is characterised by a constant density. While all real fluids are compressible, a flow problem is often considered incompressible if the density changes in the problem have a small effect on the outputs of interest. This is more likely to be true when the flow speeds are significantly lower than the speed of sound. For higher speeds, the flow would encounter significant compressibility as it comes into contact with surfaces and slows down. Incompressible aerodynamics is an account of the theory and observation of the steady flow of incompressible fluid past aerofoils, wings, and other bodies.

9.15 INCOMPRESSIBLE FLOW

Incompressible flow is the flow in which the flow speed, V, is small compared to the speed of sound, a, in the flow medium. The changes in a are very small compared to the changes in V.

When the flow Mach number is <0.3, the compressibility effects are negligibly small, and hence, the flow is called *incompressible*.

9.16 INCOMPRESSIBLE JETS

The jets with a Mach number of <0.3, up to which the compressibility effects are negligible, are called incompressible jets.

9.17 INCOMPRESSIBLE SUBSTANCE

A substance whose specific volume does not change with temperature or pressure is called an incompressible substance.

9.18 INCREASE-IN-ENTROPY PRINCIPLE

The increase-in-entropy principle for any process is expressed as

$$S_{gen} = \Delta S_{tot} = \Delta S_{sys} + \Delta S_{surr} \geq 0$$

where S_{gen} is the entropy generated, ΔS_{tot} is the total entropy, ΔS_{sys} is the entropy change in the system, and ΔS_{surr} is the entropy change in the surroundings. This is a general expression for an increase of entropy principle. It is applicable to both closed and open systems. The equality holds for reversible processes and inequality for irreversible processes. However, we know that no actual process is truly reversible, therefore, we can conclude that the net entropy change for any process that takes place is positive, and hence, the entropy of the universe, which can be considered as an adiabatic system, is increasing continuously.

9.19 INDICATED AIRSPEED (IAS)

Indicated airspeed (IAS) is the airspeed read directly from the airspeed indicator (ASI) on an aircraft, driven by the pitot-static system. It uses the difference between the total pressure and the static pressure, provided by the system, to either mechanically or electronically measure dynamic pressure. The dynamic pressure includes terms for both density and airspeed. Since the airspeed indicator cannot know the density, it is by design calibrated to assume the sea level standard atmospheric density when calculating airspeed. Since the actual density will vary considerably from this assumed value as the aircraft changes altitude, the IAS varies considerably from true airspeed (TAS), the relative velocity between the aircraft and the surrounding air mass. Calibrated airspeed (CAS) is the IAS corrected for instrument and position error.

9.20 INDUCED AIR SPEED

The induced air speed is equivalent to the induced velocity generated by a wing with assumed elliptical lift distribution which results in uniform downwash.

9.21 INDUCED DRAG

Induced drag caused by the downwash is an additional component of drag.

The trailing vortices shed from the tips of a finite wing contain energy associated with the rotational velocities. This energy is taken from the airflow over the top and bottom surfaces of the wing, that is, from the freestream velocity. Therefore, some power should be provided to compensate for the energy taken by the tip vortices to maintain the airflow at a given velocity. This power must be equal to the rate of flow of energy associated with the trailing vortices. This can be regarded as equivalent to an associated drag force on the wing, which should be added to its profile drag. This additional drag is termed induced drag.

9.22 INDUCED DRAG FACTOR

For any loading other than elliptical loading, the induced drag is given by

$$C_{D_v} = \frac{kC_L^2}{\pi A\mathbb{R}}$$

where k is a constant for a given wing and it is greater than unity. This constant k is called the induced drag factor, C_L is the lift coefficient and $A\mathbb{R}$ is the aspect ratio.

9.23 INDUCED FLOW

Induced flow is the downward vertical movement of air through the rotor system due to the production of lift, often referred to as downwash.

9.24 INDUCED VELOCITY FIELD

The velocity field, which co-exists with the vortex, is known as the induced velocity field and the velocity at any point of it is called the induced velocity.

9.25 INDUSTRIAL TUNNELS

Depending on the simulation task (i.e., small missiles or re-entry vehicles), the test section diameter should range between 500 and 1,000 mm.

9.26 INERT GAS

An inert gas is a gas that does not undergo chemical reactions under a set of given conditions.

9.27 INERTIA FORCE

Inertia force is the force due to the rate of change of momentum.

Inertial force, also called fictitious force, is any force invoked by an observer to maintain the validity of Isaac Newton's second law of motion in a reference frame that is rotating or otherwise accelerating at a constant rate.

9.28 INFINITE VORTEX

An infinite vortex is that with both ends stretching to infinity.

9.29 INFLAMMABLE

Inflammable and flammable are synonyms and mean 'able to burn', even though they look like opposites.

9.30 INFRARED THERMOGRAPHY

It is a technique, based on radiation, to assess energy losses by radiation. An important energy radiation level is that of the infrared waveband. This occurs at a wavelength between the appropriate limits of 10^{-3} to 10^{-6} metres. This is below the visual limit of red colour. Because it is not visible in the normal sense by eye detection, special detectors are required to record its presence. This has led to the development of infrared thermography.

9.31 INFRASONIC WAVES

The waves with a frequency below the human audible range are called infrasonic waves.

The term 'infrasonic' applied to sound refers to sound waves below the frequencies of audible sound and nominally includes anything under 20 Hz. Sources of infrasound in nature include volcanoes, avalanches, earthquakes and meteorites.

9.32 INK

Ink is a gel, sol or solution that contains at least one colourant, such as a dye or pigment, and is used to colour a surface to produce an image, text or design.

9.33 INLET FLOW PROCESS

The flow in the inlet essentially decelerates. In other words, the inlet duct acts as a 'diffuser'. The fluid momentum decreases and pressure increases, with no work being done, as it passes through the inlet.

9.34 INSTRUMENT LANDING SYSTEM (ILS)

In aviation, the instrument landing system (ILS) is a radio navigation system that provides short-range guidance to aircrafts to allow them to approach a runway at night or in bad weather. In its original form, it allows an aircraft to approach until it is 200 feet (61 m) above the ground, within a ½ mile of the runway. At that point, the runway should be visible to the pilot; if it is not, they perform a missed approach. Bringing the aircraft this close to the runway dramatically improves the weather conditions in which a safe landing can be made. Later versions of the system, or 'categories', have further reduced the minimum altitudes.

9.35 INTAKE

An intake (inlet) is an opening, structure or system through which a fluid is admitted to a space or machine as a consequence of a pressure differential between the outside and the inside.

9.36 INTEGRAL ANALYSIS

The analysis in which large control volumes are used to obtain the aggregate forces or transfer rates is termed integral analysis.

9.37 INTENSIVE PROPERTIES

Intensive properties are those which are independent of the size of a system. Pressure, temperature and density are intensive properties.

9.38 INTERFERENCE

Interference, in physics, is the net effect of the combination of two or more wave trains moving on intersecting or coincident paths. The effect is that of the addition of the amplitudes of the individual waves at each point affected by more than one wave.

9.39 INTERFEROMETER

Interferometer is an optical technique to visualise high-speed flows in the ranges of transonic and supersonic Mach numbers. This gives a qualitative estimate of flow density in the field.

Interferometer makes visible the optical phase changes resulting from the relative retardation of the disturbed rays.

The optical patterns given by interferometer are sensitive to the flow density.

9.40 INTERFEROMETRY

Interferometer is an optical method most suited for qualitative determination of the density field of high-speed flows. Several types of interferometers are used for the measurement of the refractive index, but the instrument most widely used for density measurements in gas streams (wind tunnels) is that attributed to Mach and Zhender.

9.41 INTERMITTENCY

The ratio of the time duration for which the flow is turbulent to the total duration is known as intermittency. For laminar flow, the intermittency=0 and it is equal to unity for fully turbulent flow. During the transition, the intermittency varies between 0 and 1.0.

9.42 INTERNAL AERODYNAMICS

Internal aerodynamics is the study of flow through passages in solid objects. For instance, internal aerodynamics encompasses the study of the airflow through a jet engine.

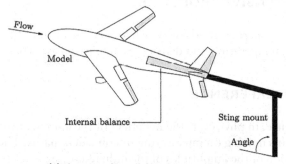

(a) An aircraft model with internal balance.

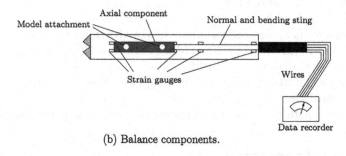

(b) Balance components.

Figure 9.3 An aircraft model mounted on a three-component sting balance.

9.43 INTERNAL BALANCE

The balance placed inside the model, as in Figure 9.3, is called internal balance.

9.44 INTERNAL ENERGY

Internal energy of a system is the sum of all the microscopic forms of energy of the system. It is the sum of all the microscopic energies such as translational kinetic energy, vibrational and rotational kinetic energy, and potential energy from intermolecular forces.

9.45 INTERNAL FLOW

An internal flow is any flow through a (circular) pipe, (non-circular) duct or (open, liquid-flow) channel where confining walls, or a free surface, guide the flow from an arbitrarily defined inlet state to an equally arbitrary outlet state.

9.46 INTERNALLY REVERSIBLE PROCESS

Internally reversible process is that for which no irreversibilities occur within the boundaries of the system during the process. That is, the system proceeds through a series of equilibrium states during an internally reversible process, and when the process is reversed, the process passes through exactly the same equilibrium states while returning to its initial state.

9.47 INTERNATIONAL SPACE STATION

The international space station is the centrepiece of our human space flight activities in low-Earth orbit.

9.48 INTERNATIONAL STANDARD ATMOSPHERE

The International Standard Atmosphere (ISA) is an atmospheric model of how the pressure, temperature, density and viscosity of the Earth's atmosphere change over a wide range of altitudes.

The performance of a flying vehicle is dependent on the physical properties, like density and temperature, of the air in which it flies. Therefore, the comparison between flying vehicles should be based on similar atmospheric conditions. To assist aircraft designs and operators in this, an agreement has been reached as an International Standard Atmosphere (ISA) intends to approximate to the atmospheric conditions prevailing for most of the year

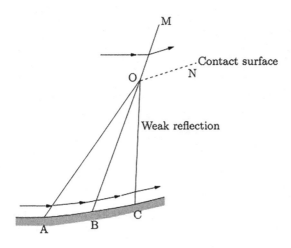

Figure 9.4 Intersection of waves of the same family.

in temperate latitudes, e.g., Europe and North America. The ISA is defined by the pressure and temperature at the mean sea level and the variation of temperature with altitude.

9.49 INTERSECTION OF SHOCKS OF THE SAME FAMILY

When a shock intersects another shock of the same family, the shocks cannot pass through, as in the case of intersection of shocks of an opposite family. The shocks will coalesce to form a single stronger shock, as shown in Figure 9.4, where successive corners in the same wall produce shocks of the same family.

9.50 INVERTED LOOP

The inverted loop or 'double bunt' is a manoeuvre in which the pilot is on the outside of the loop. The difficulty associated with an inverted loop is because in the normal loop, the climb to the top of the loop is completed while there is speed and power in hand and the engine and aerofoils are functioning in the normal fashion, whereas in the inverted loop, the climb to the top is required during the second portion of the loop when the aerofoils are in the inefficient inverted position. The path traced by an aircraft in an inverted loop is illustrated in Figure 9.5.

9.51 INVERTED MANOEUVRES

Inverted manoeuvres are essentially a flight mode in which the aircraft is in upside-down flight. To maintain the height during inverted flight, the engine must continue to run. The aircraft will be inverted; therefore, to produce an angle of attack, the fuselage will have to be in a 'tail-down' attitude. This attitude will affect the stability, although some aircrafts have been more stable when upside down than the right way up, and considerable difficulty has been experienced in restoring the normal flight.

9.52 INVERTED SPIN

The inverted spin has most of its characteristics similar to the normal spin. However, the loads on the aircraft structure are reversed, and the pilot must rely on his/her straps to hold him/her in the seat.

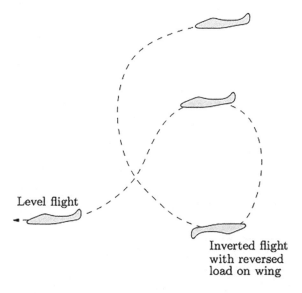

Level flight

Inverted flight
with reversed
load on wing

Figure 9.5 Inverted loop.

9.53 INVISCID FLOW

The term inviscid flow implies that the combined product of viscosity and the relevant viscosity gradient (or both normal and shear stresses) has a small effect on the flow field.

Inviscid flow is the flow of an inviscid fluid in which the viscosity of the fluid is equal to zero. Though there are limited examples of inviscid fluids, known as superfluids, the inviscid flow has many applications in fluid dynamics. The Reynolds number of inviscid flow approaches infinity as the viscosity approaches zero. When viscous forces are neglected, such as in the case of inviscid flow, the Navier–Stokes equation can be simplified to a form known as the Euler equation. This simplified equation is applicable to inviscid flow as well as flow with low viscosity and a Reynolds number much greater than one. Using the Euler equation, many fluid dynamics problems involving low viscosity are easily solved, however, the assumed negligible viscosity is no longer valid in the region of fluid near a solid boundary.

9.54 IONISATION

Ionisation is the process leading to the formation of ions by separating atoms or molecules or radicals or by adding or subtracting electrons from atoms by strong electric fields in a gas.

9.55 IONISATION GAUGE

The ionisation gauges are capable of measuring pressures as low as 10^{-12} torr.

9.56 IONISATION STATE

A molecule's charge changes whenever it gains or loses a proton, H^+. The molecule's charge is known as its ionisation state.

9.57 IRREVERSIBLE PROCESS

If $ds > 0$, the process is called an irreversible process.

9.58 IRREVERSIBILITIES

Irreversibilities are factors that cause a process to be irreversible. The well-known irreversibilities are friction, electric resistance, inelastic deformation of solids, unrestrained expansion, heat transfer through a finite temperature difference, non-quasi-equilibrium changes, and so on.

9.59 IRRIGATION

Irrigation is an artificial process of applying controlled amounts of water to land to assist in the production of crops.

9.60 IRRIGATION SPRINKLER

An irrigation sprinkler is a device used to irrigate crops, lawns, landscapes, golf courses, and other areas.

9.61 IRROTATIONAL FLOW

For irrotational flows (curl $V = 0$), a potential function Φ exists such that $V = \text{grad } \Phi$. Irrotational flow is that in which the vorticity components are zero. Inviscid flows are essentially irrotational flows. In this flow, each element of the moving fluid suffers no *net* rotation from one instant to the next concerning a given frame of reference. The classic example of irrotational motion (although not a fluid) is that of carriages on a giant wheel used for amusement rides. Each carriage describes a circular path as the wheel revolves, but does not rotate with respect to the earth. However, in an irrotational flow, a fluid element may deform providing the axes of the element to rotate equally toward or away from each other, so that there is no *net* rotation. As long as the algebraic average rotation (or angular velocity) is zero, the motion is irrotational.

9.62 ISENTROPIC COMPRESSION

Isentropic compression is the limiting case of a multi-shock compression scheme. In isentropic compression, the pressure recovery is achieved through a large number of weak shocks also termed isentropic compression waves.

A compression process can be treated as isentropic when the turning of the flow is achieved through a large number of weak oblique shocks. These kinds of compressions through a large number of weak compression waves are termed continuous compression. These kinds of corners are called continuous compression corners. Thus, the geometry of the corner should have continuous smooth turning to generate a large number of weak (isentropic) compression waves.

9.63 ISENTROPIC FLOW

Isentropic flow is a flow process which is reversible and adiabatic. It obeys the isentropic process relation:

$$\frac{p}{\rho^{\gamma}} = \text{constant}$$

where p, ρ, and γ are the pressure, density, and ratio of specific heats, respectively.

For an isentropic flow, the viscosity is assumed to be zero and the stagnation temperature is treated as a constant.

9.64 ISENTROPIC INDEX

Isentropic index is the power to the density in the flow processes equation,

$$\frac{p}{\rho^{\gamma}} = \text{constant}$$

where the process index γ is the ratio of specific heats.

9.65 ISENTROPIC PROCESS

A process during which the change in entropy is zero is called an isentropic process. It is adiabatic and reversible.

In thermodynamics, an isentropic process is an idealised thermodynamic process that is both adiabatic and reversible. The work transfers of the system are frictionless, and there is no net transfer of heat or matter.

9.66 ISENTROPIC PROCESS RELATION

The isentropic process relation is

$$\frac{p}{\rho^\gamma} = \text{constant}$$

where p is the pressure, ρ is the density and γ is the ratio of specific heats.

9.67 ISENTROPIC RELATIONS

The relations between pressure, temperature and density for an isentropic process of a perfect gas are

$$\frac{p}{p_0} = \left(\frac{\rho}{\rho_0}\right)^\gamma \quad \text{and} \quad \frac{T}{T_0} = \left(\frac{p}{p_0}\right)^{(\gamma-1)/\gamma}$$

Also, the pressure-temperature density relation of a perfect gas is

$$\frac{p}{RT} = \frac{p_0}{\rho_0 T_0} = R$$

The temperature, pressure, and density ratios as functions of the Mach number are

$$\frac{T_0}{T} = \left(1 + \frac{\gamma-1}{2} M^2\right)$$

$$\frac{p_0}{p} = \left(1 + \frac{\gamma-1}{2} M^2\right)^{\frac{\gamma}{\gamma-1}}$$

$$\frac{\rho_0}{\rho} = \left(1 + \frac{\gamma-1}{2} M^2\right)^{\frac{1}{\gamma-1}}$$

where T_0, p_0, and ρ_0 are the temperature, pressure, and density, respectively, at the stagnation state.

9.68 ISENTROPIC STAGNATION STATE

Isentropic stagnation state is the state of zero velocity reached by decelerating a flow isentropically. The state of zero velocity is called the *isentropic stagnation state*.

If the flow is assumed to be isentropic for a channel flow, all states along the channel or stream tube lie on a line of constant entropy and have the same stagnation temperature.

9.69 ISOBAR

A line on a flow field that joins places that have the same air pressure at a particular time.

9.70 ISOBARIC PROCESS

An isobaric process is a type of thermodynamic process in which the pressure of the system stays constant. The heat transferred to the system does work but also changes the internal energy of the system.

9.71 ISOCHORIC PROCESS

An isochoric process, also called a constant-volume process, an isovolumetric process, or an isometric process, is a thermodynamic process during which the volume of the closed system undergoing such a process remains constant.

9.72 ISOLATED SYSTEM

An isolated system is a thermodynamic system that cannot exchange either energy or matter outside the boundaries of the system.

9.73 ISOLATOR

Isolator is a constant area duct meant for isolating the intake from getting disturbed by the instabilities of the combustor.

9.74 ISOTHERMAL PROCESS

An isothermal process is a thermodynamic process in which the temperature of a system remains constant. The transfer of heat into or out of the system happens so slowly that thermal equilibrium is maintained.

9.75 ISOTOPES

Isotopes are atoms of the same element that have different numbers of neutrons but the same number of protons and electrons. The difference in the number of neutrons between the various isotopes of an element means that the various isotopes have different masses.

Chapter 10

Jelly to Joule–Thomson Effect

10.1 JELLY

A soft, solid brightly coloured food that shakes when it is moved. Jelly is made from sugar and fruit juice and is eaten cold at the end of a meal, especially by children.

10.2 JET

A jet may be defined as a pressure-driven shear flow that exhibits a characteristic that the width-to-axial distance is a constant. This constant assumes a value of 8 for jet Mach numbers < 0.2 and the constant decreases with an increase in the Mach number.

The included angle for low-speed jets is around $10°$, and for high-speed jets, it is < $10°$. The features of the jet flow field – the jet core, characteristic decay and fully developed zones – are shown schematically in Figure 10.1.

10.3 JET CONTROL METHODS

All types of jet controls can be broadly classified into either active or passive controls. In active control, an auxiliary power source (like microjets) is used to control the jet's characteristics. In passive control, the controlling energy is drawn directly from the flow to be controlled. Both active and passive controls mainly aim at modifying flow and noise characteristics.

10.4 JET CORE

For subsonic jets, the axial distance up to which the jet velocity along the axis is unaffected is called the jet core. In other words, the axial distance from the nozzle exit at which the jet velocity (that is, the nozzle exit velocity) begins to decrease is termed as jet core.

DOI: 10.1201/9781003348405-10

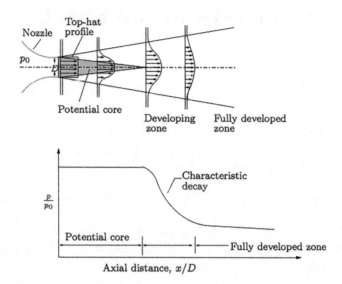

Figure 10.1 Different zones of a subsonic jet and the corresponding pitot pressure variation along the centreline.

10.5 JET ENGINE

All jet engines, which are also called gas turbines, work on the same principle. The engine sucks air with a fan. A compressor raises the pressure of
the air and it is made with many blades attached to a shaft. The blades spin
at high speed and compress or squeeze the air. The compressed air is then
sprayed with fuel and an electric spark lights the mixture. The burning
gases expand and discharge out through the nozzle at the back of the engine.
As the jet of gas shoots backward, thrust is produced. This thrust moves
the aircraft forward. As the hot air is going through the nozzle, it passes
through another group of blades called the turbine. It is attached to the
same shaft as the compressor. Spinning the turbine causes the compressor
to spin. A view of the complete gas turbine engine is shown in Figure 10.2.

10.6 JET ENGINE COMBUSTOR

In a combustor, the air is mixed with fuel and then ignited. There are as
many as 20 nozzles to spray fuel into the airstream. The mixture of air
and fuel catches fire. This provides a high-temperature, high-energy airflow. The fuel burns with the oxygen in the compressed air, producing hot
expanding gases. The inside of the combustor is often made of ceramic
materials to provide a heat-resistant chamber. The heat of the combustion
product can reach about 2,700°C.

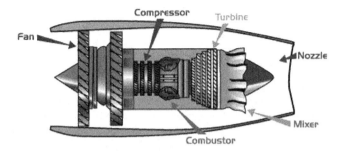

Figure 10.2 A view of complete engine.

10.7 JET ENGINE COMPONENTS

The major parts of a jet engine are a fan, compressor, combustion chamber, turbine, mixer and nozzle.

10.8 JET ENGINE COMPRESSOR

The compressor is the first component in the engine core. It is made up of fans with many blades and is attached to a shaft and squeezes the air that enters it into progressively smaller areas increasing the air pressure. This results in an increase in the energy potential of the air. The squashed air is forced into the combustion chamber.

10.9 JET ENGINE FAN

The fan is the first component in a turbofan engine. The large spinning fan sucks in a large quantity of air. Most blades of the fan are made of titanium. It then speeds up this air and splits it into two parts. One part continues to flow through the 'core' or centre of the engine, where it is acted upon by the other engine components. The second part 'bypasses' the core of the engine. It goes through a duct that surrounds the core to the back of the engine, where it produces much of the force that propels the aeroplane forward. This cooler air helps to quieten the engine in addition to adding thrust to the engine.

10.10 JET ENGINE FUEL

Jet fuel is a clear to straw-coloured fuel, based on either an unleaded kerosene (Jet A-1), or a naphtha-kerosene blend (Jet B). Similar to diesel fuel, it can be used in either compression ignition engines or turbine engines.

10.11 JET ENGINE NOZZLE

A nozzle is the exhaust duct of the engine. This is the engine part that produces the thrust for the aeroplane. The energy-depleted airflow that passed the turbine, in addition to the colder air that bypassed the engine core, produces a force when exiting the nozzle that acts to propel the engine, and therefore, the aeroplane moves forward. The combination of the hot air and cold air is expelled and produces an exhaust which causes a forward thrust. The nozzle may be preceded by a mixer, which combines the high-temperature air coming from the engine core with the lower-temperature air that was bypassed in the fan. The mixer helps to make the engine quieter.

10.12 JET ENGINE TURBINE

The high-energy airflow coming out of the combustor goes into the turbine, causing the turbine blades to rotate. The turbines are linked by a shaft to turn the blades in the compressor and spin the intake fan at the front. This rotation takes some energy from the high-energy flow that is used to drive the fan and the compressor. The gases produced in the combustion chamber move through the turbine and spin its blades. The turbines of the jet engine spin around thousands of times per second. (For example, large jet engines operate around 10,000–25,000 rpm, while micro turbines spin as fast as 500,000 rpm.) They are fixed on shafts that have several sets of ball bearings in between them.

10.13 JET FLAP

The jet flap consists of a very high-speed jet of air blown out through a narrow slit in the trailing edge of the wing, as illustrated in Figure 10.3. The jet, deflected slightly downwards, divides the upper surface flow from the

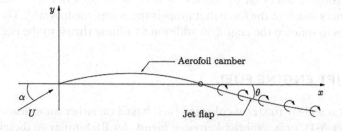

Figure 10.3 A jet flap.

lower surface flow and produces an effect on the flow over the wing similar to that produced by a very large physical trailing edge flap. There is an additional increment in lift due to the downward component of the momentum of the jet leading to a very high-lift coefficient.

Jet flap has a high-velocity sheet of air issuing from the trailing edge of an aerofoil at some downward angle to the chord line of the aerofoil. The downward deflection of the efflux produces a lifting component of reaction. The jet affects the pressure distribution on the aerofoil in a similar manner to that obtained by an addition to the circulation around the aerofoil.

10.14 JET FUEL COMPOSITION

Aviation fuels consist primarily of hydrocarbon compounds (paraffins, cycloparaffins or naphthenes, aromatics and olefins) and contain additives that are determined by the specific uses of the fuel (CRC 1984; Dukek 1978; IARC 1989). Paraffins and cycloparaffins are the major components.

10.15 JET LIFT

Jet lift and thrust vectoring aircraft is an alternative to the helicopter.

10.16 JET NOISE

The jet noise is another important parameter to be considered in the study of high-speed jets. For subsonic jets, the turbulent mixing noise is the major component of the noise. But for supersonic jets, in addition to the mixing noise, the shock-associated noise can contribute significantly to the overall jet noise. The shock-associated noise can contain broadband shock noise and screech noise. The jet noise can be studied by measuring the noise with a sound level meter.

10.17 JET PLANE

Jet plane is defined as a type of aircraft or plane, or a high-pressure stream of liquid or gas, or is a nozzle out of which a high-pressure stream comes.

10.18 JET-PROPULSION CYCLE

A jet-propulsion cycle is an open cycle on which aircraft gas turbines operate. It differs from the ideal Brayton cycle in the fact that the gases are not expanded to the ambient pressure in the turbine; instead, the gases are expanded to a pressure such that the power produced by the turbine is just sufficient to run the compressor and the auxiliary equipment. That is, the net output of a jet-propulsion cycle is zero. The thrust required to propel the aircraft is provided by a nozzle, which accelerates the gases that exit the turbine at a relatively high pressure.

10.19 JET STREAMS

Jet streams are some of the strongest winds in the atmosphere. A jet stream is a very cold, fast-moving wind found high in the atmosphere.

10.20 JET STREAM SPEED

Jet stream speeds usually range from 129 to 225 km/h, but they can reach more than 443 km/h.

10.21 JOUKOWSKI HYPOTHESIS

Joukowski postulated that 'the aerofoil generates sufficient circulation to depress the rear stagnation point from its position, in the absence of circulation, down to the sharp trailing edge'.

10.22 JOULE'S LAW

For a perfect gas, the internal energy is a function of the absolute temperature alone. This hypothesis is a generalisation of experimental results. It is known as Joule's law.

10.23 JOULE–THOMSON COEFFICIENT

The Joule–Thomson coefficient is a measure of the change in temperature, T with pressure, p, during a constant enthalpy process.
   or
   The Joule–Thomson coefficient is the ratio of the temperature decrease to the pressure drop and is expressed in terms of the thermal expansion coefficient and the heat capacity.

10.24 JOULE–THOMSON EFFECT

The Joule–Thomson effect is the cooling or heating observed during the adiabatic and isenthalpic expansion of fluids. In thermodynamics, the Joule–Thomson effect describes the temperature change of a real gas or liquid when it is forced through a valve or porous plug while keeping it insulated so that no heat is exchanged with the environment. This procedure is called a throttling process or Joule–Thomson process.

Chapter 11

Kaplan Turbine to Kutta–Joukowski Transformation

11.1 KAPLAN TURBINE

Kaplan turbine is a propeller turbine with movable blades whose pitch can be adjusted to suit existing operating conditions. The Kaplan turbine is a propeller-type water turbine that has adjustable blades, as shown in Figure 11.1. It was developed in 1913 by Austrian professor Viktor Kaplan, who combined automatically adjusted propeller blades with automatically adjusted wicket gates to achieve efficiency over a wide range of flow and water levels.

11.2 KARMAN VORTEX STREET

At low Reynolds numbers between 40 and 150, these vortices are shed in stable, well-defined patterns, which persist for long distances downstream (see Figure 11.2), and this is termed the Karman vortex street.

11.3 KELVIN SCALE

The absolute temperature scale in the SI system is the Kelvin scale. The relation between the Celsius scale and Kelvin scale is $T\ (\text{K}) = T\ (°\text{C}) + 273.15$.

11.4 KELVIN'S CIRCULATION THEOREM

This theorem states that 'in a flow of inviscid and barotropic fluid, with conservative body forces, the circulation around a closed curve (material line) moving with the fluid remains constant with time', if the motion is observed from a non-rotating frame. This theorem is also known as *Thomson's vortex theorem*.

DOI: 10.1201/9781003348405-11

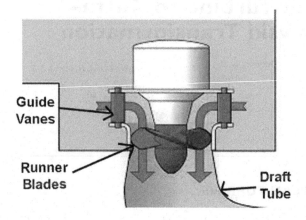

Figure 11.1 Kaplan turbine.

Figure 11.2 Karman vortex street behind a cylinder.

11.5 KELVIN-PLANCK STATEMENT

The Kelvin-Planck statement of the second law of thermodynamics is, 'it is impossible for any device that operates on a cycle to receive heat from a single reservoir and produce a net amount of work'.

The Kelvin–Planck statement can also be expressed as, 'for a power plant to operate, the working fluid must exchange heat with the environment as well as the furnace'.

or

'No heat engine can have a thermal efficiency of 100%'.

11.6 *K*-EPSILON TURBULENCE MODEL

The k -epsilon $(k - \varepsilon)$ turbulence model is the most common model used in Computational Fluid Dynamics (CFD) to simulate the mean flow characteristics for turbulent flow conditions. It is a two-equation model that gives a general description of turbulence by means of two transport equations (Partial Differential Equations, PDEs).

11.7 KEPLER'S LAWS OF PLANETARY MOTION

1. The planets move in ellipses with the sun at one focus. That is, the path of the planets about the sun is elliptical in shape, with the centre of the sun being located at one focus (The Law of Ellipses).
2. Areas swept out by the radius vector from the sun to a planet in equal times are equal. In other words, an imaginary line drawn from the centre of the sun to the centre of the planet will sweep out equal areas in equal intervals of time (The Law of Equal Areas).
3. The square of the period of revolution is proportional to the cube of the semi-major axis. That is, $T^2 = \text{constant} \times a^3$. The ratio of the squares of the periods of any two planets is equal to the ratio of the cubes of their average distances from the sun (The Law of Harmonies).

11.8 KEROSENE

Kerosene, also known as paraffin, is a combustible hydrocarbon liquid that is derived from petroleum. It is widely used as a fuel in aviation as well as in households.

11.9 KEROSENE BOILING POINT

Kerosene, which is used in heating, cooking and jet fuel, has a boiling temperature ranging from 200°C to 300°C.

11.10 KEROSENE DENSITY

Kerosene is a low viscosity, clear liquid formed from hydrocarbons obtained from the fractional distillation of petroleum between 150°C and 275°C (300°F and 525°F), resulting in a mixture with a density of 0.78 – 0.81 g/cm³ (0.45 – 0.47 oz/cu in) composed of carbon chains that typically contain between 10 and 16 carbon atoms.

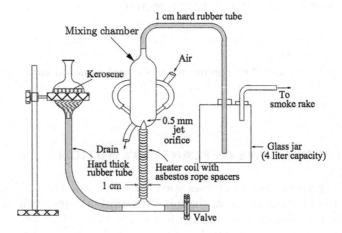

Figure 11.3 Kerosene smoke generator.

11.11 KEROSENE SMOKE GENERATOR

In this method, smoke is produced by evaporation and atomisation of kerosene in an air stream. The system is compact and electrically operated. Smoke can be generated within a few minutes. No solid deposits are formed.

The kerosene smoke generator consists of a reservoir, an electrically heated glass tube to the top of which is attached a narrow nozzle and a mixing chamber, as shown in Figure 11.3. Kerosene is heated to form vapour, which emerges through the nozzle in the form of a jet inside the mixing chamber. Two air jets impinging on the nozzle opening atomise the hot kerosene vapour forming white smoke. The smoke is taken out through the outlet provided at the top of the mixing chamber.

11.12 KIDNEYS

The kidneys are a pair of bean-shaped organs on either side of your spine, below your ribs and behind your belly. Each kidney is about 4 or 5 inches long, roughly the size of a large fist. The kidneys' job is to filter your blood.

11.13 KINEMATICS OF FLUIDS

Kinematics of fluids deals with translation, rotation and rate of deformation of fluid particles. This analysis is useful in determining the methods to describe the motion of fluid particles and analysing the flow patterns. However, the velocity and acceleration of fluid particles cannot be obtained from kinematic study alone, since the interaction of fluid particles with one another makes the fluid a disturbed medium.

11.14 KINEMATICS OF FLUID FLOW

Kinematics is the branch of physics that deals with the characteristics of motion without regard for the effects of forces or mass. In other words, kinematics is the branch of mechanics that studies the motion of a body or a system of bodies without consideration given to its mass or the forces acting on it. It describes the spatial position of bodies or systems, their velocities and their acceleration. If the effects of forces on the motion of bodies are accounted for, the subject is termed dynamics. Kinematics differs from dynamics in that the latter takes these forces into account.

11.15 KINEMATIC VISCOSITY COEFFICIENT

Kinematic viscosity coefficient is a convenient form of expressing the viscosity of a fluid. It is the ratio of the density ρ and the absolute coefficient of viscosity μ,

$$v = \frac{\mu}{\rho}$$

The kinematic viscosity coefficient v is expressed as m^2/s and $1\,cm^2/s$ is known as *stoke*.

The kinematic viscosity coefficient is a measure of the relative magnitudes of viscosity and inertia of the fluid. Both dynamic viscosity coefficient μ and kinematic viscosity coefficient v are functions of temperature. For liquids, μ decreases with increase of temperature, whereas for gases, μ increases with increase of temperature. This is one of the fundamental differences between the behaviour of gases and liquids. The viscosity is practically unaffected by the pressure.

11.16 KINETIC DIAMETER

Kinetic diameter is a measure applied to atoms and molecules that expresses the likelihood that a molecule in a gas will collide with another molecule.

11.17 KINETIC ENERGY

The energy that a system possesses as a result of its motion relative to some reference frame is called kinetic energy.

II.18 KINETIC ENERGY OF GAS

The average kinetic energy of gas particles is proportional to the absolute temperature of the gas, and all gases at the same temperature have the same average kinetic energy.

II.19 KINETIC ENERGY THICKNESS

The kinetic energy thickness may be defined as the distance through which the surface would have to be displaced in order that, with no boundary layer, the total flow of kinetic energy at the station considered would be the same as that which would occur. This quantity is defined with reference to the kinetic energy of the fluid in a manner comparable with the momentum thickness.

II.20 KINETIC THEORY OF GASES

Kinetic theory of gases is a theory based on a simplified molecular or particle description of a gas from which many gross properties of gas can be derived.

II.21 KINETIC THEORY EQUIVALENT OF THE PERFECT GAS STATE EQUATION

The kinetic theory equivalent of the perfect gas state equation is

$$\frac{p}{\rho} = \frac{1}{3}\overline{C^2}$$

where $\overline{C^2} = 3\,RT$.

II.22 KING'S RELATION

It is the relation meant for analysing the behaviour of hot-wire systems.

II.23 KNOT

The knot is a unit of speed equal to one nautical mile per hour exactly 1.852 km/h (~1.151 mph or 0.514 m/s).

11.24 KNUDSEN DIFFUSION

Knudsen diffusion is a means of diffusion that occurs when the scale length of a system is comparable to or smaller than the mean free path of the particles involved.

11.25 KNUDSEN EFFECT

When the pore diameter of the material becomes less than the average free length of the path of gas molecules, the molecules will only collide with the pore surfaces without transferring energy. This is known as the Knudsen effect.

11.26 KNUDSEN GAUGE

The Knudsen gauge is capable of measuring low pressures in the range between 10^{-5} m and $10\,\mu$m of Hg. This is used as a calibration device for other gauges meant for pressure measurement in this range.

11.27 KNUDSEN NUMBER

The ratio of the mean free path λ to the characteristic dimension L defines the Knudsen number (Kn). Knudsen number is a non-dimensional parameter obtained by dividing the mean free path, λ in gas to a characteristic dimension, L.

$$Kn = \frac{\lambda}{L}$$

It plays a dominant role in the field of rarefied gas dynamics.

The widely accepted classification of flow regimes based on the Knudsen number is as follows:

(i) $Kn < 0.01$ (continuum flow)
(ii) $0.01 < Kn < 0.1$ (slip flow)
(iii) $0.1 < Kn < 10$ (transition flow)
(iv) $Kn > 10$ (free molecule flow)

11.28 KUTTA CONDITION

The Kutta condition can be stated as follows:

'A body with a sharp trailing edge moving through a fluid will create about itself a circulation of sufficient strength to hold the rear stagnation point at the trailing edge'.

In fluid flow around a body with a sharp corner, the Kutta condition refers to the flow pattern in which fluid approaches the corner from both directions, meets at the corner, and then, flows away from the body. None of the fluid flows around the corner remaining attached to the body. The Kutta condition is significant when using the Kutta–Joukowski theorem to calculate the lift generated by an aerofoil.

Actual wing profiles are with a rounded trailing edge of finite thickness. Because of the rounded trailing edge of the wings, in actual flow, where viscous boundary layer and wake exist, the position of the rear stagnation point may differ from the location predicted by potential flow theory and the full Joukowski circulation, may not be established. This is because for realising full Joukowski circulation, the trailing edge should be of zero thickness and without any wake. This condition of realising full Joukowski circulation, resulting in flow without wake is known as the Kutta condition.

The Kutta condition is a principle in steady-flow fluid dynamics, especially aerodynamics, that is applicable to solid bodies with sharp corners, such as the trailing edges of aerofoils. It is named after the German mathematician and aerodynamicist Martin Kutta.

II.29 KUTTA–ZHUKOVSKY LIFT THEOREM

Lift force proportional to circulation is known as the Kutta–Zhukovsky lift theorem.

The Kutta–Joukowski theorem states that 'the force per unit length acting on a right cylinder of any cross section whatsoever is equal to $\rho_\infty V_\infty \Gamma_\infty$, and is perpendicular to the direction of V_∞'.

or

The circulation of the vortex determines the lift, and the lift formula, which gives the relation between circulation, Γ, and lift per unit width, l in inviscid flow, is the Kutta–Joukowski theorem, namely

$$l = \rho\, \Gamma\, V_\infty$$

where l is the lift per unit span of the wing, Γ is the circulation around the wing, V_∞ is the freestream velocity, and ρ is the density of the flow.

The Kutta–Joukowski theorem is a fundamental theorem in aerodynamics used for the calculation of the lift of an aerofoil and any two-dimensional bodies including circular cylinders translating into a uniform fluid at a constant speed large enough so that the flow seen in the body-fixed frame is steady and unseparated. The theorem relates the lift generated by an aerofoil to the speed of the aerofoil through the fluid, the density of the fluid and the circulation around the aerofoil. The circulation is defined as the line integral around a closed loop enclosing the aerofoil of the component of the velocity of the fluid tangent to the loop. It is named after Martin Kutta and Nikolai Zhukovsky (or Joukowski) who first developed its key ideas in the early twentieth century. The Kutta–Joukowski theorem is an inviscid theory, but it is a good approximation for real viscous flow in typical aerodynamic applications.

11.30 KUTTA–JOUKOWSKI TRANSFORMATION

A transformation, which generates a family of aerofoil-shaped curves along with their associated flow patterns by applying a certain transformation is the Kutta–Joukowski transformation.

Kutta–Joukowski transformation is the simplest of all transformations developed for generating aerofoil-shaped contours. Kutta used this transformation to study circular-arc wing sections, while Joukowski showed how this transformation could be extended to produce wing sections with thickness t as well as camber.

Chapter 12

Lactic Acid to Lungs

12.1 LACTIC ACID

Lactic acid is an alpha-hydroxy acid (AHA) due to the presence of a hydroxyl group adjacent to the carboxyl group.

12.2 LACTO CALAMINE

Lacto calamine lotion is a paraben-free formula, perfect for giving you a clear, matte face. The lotion absorbs excess oil that helps you keep away problems like pimples, acne, dark spots, blackheads, whiteheads and patchy skin.

12.3 LACTOMETER

Lactometer is a little glass instrument used to measure the amount of water in the milk you drink.

12.4 LAGRANGE POINTS

The Lagrange (sometimes called liberation) points are positions of equilibrium for a body in a two-body system.

12.5 LAGRANGIAN DESCRIPTION

The Lagrangian method describes the motion of each particle of the flow field in a separate and discrete manner. This approach of identifying material points and following them along is also termed the *particle* or *material description*. This approach is usually preferred in the description of low-density flow fields (also called rarefied flows), moving solids, like in describing the motion of a projectile, and so on. However, in a deformable system, like a continuum fluid,

DOI: 10.1201/9781003348405-12

Figure 12.1 Lahar.

there are infinite number of fluid elements whose motion must be described. For this case the Lagrangian approach becomes unmanageable.

12.6 LAHAR

A lahar is a violent type of mudflow or debris flow composed of a slurry of pyroclastic material, rocky debris and water. The material flows down from a volcano, typically along a river valley, as shown in Figure 12.1.

12.7 LAMINAR AEROFOILS

Aerofoils maintaining laminar boundary layer conditions all along the surface are known as laminar aerofoils. This is the most suitable arrangement to keep skin friction low. Though such aerofoils have been designed, they have many limitations.

12.8 LAMINAR FLOW

Laminar flow is an orderly flow in which the fluid elements move in an orderly manner such that the transverse exchange of momentum is insignificant. The laminar flow may be described as 'a well orderly pattern where fluid layers are assumed to slide over one another', that is, in laminar flow, the fluid moves in layers, or laminas, one layer gliding over an adjacent layer with an interchange of momentum only at the molecular level.

12.9 LAMINAR SEPARATION BUBBLE

A laminar separation bubble is formed when the previously attached laminar boundary layer encounters an adverse pressure gradient of sufficient magnitude to cause the flow to separate.

12.10 LAMINAR SUBLAYER

The laminar sublayer is that zone adjacent to the boundary where the turbulence is suppressed to such a degree that only the laminar effects prevail.

Laminar sublayer or viscous sublayer is a thin layer within a boundary layer just adjacent to the surface over which the boundary layer is formed.

In the turbulent boundary layer, as noted, large Reynolds stresses set up due to mass transport in the direction perpendicular to the surface, so that energy from the mainstream may easily penetrate to fluid layers quite close to the surface. Because of this, these layers have a velocity that is not much less than that of the mainstream. However, in layers that are very close to the surface, it is not possible for the velocity to exist perpendicular to the surface so in a very thin region immediately adjacent to the surface, the flow approximates laminar flow. This thin layer adjacent to the surface is termed as laminar sublayer or viscous sublayer.

12.11 LANCASTER–PRANDTL LIFTING-LINE THEORY

The Prandtl lifting-line theory is a mathematical model that predicts lift distribution over a three-dimensional wing based on its geometry. It is also known as the Lanchester–Prandtl wing theory.

It is a representation to improve on the accuracy of the horseshoe vortex system. In the lifting-line theory, the bound vortex is assumed to lie on a straight line joining the wing tips (known as the lifting line). Now, the vorticity is allowed to vary along the line. The lifting line is generally taken to lie along the line joining the section quarter-chord points of the wing. The results obtained using this representation are generally good provided that the aspect ratio of the wing is moderate or large, generally not less than 4.

12.12 LANDING

Landing of an aircraft is a flight phase of bringing the aircraft in contact with the ground at the lowest possible vertical velocity and, at the same time, somewhere near the lowest possible horizontal velocity relative to the ground.

12.13 LANDING GEAR

Landing gear is the undercarriage of an aircraft or spacecraft and may be used for either take-off or landing. For aircraft, it is generally needed for both. It was also formerly called alighting gear by some manufacturers, such as the Glenn L. Martin Company. For aircraft, Stinton makes the terminology distinction undercarriage (British)=landing gear (US).

12.14 LANDSLIDE

A landslide is the movement of rock, earth or debris down a sloped section of land, as shown in Figure 12.2. Rain, earthquakes, volcanoes or other factors that make the slope unstable cause landslides. Geologists – the scientists who study the physical formations of the Earth – sometimes describe landslides as one type of mass wasting.

12.15 LAPLACE EQUATION

Laplace equation is the governing equation for potential flows. Potential flow is based on the concept that the flow field can be represented by a potential function ϕ such that

$$\nabla^2 \phi = 0$$

This linear partial differential equation is popularly known as the *Laplace equation*. Derivatives of ϕ give velocities. Unlike the stream function ϕ, the potential function can exist only if the flow is *irrotational*, that is, when viscous effects are absent.

Figure 12.2 Landslide.

12.16 LAPSE RATE

Lapse rate λ is the rate at which the atmospheric temperature in the stratosphere decreases with altitude increase. For the International Standard Atmosphere (ISA), the value of the lapse rate λ is 0.0065 K/m.

12.17 LARGE EDDY SIMULATION

Large eddy simulation (LES) is a mathematical model for turbulence used in computational fluid dynamics.

12.18 LASER

The word LASER stands for Light Amplification by Stimulated Emission of Radiation.

A laser is a device that emits light through a process of optical amplification based on the stimulated emission of electromagnetic radiation. The word 'laser' is an acronym for 'light amplification by stimulated emission of radiation'.

12.19 LASER DOPPLER ANEMOMETER

The novel method of measuring velocities in fluid flows using the laser Doppler technique has been subjected to rapid development in the past few decades. The usefulness and reliability of this instrument in the measurement of low as well as high-speed flows have been successfully established. This technique is commonly referred to as Laser Doppler Anemometry or Laser Doppler Velocimetry. In short, it is termed as LDA or LDV.

12.20 LATENT ENERGY

The internal energy associated with the phase of a system is called latent energy.

12.21 LATENT HEAT

Latent heat is defined as the heat or energy that is absorbed or released during a phase change of a substance.

12.22 LATENT HEAT OF FUSION

The enthalpy of fusion of a substance, also known as (latent) heat of fusion, is the change in its enthalpy resulting from providing energy; typically heat, to a specific quantity of the substance to change its state from a solid to a liquid, at constant pressure.

12.23 LATENT HEAT OF VAPORISATION

Latent heat of vaporisation is the amount of energy required to vaporise a unit mass of saturated liquid at a given temperature or pressure.

12.24 LATERAL ACCELERATION

Lateral acceleration is acceleration to the side.

12.25 LATERAL STABILITY

Stability or control concerning rolling about the longitudinal axis is called lateral stability or control, respectively.

12.26 LAUNCH SYSTEM

A launch system includes the launch vehicle, launch pad, vehicle assembly and fuelling systems, range safety, and other related infrastructure.

12.27 LAUNCH VEHICLE

Launch vehicle is a rocket-propelled vehicle used to carry a payload from the Earth's surface to space, usually to the Earth's orbit or beyond.

12.28 LAVA FLOW

A lava flow is an outpouring of lava created during an effusive eruption, as shown in Figure 12.3. Explosive eruptions produce a mixture of volcanic ash and other fragments called tephra, rather than lava flows.

Figure 12.3 Lava flow.

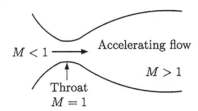

Figure 12.4 Laval nozzle.

12.29 LAVAL NOZZLE

A de Laval nozzle (or convergent-divergent nozzle, C-D nozzle) is a tube that is pinched in the middle, making a carefully balanced, asymmetric hour-glass shape, as shown in Figure 12.4. It is used to accelerate pressurised gas passing through it to a higher supersonic speed in the axial (thrust) direc-tion by converting the heat energy of the flow into kinetic energy. Because of this, the nozzle is widely used in some types of steam turbines and rocket engine nozzles. It also sees use in supersonic jet engines.

12.30 LAW OF CONSERVATION OF ENERGY

The law of conservation of energy states that energy can neither be cre-ated nor destroyed – it can only be converted from one form of energy to another.

12.31 LAW OF DIMENSIONAL HOMOGENEITY

The law of dimensional homogeneity states that 'an analytically derived equation representing a physical phenomenon must be valid for all systems of units'.

12.32 LAW OF FLOATATION

When a body floats in a liquid, the weight of the liquid displaced by its immersed part is equal to the total weight of the body. This is the law of floatation.

12.33 LAW OF INTERMEDIATE METALS

The law of intermediate metals states that a third metal (here iron) inserted between two dissimilar metals of a thermocouple junction will have no effect on the output voltage as long as the two junctions formed by the additional metal are at the same temperature.

12.34 LAW OF THE WALL

The wall shear stress in a turbulent boundary layer can also be estimated using the law of the wall.

12.35 LAYERS IN THE ISA

Different layers in the standard atmosphere are Troposphere, Tropopause, Stratosphere, Stratopause, Mesosphere, and Mesopause.

12.36 LAYERS OF THE EARTH

Starting at the centre, the Earth is composed of four distinct layers. They are, from deepest to shallowest, the inner core, the outer core, the mantle and the crust. Except for the crust, no one has ever explored these layers in person. In fact, the deepest humans have ever drilled is just over 12 km (7.5 mi) and even that took 20 years.

12.37 LDA PRINCIPLE

The principle underlying the laser Doppler anemometer is that a moving particle illuminated by a light beam scatters light at a frequency different from that of the original incident beam. This difference in frequency is known as Doppler shift and it is proportional to the velocity of the particle.

12.38 LEACHING

Leaching is the loss or extraction of certain materials from a carrier into a liquid (usually, but not always a solvent) and may refer to leaching (agriculture), the loss of water-soluble plant nutrients from the soil; or applying a small amount of excess irrigation to avoid soil salinity.

12.39 LEAD-LAGGING

Lead-lagging is the movement of a blade forward or aft in the plane of rotation.

12.40 LEADING-EDGE FLAP

The leading-edge flap is another device capable of producing lift increment by preventing flow separation from the leading edge of the wing. The leading-edge flap is also referred to as a droop snoot. High-speed aerofoil sections usually have a sharp leading edge and very little camber.

Such wings tend to stall early because, even at moderate incidence, the flow around the leading edge must negotiate a sharp bend. This gives rise to a region of separated flow known as a separation bubble behind the leading edge, as shown in Figure 12.5.

12.41 LEADING-EDGE RADIUS

The leading edge of aerofoils used in subsonic applications is rounded with a radius of about 1% of the chord length. The leading edge of an aerofoil is the radius of a circle with its centre on a line tangential to the leading-edge camber connecting tangency points of the upper and lower surfaces with

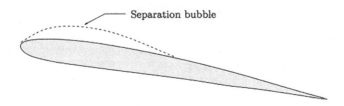

Figure 12.5 A separation bubble just behind the leading edge of an aerofoil.

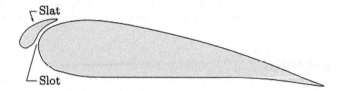

Figure 12.6 Sectional view of a wing with leading-edge slat.

the leading edge. The magnitude of the leading-edge radius has a significant effect on the stall characteristics of the aerofoil section.

12.42 LEADING-EDGE SLAT

The leading-edge slat is an auxiliary aerofoil or slat, mounted in a fixed position ahead of the leading edge of the main aerofoil (that is, wing), with a carefully designed gap, or slot, between them, as shown in Figure 12.6.

The leading-edge slat is essentially a device meant for boundary layer control. It operates by allowing the passage through the slot of air from the higher-pressure region below the wing to the lower-pressure region above it. Thus, energy is added to the boundary layer on the upper surface and any tendency to separation of the flow is much reduced.

12.43 LEAK

A small hole or crack through which liquid or gas can get through.

12.44 LEAN MIXTURE

Lean mixture is approximately the correct mixture to burn the fuel efficiently.
or
A lean mixture is a fuel/air mixture containing a relatively low proportion of fuel. A lean mixture can be caused by too little fuel or too much air. Too much oxygen indicates a lean mixture and the need for less fuel.

12.45 LEFT-RUNNING WAVE

For an observer looking in the direction of flow towards the disturbance, the wave to the left is called the left-running wave.

Shocks are referred to as left-running and right-running depending on whether they run to the left or right when viewed in the flow direction. Left-running waves constitute one family and right-running waves constitute the opposite family.

12.46 LEMON JUICE

The juice of a lemon is about 5%–6% citric acid, with a pH of around 2.2, giving it a sour taste. The distinctive sour taste of lemon juice makes it a key ingredient in drinks and foods such as lemonade and lemon meringue pie.

12.47 LIFT

The aerodynamic force F_{ad} can be resolved into two component forces, one at right angles to freestream velocity V and the other opposite to V. The force component normal to V is called lift L.

Lift is the component of the aerodynamic force perpendicular to the direction of motion.

12.48 LIFT COEFFICIENT

The lift force, L, can be expressed as a dimensionless number, popularly known as lift coefficient C_L, given by

$$C_L = \frac{L}{\frac{1}{2}\rho V^2 S}$$

where ρ, V, and S are the flow density, flight speed, and wing area, respectively.

12.49 LIFT CURVE

Lift curve is a plot of lift coefficient variation with the angle of incidence.

12.50 LIFT-CURVE SLOPE

Differentiating L with respect to the angle of attack, α, we have

$$\frac{dC_L}{d\alpha} = a$$

This derivative is called the lift-curve slope. The theoretical value of the lift-curve slope is 2π, but the experimental results show that the value for a two-dimensional aerofoil is about 5.7 per radian, which is about 0.1 per degree.

12.51 LIFT-INDUCED DRAG

In aerodynamics, lift-induced drag, induced drag, vortex drag or sometimes drag due to lift, is an aerodynamic drag force that occurs whenever a moving object redirects the airflow coming at it. This drag force occurs in aeroplanes due to wings or a lifting body redirecting air to cause lift and in cars with aerofoil wings that redirect air to cause a downforce.

12.52 LIFTING-LINE THEORY

The Prandtl lifting-line theory is a mathematical model that predicts lift distribution over a three-dimensional wing based on its geometry. It is also known as the Lanchester–Prandtl wing theory. Frederick W. Lanchester expressed the theory independently in 1907 and Ludwig Prandtl in 1918–1919 after working with Albert Betz and Max Munk.

12.53 LIFTING SURFACE THEORY

Lifting surface theory is a method that treats the aerofoil as a vortex sheet over which vorticity is spread at a given rate. In other words, the aerofoil is regarded as a surface composed of lifting elements. This is different from the lifting-line theory. The essential difference between the lifting surface theory and lifting-line theory is that in the former, the aerofoil is treated as a vortex sheet, whereas in the latter, the aerofoil is represented by a straight line joining the wing tips, over which the vorticity is distributed.

12.54 LIGHT-YEAR

The light-year is a unit of length used to express astronomical distances and is equivalent to about 9.46 trillion kilometres (9.46×10^{12} km) or 5.88 trillion miles (5.88×10^{12} mi).

12.55 LIGHTNING

Lightning is a brilliant electric spark discharge in the atmosphere, occurring within a thundercloud, between clouds, or between a cloud and the ground, as shown in Figure 12.7.

Figure 12.7 Lightning.

12.56 LIKE REFLECTION

The reflection of an incident wave from a solid boundary is termed like reflection, that is, a shock wave reflects as a shock wave and an expansion wave reflects as an expansion wave.

12.57 LIMING

Liming is one of the main steps carried out during leather production in the tannery. The main purpose of liming is to separate the hair from the hides.

12.58 LIMITATIONS ON AIR AS A PERFECT GAS

When the temperature is less than 500 K, air can be treated as a perfect gas, and the ratio of specific heats, γ, takes a constant value of 1.4.

When the temperature lies between 500 and 2,000 K, air is only thermally perfect (but calorically imperfect), and the state equation $p = \rho RT$ is valid, but c_p and c_v become functions of temperature, $c_p = c_p(T)$ and $c_v = c_v(T)$. Even though c_p and c_v are functions of temperature, their ratio γ continues to be independent of temperature. That is, c_p and c_v vary with temperature in such a manner that their ratio continues to be the same constant as in temperatures below 500 K. For temperatures of more than 2,000 K, air becomes both thermally and calorically imperfect. That is, c_p, c_v as well as γ become functions of temperature.

12.59 LIMITATIONS OF HOT-WIRE ANEMOMETER

Even though the hot-wire anemometer is considered to be one of the most useful instruments for measuring turbulence, it has many limitations. Some of these limitations have become apparent in the course of our discussions in the preceding sections. These involve:

1. A nonlinear character of heat transfer with respect to velocity and temperature.
2. An onset of practical limitations due to the complex nature of the heat transfer between the wire and fluid in compressible flows.
3. A limitation set by the resolution power in space, that is, in the direction of the wire due to its finite length.
4. A limitation set by resolution power in time in the flow direction due to the finite time constant of the hot-wire.

12.60 LIMITATIONS OF HYDRAULIC ANALOGY

The analogy is valid for one-dimensional and two-dimensional flows only. But the two-dimensionality limitation is not a serious drawback since most of the problems in gas dynamics are of two-dimensional nature and even flows that are not strictly two-dimensional can be approximated as two-dimensional flows without introducing significant error.

The hydraulic analogy is well established for gas with $\gamma = 2$. The influence of the γ value, with reference to the analogy, varies considerably according to the type of the flow. For moderate subsonic speeds or uniform streams with small perturbations, the pressure distribution as given by various linearised theories, like the Prandtl–Glauert rule, Karman–Tsien theory, etc. is independent of the specific heat ratio γ. Hence, the influence of γ does not come into the picture at all. Even for high-speed flows or for large changes in velocity, the effect of γ is not so large as to change the order of magnitude of the numerical results. For supersonic flows, the effect of γ varies with the shock strength. In transonic flows, the effect of γ can be taken care of by making use of the transonic similarity laws.

However, for one-dimensional flows, it is well established that by altering the water flow channel cross-section, flow conditions corresponding to gases with different values of γ can be obtained. The flow in a rectangular section corresponds to the flow of gas with $\gamma = 2$. The flow in a triangular section corresponds to $\gamma = 1.5$, and the flow in a parabolic section corresponds to $\gamma = 1.4$.

12.61 LINE VORTEX

A line vortex is a string of rotating particles. In a line vortex, a chain of fluid particles are spinning about their common axis and carrying around with them a swirl of fluid particles that flow around in circles. A cross-section of such a string of particles and the associated flow show a spinning point outside of which the flow streamlines are concentric circles.

12.62 LINEAR FLIGHT

When the aircraft velocity V is in a fixed straight line, the flight is termed linear. There are three types of linear symmetrical flight: gliding, horizontal and climbing. Among these, gliding is the only flight possible without the use of an engine.

12.63 LINEAR VARIABLE DIFFERENTIAL TRANSFORMER (LVDT)

Linear differential transformer works on the induction principle, like any other transformer, but with a movable core fixed to the diaphragm. The LVDT consists of a primary coil, two secondary coils and a movable iron core. The secondary coils are connected in opposition. When the core is in the null position between the two secondary coils, the voltage induced gets cancelled. Any movement of the core on either side gives an output proportional to the displacement from the null point. This instrument is capable of detecting displacements of the order of a few microns. Applying known pressures, the associated displacements of the core and the corresponding outputs of the LVDT can be noted to make the calibration chart.

12.64 LIQUID

In the liquid phase, the molecular spacing is not much different from that of the solid phase, except that the molecules are no longer at a fixed position relative to each other.

A liquid is a nearly incompressible fluid that conforms to the shape of its container but retains a (nearly) constant volume independent of pressure. As such, it is one of the four fundamental states of matter (the others being solid, gas and plasma), and is the only state with a definite volume but no fixed shape.

12.65 LIQUID HYDROGEN

Liquid hydrogen (LH_2) is the liquid state of the element hydrogen. Hydrogen is found naturally in the molecular H_2 form. To exist as a liquid, H_2 must be cooled below its critical point of 33 K.

12.66 LIQUID MANOMETERS

Liquid manometers measure differential pressure by balancing the weight of a liquid between two pressures. Light liquids such as water can measure small pressure differences; mercury or other heavy liquids are used for large pressure differences.

12.67 LIQUID NITROGEN

Liquid nitrogen is nitrogen in a liquid state at a low temperature of –196°C (77 K). It is produced industrially by fractional distillation of liquid air. It is a colourless, low-viscosity liquid that is widely used as a coolant.

12.68 LIQUID OXYGEN

Liquid oxygen is a form of molecular oxygen. It is a cryogenic liquid. Cryogenic liquids are liquefied gases that have a normal boiling point below –90°C.

12.69 LIQUID-PROPELLANT ROCKET

In a liquid-propellant rocket, the fuel and the oxidiser are stored in the rocket in liquid form and pumped into the combustion chamber.

12.70 LIQUID PARAFFIN

Liquid paraffin, also known as *paraffinum liquidum* or Russian mineral oil, is a very highly refined mineral oil used in cosmetics and medicines. Cosmetic or medicinal liquid paraffin should not be confused with the paraffin (or kerosene) used as a fuel.

12.71 LIQUEFACTION

Liquefaction is a phenomenon in which the strength and stiffness of soil are reduced by earthquake shaking or other rapid loadings. Liquefaction and the related phenomena have been responsible for tremendous amounts of damage in historical earthquakes around the world.

In materials science, liquefaction is a process that generates a liquid from a solid or a gas or that generates a non-liquid phase that behaves as per fluid dynamics. It occurs both naturally and artificially.

12.72 LIQUEFIED PETROLEUM GAS

Liquefied petroleum gas (LPG, LP gas or condensate) is a flammable mixture of hydrocarbon gases such as propane and butane.

12.73 LITMUS

Litmus is a water-soluble mixture of different dyes extracted from lichens. It is often absorbed onto a filter paper to produce one of the oldest forms of pH indicator used to test materials for acidity.

12.74 LOAD FACTOR

The load factor, n, for an aircraft is defined as the ratio of the lift, L, to the weight, W:

$$n = \frac{L}{W}$$

In a straight and level flight, $L = W$, so that $n = 1$. In manoeuvres, $L = n\,W$.

12.75 LOBED-IMPELLER METER

The lobed-impeller meter, shown in Figure 12.8, is a positive-displacement metre meant for mass flow measurement of gas and liquid flows.

Accurate fit of the casing and impeller are ensured in these metres. In this way, the incoming fluid is always trapped between the rotors and is

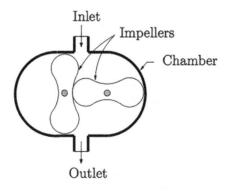

Figure 12.8 Lobed-impeller meter.

conveyed to the outlet by the rotation of the rotors. The number of revolutions per unit time of the rotors is a measure of the volumetric flow rate. The accuracy of the displacement meters can be very good even at the low end of the mass/volume range.

12.76 LOCAL CHEMICAL EQUILIBRIUM

A flow is said to be in local chemical equilibrium if the local chemical composition at each point in the flow is the same as that determined by the chemical equilibrium calculations.

12.77 LOCAL RATE OF CHANGE

The rate of change of properties measured by probes at fixed locations is referred to as local rates of change.

12.78 LOCAL THERMODYNAMIC EQUILIBRIUM

A flow is said to be in local thermodynamic equilibrium if a local Boltzmann distribution exists at each point in the flow at the local temperature T.

12.79 LONGITUDINAL AXIS

The longitudinal axis is a straight line running from the foremost point to the rearmost point and passing through the centre of gravity and is horizontal when the aircraft is in level flight.

12.80 LONGITUDINAL DIHEDRAL

The angle between the tailplane chord and wing chord is called longitudinal dihedral. The longitudinal dihedral is regarded as an important characteristic dictating the longitudinal stability of most types of aircraft.

12.81 LONGITUDINAL STABILITY

The stability or control of an aircraft concerning pitching about the lateral axis is called longitudinal stability or control, respectively.

12.82 LONGITUDINAL WAVES

Longitudinal waves are waves in which the displacement of the medium is in the same (or opposite) direction of the wave propagation. Sound travels through longitudinal waves.

12.83 LOSSES IN DIFFUSER

The loss of energy in the diffuser is due to (a) skin friction and (b) expansion.

12.84 LOTION

A lotion is a low-viscosity topical preparation intended for application to the skin. By contrast, creams and gels have a higher viscosity, typically due to lower water content.

12.85 LOUDSPEAKER

A loudspeaker (or speaker driver, or most frequently just speaker) is an electroacoustic transducer, that is, a device that converts an electrical audio signal into a corresponding sound. A speaker system, also often simply referred to as a 'speaker' or 'loudspeaker', comprises one or more such speaker drivers (above definition), an enclosure and electrical connections possibly including a crossover network.

12.86 LOW EARTH ORBIT

A low Earth orbit (LEO) is an orbit that is relatively close to the Earth's surface. It is normally at an altitude of <1,000 km but could be as low as 160 km above Earth – which is low compared to other orbits, but still very far above the Earth's surface.

12.87 LOW-SPEED FLOW

Flows with Mach number <0.5 are called low-speed flows. For low-speed flow problems, thermodynamic considerations are not needed because the heat content of the fluid flow is so large, compared to the kinetic energy of the flow, that the temperature remains nearly constant even if the whole kinetic energy is transformed into heat.

The difference between the static and stagnation temperatures is not significant in low-speed flows.

12.88 LOW-SPEED WIND TUNNELS

Low-speed tunnels are those with a test section speed of <650 kmph.

12.89 LOW WING

Low wing means the wing's lower surface is level with (or below) the bottom of the fuselage.

12.90 LOWER CRITICAL REYNOLDS NUMBER

Lower critical Reynolds number is the Reynolds number below which the entire flow is laminar.

12.91 LUBRICANT

A lubricant is a substance that helps to reduce friction between surfaces in mutual contact, which ultimately reduces the heat generated when the surfaces move.

Typically, lubricants contain 90% base oil (most often petroleum fractions called mineral oils) and <10% additives. Vegetable oils or synthetic liquids such as hydrogenated polyolefins and esters, silicones, fluorocarbons and many others are sometimes used as base oils.

12.92 LUBRICATION

Lubrication is the control of friction and wear by the introduction of a friction-reducing film between moving surfaces in contact.

12.93 LUDWIEG TUBE

Basically Ludwieg tube tunnel is a blowdown-type hypersonic tunnel. Due to its special fluid dynamic features, no devices are necessary to control pressure or temperature during the run. It thus can be regarded as an 'intelligent blowdown facility'. The test gas storage occurs in a long charge tube, which is by a fast-acting valve separated from the nozzle, test section and the discharge vacuum tank.

Hubert Ludwieg, a German scientist, first proposed the concept of the Ludwieg tube in 1955. The beauty of the Ludwieg tube is that it provides a clean supersonic/hypersonic flow with relative ease and low cost.

12.94 LUDWIEG TUBE ADVANTAGES

Advantages of the Ludwieg tube tunnel compared to standard blowdown tunnels are the following:

1. The Ludwieg tube tunnel requires an extremely short start and shut-off time.
2. No regulation of temperature and pressure during the run is necessary. No throttle valve upstream of the nozzle is necessary. The gas dynamic principle regulates the temperature and pressure.
3. From points 1 and 2 above follows an impressive economy, since there exists no wastage of mass and energy of flow during tunnel start and shutoff.
4. Due to the elimination of the pressure regulating valve, the entrance to the nozzle can be kept clean. This ensures a low turbulence level in the test section.
5. The Ludwieg tube is well suited for transient heat transfer tests.
6. There is no unit Reynolds number effect in the Ludwieg tube as in the case of many facilities.

12.95 LUMINOSITY

Luminosity is an absolute measure of radiated electromagnetic power – the radiant power emitted by a light-emitting object. In astronomy, luminosity is the total amount of electromagnetic energy emitted per unit of time by a star, galaxy or any other astronomical object.

12.96 LUNGS

The lungs are a pair of spongy, air-filled organs located on either side of the chest. The windpipe conducts the inhaled air into the lungs through its tubular branches called bronchi.

Chapter 13

Mach to Munk's Theorem of Stagger

13.1 MACH

Mach (Figure 13.1) was the first scientist who experimentally observed the phenomenon of shock.

13.2 MACH ANGLE

The angle between the Mach line and the direction of motion of the body (flow direction) is called the Mach angle μ.

13.3 MACH ANGLE - MACH NUMBER RELATION

The angle μ is simply a characteristic angle associated with the Mach number M by the relation

$$\sin \mu = \frac{1}{M}$$

This is called the Mach angle-Mach number relation. The lines, which may be drawn at any point in the flow field with inclination μ, are called Mach lines or Mach waves. It is essential to understand the difference between the Mach waves and Mach lines. Mach waves are the weakest isentropic waves in a supersonic flow field, and the flow through them will experience only negligible changes in flow properties. Thus, a flow traversed by the Mach waves does not experience a change in Mach number, whereas the Mach lines, even though are weak isentropic waves, will cause small but finite changes to the properties of a flow passing through them. In uniform supersonic flows, the Mach waves and Mach lines are linear and inclined at an angle given by

$$\mu = \sin^{-1}\left(\frac{1}{M}\right)$$

DOI: 10.1201/9781003348405-13

Figure 13.1 Ernst Waldfried Josef Wenzel Mach (18 February 1838 to 19 February 1916).

But in nonuniform supersonic flows, the flow Mach number M varies from point to point, and hence, the Mach angle μ, being a function of the flow Mach number, varies with M and the Mach lines are curved.

13.4 MACH CONE

Mach cone is the conical pressure wave front produced by a body moving at a speed greater than that of sound.

13.5 MACH LINE

Mach line is a theoretical line representing the back-sweep of a cone-shaped shock wave made by an assumed infinitesimally small particle moving at the same speed and along the same flight path as that of an actual body or particle. Propagation of disturbance waves created by an object moving with velocity (a) V=0, (b) V=a/2, (c) V=a and (d) V>a are illustrated in Figure 13.2.

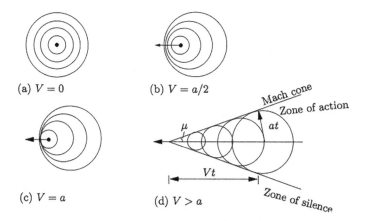

(a) $V = 0$

(b) $V = a/2$

(c) $V = a$

(d) $V > a$

Figure 13.2 Propagation of wave, Mach cone, zone of action and zone of silence.

13.6 MACH NUMBER

Mach number M is a dimensionless parameter expressed as the ratio between the magnitudes of local flow velocity and local velocity of sound, i.e.,

$$M = \frac{\text{Local flow velocity}}{\text{Local velocity of sound}} = \frac{V}{a}$$

Mach number is the ratio of local flow speed to the local speed of sound or the ratio of inertial force to elastic force. It is a measure of compressibility. For an incompressible fluid, $M = 0$.

13.7 MACH NUMBER INDEPENDENCE PRINCIPLE

For slender configurations, such as sharp cones and wedges, the strong shock assumption is

$$M_\infty \sin\theta_b \gg 1$$

where M_∞ is the freestream Mach number and θ_b is the semi-angle for the nose. The concept termed the Mach number independence principle depends on this assumption. Oswatitsch derived the Mach number independence principle for inviscid flow. The pressure forces are much larger than the viscous forces for blunt bodies or slender bodies at large angles of incidence when the Reynolds number exceeds 10^5; therefore, one would expect the Mach number independence principle to hold at these conditions.

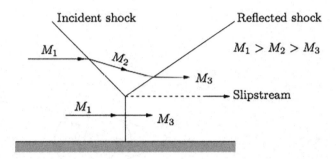

Figure 13.3 Mach reflection.

13.8 MACH REFLECTION

Mach reflection is a supersonic fluid dynamics effect named after Ernst Mach and is a shock wave reflection pattern involving three shocks.

The intersection of normal shock and the right-running oblique shock gives rise to a reflected left-running oblique shock to bring the flow into the original direction, as illustrated in Figure 13.3. This kind of reflection is termed Mach reflection. The left-running shock must have lesser strength compared to the right-running shock because of the flow deflection angle θ involved in the compression process associated with the shock, but $M_1 > M_2$.

13.9 MACH STEM

The reflected blast wave merges with the incident shock wave to form a single wave known as the Mach stem.

13.10 MACH TUCK

Mach tuck is an aerodynamic effect whereby the nose of an aircraft tends to pitch downwards as the airflow around the wing reaches supersonic speeds. This diving tendency is also known as tuck under. The aircraft will first experience this effect at significantly below Mach 1.

13.11 MACH WAVE

Mach wave is the weakest isentropic wave in a supersonic flow. Neither the flow properties nor the flow direction will be influenced by the Mach wave. Another important aspect is that the Mach waves will be present even in uniform and unidirectional supersonic flow.

13.12 MACH-WEDGE

For supersonic flow over two-dimensional objects, we will have a 'Mach-wedge' instead of a Mach cone.

13.13 MACHMETER

A Machmeter is an aircraft pitot-static system that shows the ratio of the true airspeed to the speed of sound – a dimensionless quantity called the Mach number. This is shown on a Machmeter as a decimal fraction. An aircraft flying at the speed of sound is flying at a Mach number of one expressed as Mach 1.

13.14 MACROSTATE

A macrostate of a system is a state that is observed by the experimenter. Its description need not be given to the ultimate limits in detail, but only to the desired degree of accuracy of laboratory experiment.

13.15 MAGNETOFLUIDMECHANICS

Magnetofluidmechanics is an extension of fluid mechanics with thermodynamics, mechanics, materials and the electrical sciences. Astrophysicists initiated this branch. Other names, which are used to refer to this discipline, are magnetogasdynamics and hydromagnetics.

Magnetofluidmechanics is the science of the motion of an electrically charged conducting fluid in the presence of a magnetic field. The motion of the electrically conducting fluid in the magnetic field will induce electric currents in the fluid, thereby, modifying the field. The mechanical forces produced by it will also modify the flow field. The interaction between the field and the motion makes the magnetofluid dynamic analysis difficult.

A gas at normal and moderately high temperatures is a nonconductor. But at very high temperatures of the order of 10,000 K and above, thermal excitation sets in. This leads to dissociation and ionisation. Ionised gas is called plasma, which is an electrically conducting medium. Electrically conducting fluids are encountered in engineering problems like re-entry of missiles and spacecraft, plasma jet, controlled fusion research and magnetohydrodynamic generator.

13.16 MAGNUS EFFECT

The lateral force experienced by rotating bodies is called the Magnus effect. It is the generation of a sidewise force on a spinning cylindrical or spherical solid immersed in a fluid (liquid or gas) when there is relative motion between the spinning body and the fluid.

13.17 MALLEABLE

Malleable means capable of being shaped, bent or drawn out by hammering or pressure.

13.18 MALTOSE

Maltose is a sugar made out of two glucose molecules bound together. It's created in seeds and other parts of plants as they break down their stored energy in order to sprout. Thus, foods like cereals, certain fruits and sweet potatoes contain naturally high amounts of this sugar.

13.19 MANOEUVRES OF AN AIRCRAFT

The manoeuvres of an aircraft are (a) Movement forward or backwards, (b) Movement up or down, (c) Movement sideways, to right or left, (d) Rolling, (e) Yawing and (f) Pitching.

13.20 MANOMETERS

Manometers measure the difference between a known and an unknown pressure by observing the difference in the heights of two fluid columns. Two popular types of manometers are illustrated in Figure 13.4.

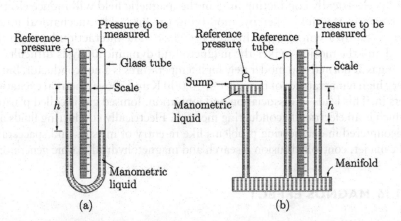

Figure 13.4 Manometers: (a) U-tube and (b) multitube.

13.21 MANOMETRY

Manometry is a science of pressure measurement using liquid column height. The pressure measuring devices meant for measurements in fluid flow may broadly be grouped into manometers and pressure transducers. Various types of liquid manometers are employed depending upon the range of pressures to be measured and the degree of precision required. The U-tube manometers, multitube manometers and Betz-type manometers are some of the popular liquid manometers used for pressure measurements in fluid flow.

The basic measuring principle of the liquid manometer is that 'the pressure applied is balanced by the weight of a liquid column'.

13.22 MARINE ENGINEERING

Marine engineering includes the engineering of boats, ships, oil rigs and any other marine vessel or structure as well as oceanographic engineering, oceanic engineering or nautical engineering. Specifically, marine engineering is the discipline of applying engineering sciences, including mechanical engineering, electrical engineering, electronic engineering and computer science, to the development, design, operation and maintenance of watercraft propulsion and on-board systems and oceanographic technology. It includes, but is not limited, to power and propulsion plants, machinery, piping, automation and control systems for marine vehicles of any kind, such as surface ships and submarines.

13.23 MARSH

A marsh is a wetland that is dominated by herbaceous rather than woody plant species. Marshes can often be found at the edges of lakes and streams, where they form a transition between the aquatic and terrestrial ecosystems. Grasses, rushes or reeds often dominate them.

13.24 MASS DIFFUSIVITY

Diffusivity, mass diffusivity or diffusion coefficient is the proportionality constant between the molar flux due to molecular diffusion and the gradient in the concentration of the species (or the driving force for diffusion). Diffusivity is encountered in Fick's law and numerous other equations of physical chemistry.

13.25 MASS FLOW RATE

Mass flow rate is the mass of a substance that passes per unit of time.

13.26 MASS FRACTION

It is the ratio of the mass of a component to the mass of the mixture.

13.27 MASS-MOTION VELOCITY

The fluid velocity behind the moving shock is called mass-motion velocity.

13.28 MASS TRANSFER

Mass transfer is the net movement of mass from one location, usually meaning stream, phase, fraction or component, to another. Mass transfer occurs in many processes, such as absorption, evaporation, drying, precipitation, membrane filtration and distillation.

13.29 MATERIAL DERIVATIVE

The time rate change of any property ϕ experienced by a fluid element (Lagrangian perspective).

$$\frac{D\phi}{Dt} = \frac{\partial \phi}{\partial t} + u \cdot \nabla \phi$$

13.30 MATERIAL DESCRIPTION

Material description or the Lagrangian method describes the motion of each particle of the flow field in a separate and discrete manner. For example, the velocity of the nth particle of an aggregate of particles moving in space can be specified by the scalar equations:

$$(V_x)_n = f_n(t)$$

$$(V_y)_n = g_n(t)$$

$$(V_z)_n = h_n(t)$$

where V_x, V_y, V_z are the velocity components in x, y, z directions, respectively.

13.31 MATERIALS SCIENCE

The interdisciplinary field of materials science, also commonly termed materials science and engineering, covers the design and discovery of new materials, particularly solids. The intellectual origins of materials science stem from the enlightenment when researchers began to use analytical thinking from chemistry, physics and engineering to understand ancient, phenomenological observations in metallurgy and mineralogy. Materials science still incorporates elements of physics, chemistry and engineering. As such, the field was long considered by academic institutions as a sub-field of these related fields. Beginning in the 1940s, materials science began to be more widely recognised as a specific and distinct field of science and engineering, and major technical universities around the world created dedicated schools for its study.

13.32 MATHEMATICS

Mathematics includes the study of such topics as quantity (number theory), structure (algebra), space (geometry) and change (analysis). It has no generally accepted definition.

13.33 MATTER

Matter is everything around you. Atoms and compounds are all made of very small parts of matter. These atoms go on to build the things you see and touch every day. Matter is defined as anything that has mass and takes up space (it has volume).

13.34 MAXWELL'S DEFINITION OF VISCOSITY

Maxwell's definition of viscosity states that 'the coefficient of viscosity is the tangential force per unit area on either of two parallel plates at a unit distance apart, one fixed and the other moving with unit velocity'.

13.35 MAXWELL RELATIONS

Maxwell relations are equations that relate the partial derivatives of pressure p, specific volume v, temperature T and entropy s of a simple compressible substance to each other. They are obtained from the four Gibbs equations by exploiting the exactness of the differentials of thermodynamic properties.

13.36 MAXWELL–BOLTZMANN STATISTICS

In statistical mechanics, Maxwell–Boltzmann statistics describe the average distribution of non-interacting material particles over various energy states in thermal equilibrium and is applicable when the temperature is high enough or the particle density is low enough to render quantum effects to be negligible.

13.37 MAYER'S RELATION

$$c_p - c_v = R$$

This relation is popularly known as Mayer's relation in honour of Julius Robert von Mayer (25 November 1814 to 20 March 1878), a German physician and physicist and one of the founders of thermodynamics. He is best known for enunciating in 1841 one of the original statements of the conservation of energy or what is now known as one of the first versions of the first law of thermodynamics, namely 'energy can be neither created nor destroyed'.

13.38 MEAN AERODYNAMIC CHORD

Mean aerodynamic chord is an average chord that when multiplied by the product of the average section moment coefficient, the dynamic pressure, and the wing area, gives the moment for the entire wing.

   The mean aerodynamic, mac, chord is given by

$$\text{mac} = \frac{1}{S} \int_{-b}^{b} \left[c(y) \right]^2 dy$$

where S is the wing area, b is the semi-span of the wing , $c(y)$ is the local chord, and y is the transverse axis.

13.39 MEAN CAMBER LINE

Mean camber line is the locus of the points midway between the upper and lower surfaces of the aerofoil. In other words, mean camber line is the bisector of the aerofoil thickness. The shape of the mean camber line plays an important role in the determination of the aerodynamic characteristics of the aerofoil section.

13.40 MEAN EFFECTIVE PRESSURE

The mean effective pressure (MEP) is an important term associated with reciprocating engines. It is a fictitious pressure and is defined as that pressure which when acting on the piston during the entire power stroke would produce the same amount of net work as that produced during the actual cycle.

13.41 MEAN FREE PATH

Mean free path is the distance travelled by a molecule between two successive collisions.

13.42 MEAN FREE PATH LENGTH

The mean value of mean free path of all molecules in a gas is called the molecular mean free path length.

13.43 MEASURING PRINCIPLE OF THE LIQUID MANOMETER

The basic measuring principle of the liquid manometer is that the pressure applied is balanced by the weight of a liquid column.

13.44 MECHANICAL EQUILIBRIUM

Mechanical equilibrium is related to pressure, and a system is said to be in mechanical equilibrium if there is no change in pressure with time at any point in the system.

13.45 MECHANICS OF FLUIDS

The science of fluid motion is referred to as the mechanics of fluids, an allied subject of the mechanics of solids and engineering materials and built on the same fundamental laws of motion. Therefore, unlike empirical hydraulics, it is based on the physical principles and has a close correlation with experimental studies that both compliment and substantiate the fundamental analysis. Unlike the classical hydrodynamics that is based purely on mathematical treatment.

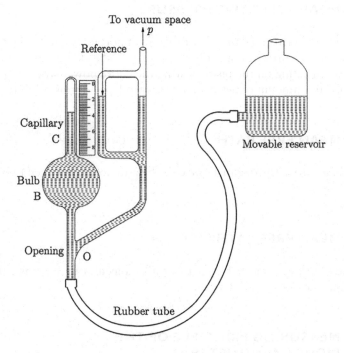

Figure 13.5 The McLeod gauge.

13.46 McLEOD GAUGE

The McLeod gauge is used as a standard for measuring low vacuum pressures. It is basically a mercury manometer. A typical McLeod gauge is shown schematically in Figure 13.5. The measuring procedure of a McLeod gauge is as follows. The mercury reservoir is lowered until the mercury column drops below the opening O. Now, the bulb B and the capillary C are at the same pressure as the vacuum space p. The reservoir is raised until the mercury fills the bulb and rises in the capillary up to a level at which the mercury in the reference capillary is at the zero level marked on it.

13.47 MEDICAL OXYGEN

Medical oxygen is high purity oxygen that is used for medical treatments and is developed for use in the human body. Medical oxygen comprises minimum 90% oxygen with 5% nitrogen and 5% argon. There are different ways of making oxygen, but all the techniques eliminate nitrogen, moisture, hydrocarbons and CO_2 leaving behind only oxygen.

13.48 MELTING

Melting, or fusion, is a physical process that results in the phase transition of a substance from a solid to a liquid. This occurs when the internal energy of the solid increases, typically by the application of heat or pressure, which increases the substance's temperature to the melting point.

13.49 MELTING POINT

The melting point is the temperature at which a solid turns into a liquid. For ice, the melting point is 0°C or 273 K. The chemical element with the highest melting point is tungsten (3,410°C), which is used for making filaments in light bulbs.

13.50 MELTING POINT OF WATER

Pure water transitions between the solid and liquid states at 0°C at sea level. This temperature is referred to as the melting point when rising temperatures are causing ice to melt and change its state from a solid to a liquid (water).

13.51 MERCURY

Mercury is a heavy, silver-white, highly toxic metallic element and the only common metal that is liquid at room temperature. It is also called quicksilver. It is used in barometers, thermometers, pesticides, pharmaceutical preparations, reflecting surfaces of mirrors, and dental fillings'; in certain switches, lamps, and other electric apparatus, and as a laboratory catalyst. Its symbol is Hg; atomic weight: 200.59; atomic number: 80; specific gravity: 13.546 at 20°C; freezing point: −38.9°C; and boiling point: 357°C.

13.52 METAL FILM RTDs

In modern construction techniques, a platinum or metal-glass slurry film is deposited as a screen onto a small ceramic substrate, etched with a laser-trimming system and sealed. The film RTD offers a substantial reduction in assembly time and has further advantage of increased resistance for a given size. Due to the manufacturing technology, the device size itself is small, which makes it respond quickly to changes in temperature. Film RTDs are less stable than their hardware counterparts, but they are becoming more popular because of their advantages in size and production cost.

13.53 METEOROLOGY

Meteorology is a branch of the atmospheric sciences (which includes atmospheric chemistry and atmospheric physics) with a major focus on weather forecasting. The study of meteorology dates to thousands of years,, though significant progress in meteorology did not begin until the eighteenth century. The nineteenth century saw a modest progress in the field after weather observation networks were formed across broad regions. Prior attempts at prediction of weather depended on historical data. It was not until after the elucidation of the laws of physics, and more particularly, the development of the computer, allowing for the automated solution of a great many equations that model the weather, in the latter half of the twentieth century that significant breakthroughs in weather forecasting were achieved. An important branch of weather forecasting is marine weather forecasting as it relates to maritime and coastal safety in which weather effects also include atmospheric interactions with large bodies of water.

13.54 METHANOL

Methanol, sometimes called 'wood alcohol', is a clear liquid with the chemical formula CH_3OH.

13.55 METHOD OF CHARACTERISTICS

Method of characteristics is a numerical technique used to design contoured nozzles to generate uniform and unidirectional supersonic flows.

The mathematical concept of characteristics (taken as identical to the Mach lines), even though not physical, forms the basis for the numerical method termed method of characteristics.

13.56 MICRO MANOMETER

Micro manometer is basically a precision U-tube manometer. It is a rugged instrument capable of measuring liquid column height with an accuracy of ± 0.1 mm. A typical micro manometer is shown in Figure 13.6. From the figure, it is seen that the micro manometer is a simple U-tube manometer with a vernier scale for precise measurement of liquid level. The pressure p_2 to be measured is applied to the manometer tube and a reference pressure (usually the ambient atmospheric pressure), p_1, is applied to the reservoir.

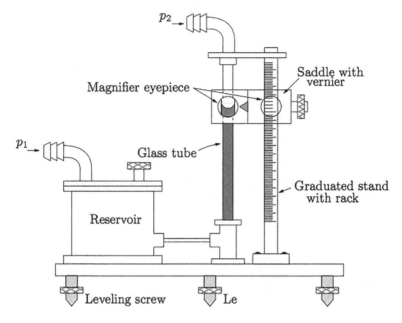

Figure 13.6 Micro manometer.

13.57 MICROPHONE

A microphone, colloquially called a mic or mike, is a device – a transducer –
that converts sound into an electrical signal. Microphones are used in many
applications such as telephones, hearing aids, public address systems for concert halls and public events, motion picture production, live and recorded
audio engineering, sound recording, two-way radios, megaphones and radio
and television broadcasting. They are also used in computers for recording
voice, speech recognition, VoIP (Voice over Internet Protocol) and for non-
acoustic purposes such as ultrasonic sensors or knock sensors.

13.58 MICROSTATE

A microstate of a system species is the state of the system to the ultimate
limit of detail (a complete specification of the coordinates and molecules
comprising the system would define a microstate of the system).

13.59 MID-WING

The wing is mounted midway up the fuselage.

13.60 MILK

Milk is a nutrient-rich liquid food produced by the mammary glands of mammals. It is the primary source of nutrition for young mammals, including breastfed human infants before they are able to digest solid food.

13.61 MILKSHAKE

A milkshake is a sweet drink made by blending milk, ice cream and flavourings or sweeteners such as butterscotch, caramel sauce, chocolate syrup, fruit syrup or whole fruit into a thick, sweet and cold mixture.

13.62 MINOR LOSSES

In many circuits, the pipe flow may be required to pass through fittings, such as bends, nozzles and diffusers involving abrupt changes in the area. Because of these passages, additional losses are encountered. The primary loss associated with these passages is separation losses, in which energy is dissipated due to the violent mixing in the separation zones. Usually, these losses are considerably smaller than the frictional loss, and hence, termed minor losses.

13.63 MISSILE

Missile (shown in Figure 13.7) is a rocket-propelled weapon designed to deliver an explosive warhead with great accuracy at high speed.

13.64 MISSILE TYPES

There are five types of missiles: air-to-air, air-to-surface, surface-to-air, antiship and antitank or assault. Ballistic missiles are most often categorised as

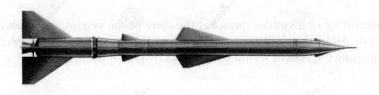

Figure 13.7 Missile.

short-range, medium-range, intermediate-range and intercontinental ballistic missiles (SRBMs, MRBMs, IRBMs and ICBMs).

13.65 MIST

Mist is a phenomenon caused by small droplets of water suspended in the air.

13.66 MIST FLOW

Mist flow is said to occur when a significant amount of liquid is transferred
from the annular film to the gas core; at high gas flow rates, nearly all the
liquid is entrained in the gas.

Mist flow is a flow regime of two-phase gas-liquid flow, as shown in
Figure 13.8. It occurs at very high flow rates and has very high flow quality.
Due to this condition, the liquid film flowing on the channel wall is thinned
by the shearing of the gas core at the interface until it becomes unstable and
is destroyed.

13.67 MIXED-FLOW PUMP

A mixed-flow pump is a centrifugal pump with a mixed-flow impeller.

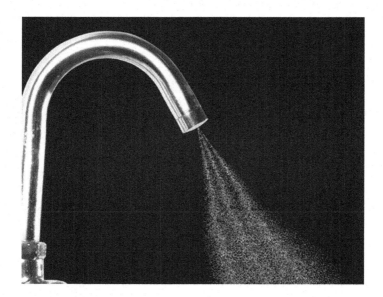

Figure 13.8 Mist flow.

13.68 MIXED-FLOW TURBINE

A mixed-flow turbine of the Deriaz type uses swivelled, variable-pitch runner blades that allow for improved efficiency at part loads in medium-sized machines.

13.69 MIXING CHAMBER

In a flow system, the section where the mixing process takes place is commonly referred to as the mixing chamber. It need not have to be a distinct 'chamber'. An ordinary T-elbow or a Y-elbow in a shower, for example, serves as a mixing chamber for the cold and hot water streams.

13.70 MIXING LENGTH

Mixing length is defined as that distance in the transverse direction which must be covered by a lump of fluid particle travelling with its original mean velocity to make the difference between its velocity and the velocity of the new layer equal to the mean transverse fluctuation in the turbulent flow.

13.71 MIXTURE

A mixture is the physical combination of two or more substances in which the identities are retained and are mixed in the form of solutions, suspensions or colloids.

13.72 MOBILITY

Mobility is the ability to move or be moved freely and easily.

13.73 MOISTURE

Moisture is the presence of a liquid, especially water, often in trace amounts. Small amounts of water may be found, for example, in the air, food and some commercial products.

13.74 MOISTURE AND HUMIDITY

Moisture is associated with the water content in the liquid phase present in any substance. Liquid water present in steam is called moisture. Humidity is associated with the water content in the gaseous phase present in the air. The air always holds a certain amount of water in the gaseous phase, which is called water vapour.

13.75 MOLAR MASS

Molar mass is defined as the mass of one mole of a substance in grams or the mass of one kmol in kilograms.

13.76 MOLASSES

Molasses or black treacle is a viscous product resulting from refining sugarcane or sugar beets into sugar.

13.77 MOLE

The mole is the unit of measurement for the amount of substance in the International System of Units (SI). It is defined as exactly $6.02214076 \times 10^{23}$, which may be atoms, molecules, ions or electrons.

13.78 MOLE FRACTION

This is the ratio of the mole number of a component to the mole number of the mixture.

13.79 MOLECULE

A molecule is a group of two or more atoms that form the smallest identifiable unit into which a pure substance can be divided and still retain the composition and chemical properties of that substance. It is an electrically neutral group of two or more atoms held together by chemical bonds. Molecules are distinguished from ions by their lack of electrical charge. The molecular sizes of oxygen, nitrogen and argon are 0.299, 0.305 and 0.363 nm.

13.80 MOLECULAR DIAMETER

Molecular diameters are an important property of gases for numerous scientific and technical disciplines. The molecular sizes of oxygen, nitrogen and argon are 0.299, 0.305 and 0.363 nm.

13.81 MOLECULAR GAS DYNAMICS

Molecular gas dynamics is the branch of gas dynamics that studies the movement of gases from the molecular-kinetic point of view.

13.82 MOLECULAR WEIGHT

The molecular weight or molar mass is defined as the mass of one mole of a substance in kilograms (or grams). When we say that the molecular mass of oxygen is 32, it means that the mass of 1 kmole of oxygen is 32 kg or 1 gmole of oxygen is 32 g.

13.83 MOLLIER DIAGRAM

The Mollier diagram (refer Figure 13.9) is a graphical representation of the relationship between air temperature, moisture content and enthalpy and is a basic design tool for building engineers and designers.

13.84 MOMENTUM DIFFUSION

Momentum diffusion most commonly refers to the diffusion, or spread of momentum between particles (atoms or molecules) of matter, often in the fluid state. This transport of momentum can occur in any direction of the fluid flow. Momentum diffusion can be attributed to either external pressure or shear stress or both.

13.85 MOMENTUM EQUATION

The momentum equation is the representation of Newton's law that states that the 'sum of the forces acting on a system of fluid particles is equal to the time rate of change of linear momentum'.

The momentum equation is based on Newton's second law and represents the balance between various forces acting on a fluid element. For gaseous medium, body forces are negligibly small as compared to other forces, and hence, can be neglected. For steady incompressible flows, the momentum equation can be written as

$$V_x \frac{\partial V_x}{\partial x} + V_y \frac{\partial V_x}{\partial y} + V_z \frac{\partial V_x}{\partial z} = -\frac{1}{\rho} \frac{\partial p}{\partial x} + v \left(\frac{\partial^2 V_x}{\partial x^2} + \frac{\partial^2 V_x}{\partial y^2} + \frac{\partial^2 V_x}{\partial z^2} \right) \quad (13.1)$$

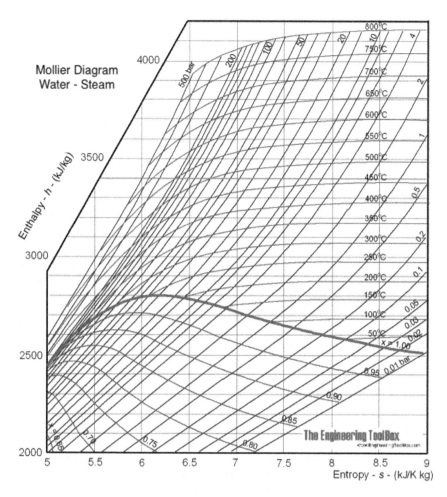

Figure 13.9 Mollier diagram: water-steam.

$$V_x \frac{\partial V_y}{\partial x} + V_y \frac{\partial V_y}{\partial y} + V_z \frac{\partial V_y}{\partial z} = -\frac{1}{\rho} \frac{\partial p}{\partial y} + v\left(\frac{\partial^2 V_y}{\partial x^2} + \frac{\partial^2 V_y}{\partial y^2} + \frac{\partial^2 V_y}{\partial z^2} \right) \qquad (13.2)$$

$$V_x \frac{\partial V_z}{\partial x} + V_y \frac{\partial V_z}{\partial y} + V_z \frac{\partial V_z}{\partial z} = -\frac{1}{\rho} \frac{\partial p}{\partial z} + v\left(\frac{\partial^2 V_z}{\partial x^2} + \frac{\partial^2 V_z}{\partial y^2} + \frac{\partial^2 V_z}{\partial z^2} \right) \qquad (13.3)$$

Equations (13.1), (13.2), and (13.3) are, respectively, the x, y, and z components of the momentum equation. These equations are generally known as *Navier-Stokes equations*.

13.86 MOMENTUM THICKNESS

The momentum thickness is defined as the thickness of a layer of fluid of velocity U_∞ for which the momentum flux is equal to the deficit of momentum flux through the boundary layer.

Momentum thickness is the distance by which the boundary, over which the boundary layer prevails, has to be hypothetically shifted so that the momentum associated with the mass passing through the actual thickness (distance) and the hypothetical thickness will be the same.

13.87 MONOPLANE AIRCRAFT

A monoplane is a fixed-wing aircraft with one main set of wing surfaces in contrast to a biplane or triplane which have multiple planes. A monoplane has inherently the highest efficiency and the lowest drag of any wing configuration and is the simplest to build. However, during the early years of flight, these advantages were offset by its greater weight and lower manoeuvrability, making it relatively rare until the 1930s. Since then, the monoplane has been the most common form of a fixed-wing aircraft.

13.88 MONSOON

Monsoon is traditionally a seasonal reversing wind accompanied by corresponding changes in precipitation. Usually, the term monsoon is used to refer to the rainy phase of a seasonally changing pattern, although technically, there is a dry phase also. The term is also sometimes used to describe locally heavy but short-term rains.

13.89 MONSOON SEASON

Monsoon season in South Asia typically occurs between June to September leading to heavy rainfall. Monsoons typically occur in tropical areas and various atmospheric conditions influence the monsoon winds.

13.90 MONTE CARLO METHODS

Monte Carlo methods, or Monte Carlo experiments, are a broad class of computational algorithms that rely on repeated random sampling to obtain

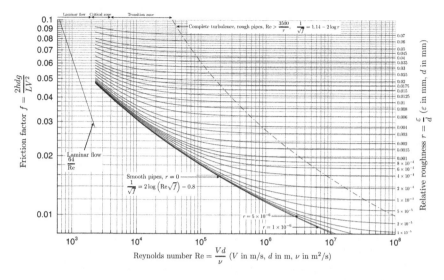

Figure 13.10 Moody's diagram.

numerical results. The underlying concept is to use randomness to solve problems that might be deterministic in principle.

or

Monte Carlo simulation, also known as the Monte Carlo method or a multiple probability simulation, is a mathematical technique, which is used to estimate the possible outcomes of an uncertain event.

13.91 MOODY CHART

Moody chart or Moody diagram (shown in Figure 13.10) is a graph in non-dimensional form that relates the Darcy–Weisbach friction factor f, Reynolds number Re, and surface roughness for fully developed flow in a circular pipe. It can be used to predict pressure drop or flow rate down such a pipe.

13.92 MOST PROBABLE MACROSTATE

The most probable macrostate has the maximum number of microstates. It is the macrostate which occurs when the system is in thermodynamic equilibrium.

This plays a dominant role in the study of high-enthalpy gas dynamics, because at temperatures above 800 K, the vibration excitation of the molecules becomes active; beyond 2,000 K, the molecules in a gas dissociate to become atoms; and beyond 3,000 K, the atoms themselves become active and get ionised, heading towards a plasma state.

13.93 MOVEMENT OF CENTRE OF PRESSURE

At low values of incidence, the centre of pressure cp moves forward with increasing incidence. This is due to the increase of suction (that is, increase in the negative magnitude of pressure coefficient) and the forward movement of its position on the upper surface of the aerofoil with increase of incidence angle α.

13.94 MOVING BOUNDARY WORK

One form of mechanical work frequently encountered in practice is associated with the compression or expansion of a gas in a piston-cylinder device. This work is called moving boundary work or simply boundary work.

13.95 MULTIPHASE FLOW

Multiphase flow is the simultaneous flow of materials with two or more thermodynamic phases.

13.96 MULTIPLE SLOTTED FLAPS

Multiple slotted flaps use the same concept as the single slotted flap but have increased effectiveness.

13.97 MUNK'S THEOREM OF STAGGER

Munk's theorem of stagger states that 'the total drag of a multi-plane system does not change when the elements are translated parallel to the direction of the wind, provided that the circulations are left unchanged'. Thus, the total induced drag depends only on the frontal aspect.

Chapter 14

Natural Convection to Nusselt Number

14.1 NATURAL CONVECTION

Natural convection is a type of flow of motion of a liquid, such as water, or gas, such as air (see Figure 14.1), in which the fluid motion is not generated by any external source (like a pump, fan, suction device, etc.) but by some parts of the fluid which are heavier than the other parts. In most cases, this leads to natural circulation – the ability of a fluid in a system to circulate continuously with gravity and possible changes in heat energy. The driving force for natural convection is gravity.

14.2 NATURAL GAS

Natural gas (also called fossil gas; sometimes just gas) is a naturally occurring hydrocarbon gas mixture consisting primarily of methane, but commonly including varying amounts of other higher alkanes, and sometimes, a small percentage of carbon dioxide, nitrogen, hydrogen sulphide, or helium.

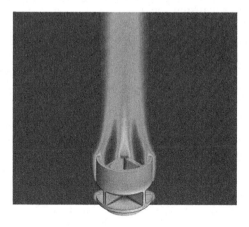

Figure 14.1 Natural convection.

DOI: 10.1201/9781003348405-14

14.3 NATURE OF HIGH-ENTHALPY FLOWS

There are two major physical characteristics that cause a high-enthalpy flow to deviate from calorically perfect gas behaviour:

1. At high temperatures, the vibrational excitation of the gas molecules becomes important, absorbing some of the energy, which, at normal temperatures, would go into the translational and rotational motion. The excitation of vibrational energy causes the specific heat of the gas to become a function of temperature, causing the gas to become calorically imperfect.
2. With further increase in temperature, the molecules begin to dissociate and even ionise. Under these conditions, the gas becomes chemically reacting, and the specific heats become functions of both temperature and pressure.

14.4 NAVIER–STOKES EQUATIONS

Navier–Stokes equations are mathematical states of conservation of momentum of a flow system.

The momentum equation that is based on Newton's second law represents the balance between various forces acting on a fluid element, namely,

1. Force due to rate of change of momentum, generally referred to as inertia force.
2. Body forces such as buoyancy force, magnetic force, and electrostatic force.
3. Pressure force.
4. Viscous forces (causing shear stress).

For a fluid element under equilibrium by Newton's second law, we have the momentum equation as

Inertia force + Body force + Pressure force + Viscous force = 0

For a gaseous medium, body forces are negligibly small compared to other forces, and hence, can be neglected. For steady incompressible flows, the momentum equation can be written as

$$V_x \frac{\partial V_x}{\partial x} + V_y \frac{\partial V_x}{\partial y} + V_z \frac{\partial V_x}{\partial z} = -\frac{1}{\rho}\frac{\partial p}{\partial x} + v\left(\frac{\partial^2 V_x}{\partial x^2} + \frac{\partial^2 V_x}{\partial y^2} + \frac{\partial^2 V_x}{\partial z^2} \right)$$

$$V_x \frac{\partial V_y}{\partial x} + V_y \frac{\partial V_y}{\partial y} + V_z \frac{\partial V_y}{\partial z} = -\frac{1}{\rho} \frac{\partial p}{\partial y} + v \left(\frac{\partial^2 V_y}{\partial x^2} + \frac{\partial^2 V_y}{\partial y^2} + \frac{\partial^2 V_y}{\partial z^2} \right)$$

$$V_x \frac{\partial V_z}{\partial x} + V_y \frac{\partial V_z}{\partial y} + V_z \frac{\partial V_z}{\partial z} = -\frac{1}{\rho} \frac{\partial p}{\partial z} + v \left(\frac{\partial^2 V_z}{\partial x^2} + \frac{\partial^2 V_z}{\partial y^2} + \frac{\partial^2 V_z}{\partial z^2} \right)$$

These equations are the x, y, z components of the momentum equation, respectively. These equations are generally known as Navier–Stokes equations. They are nonlinear partial differential equations and there exists no known analytical method to solve them.

In physics, the Navier–Stokes equations are certain partial differential equations, which describe the motion of viscous fluid substances, named after the French engineer and physicist Claude-Louis Navier and Anglo-Irish physicist and mathematician George Gabriel Stokes.

14.5 NAVIGATION

Navigation is a field of study that focuses on the process of monitoring and controlling the movement of a craft or vehicle from one place to another. The field of navigation includes four general categories: land navigation, marine navigation, aeronautic navigation, and space navigation.

14.6 NEGATIVE ANGLE OF ATTACK

A symmetric aerofoil will generate no lift at zero angle of attack and negative lift at a negative angle of attack. However, cambered aerofoils are curved such that they will generate lift at small negative angles of attack.

14.7 NEON

Neon is a chemical element with the symbol Ne and atomic number 10. It is a noble gas. Neon is a colourless, odourless, inert monatomic gas under standard conditions with about two-thirds the density of air.

14.8 NET HEAD

Net head or effective head on the turbine (including nozzle) is the static head minus the pipe friction losses.

14.9 NET WING AREA

The wing area that does not include the fuselage is called the net wing area.

14.10 NEWTONIAN FLOW MODEL

When the density ratio across the shock wave becomes small, the shock layer becomes very thin. For this kind of flow situation, we can assume that the speed and direction of the gas particles in the freestream remain unchanged until they strike the solid surface exposed to the flow. This flow model is termed the Newtonian flow model. For the Newtonian flow model, the normal component of momentum of the impinging fluid particle is wiped out, whereas the tangential component of momentum is conserved.

14.11 NEWTON'S LAW OF GRAVITATION

Newton's law of gravitation states that any particle of matter in the universe attracts any other with a force varying directly as the product of the masses and inversely as the square of the distance between them.

14.12 NEWTON'S LAWS OF MOTION

1. Everybody continues in its state of rest or of uniform motion in a straight line except insofar as it is compelled to change that state by an external impressed force.
2. The rate of change of momentum of the body is proportional to the impressed force and takes place in the direction in which the force acts.
3. To every action there is an equal and opposite reaction.

14.13 NEWTON'S LAW OF VISCOSITY

Newton's law of viscosity states that 'the stresses which oppose the shearing of a fluid are proportional to the rate of shear strain', that is, the shear stress τ is given by

$$\tau = \mu \frac{\partial u}{\partial y}$$

where μ is the absolute coefficient of viscosity and $\partial u / \partial y$ is the velocity gradient.

14.14 NEWTON'S SECOND LAW

Newton's second law is a statement of conservation of momentum, which states that 'the forces acting on a system is equal to the rate of change of momentum of the system'.

14.15 NEWTONIAN FLUID

Newtonian fluids are those that obey Newton's law of viscosity, which states that 'the stresses which oppose the shearing of a fluid are proportional to the rate of shear strain', i.e., the shear stress τ is given by

$$\tau = \mu \frac{\partial u}{\partial y}$$

where μ is the absolute coefficient of viscosity and $\partial u / \partial y$ is the velocity gradient. The viscosity μ is a property of the fluid.

14.16 NITRIC ACID

Nitric acid (HNO_3), also known as *aqua fortis* (Latin for 'strong water') and spirit of niter, is a highly corrosive mineral acid. The pure compound is colourless but older samples tend to acquire a yellow cast due to decomposition into oxides of nitrogen and water.

14.17 NITROGEN

Nitrogen is the chemical element with the symbol N and atomic number 7.

14.18 NITROGEN GAS

Nitrogen gas makes up about 80% of the air we breathe.

14.19 NITROUS OXIDE

Nitrous oxide, commonly known as *laughing gas* or *happy gas*, is a colourless, non-flammable gas. This gas is used in medical and dental procedures as a sedative. It helps to relieve anxiety before the procedure and allows the patient to relax.

14.20 NOBLE GASES

Noble gases are odourless, colourless, non-flammable, and monoatomic gases that have low chemical reactivity. The noble gases are the chemical elements in group 18 of the periodic table. They are the most stable due to having the maximum number of valence electrons.

14.21 NO-SLIP CONDITION

In fluid dynamics, the no-slip condition for viscous fluids assumes that at a solid boundary, the fluid will have zero velocity relative to the boundary. The fluid velocity at all fluid–solid boundaries is equal to that of the solid boundary.

14.22 NOISE

Unpleasant or unwanted sound is termed noise.

14.23 NOISE POLLUTION

Noise pollution, also known as environmental noise or sound pollution, is the propagation of noise with ranging impacts on the activity of human or animal life, most of them harmful to a degree. Machines, transport, and propagation systems mainly cause the source of outdoor noise worldwide.

14.24 NOISE POLLUTION EFFECTS

Noise pollution impacts millions of people on a daily basis. The most common health problem it causes is Noise-Induced Hearing Loss (NIHL). Exposure to loud noise can also cause high blood pressure, heart disease, sleep disturbances, and stress.

14.25 NOISE OF SUPERSONIC JET

The noise of a supersonic jet comprises three basic components: the turbulent mixing noise, the broadband shock-associated noise, and the screech tones.

The appearance of a screeching tone is usually accompanied by its harmonics. Sometimes, even the fourth or fifth harmonics can be detected. The relative magnitude of this noise intensity is a strong function of the direction of observation. In the downstream direction of the jet, turbulent mixing noise is the most dominant noise component. In the upstream direction, the broadband shock-associated noise is more intense. For circular jets, the screech tones radiate primarily in the upstream direction.

14.26 NON-CHEMICAL ROCKET

In a non-chemical rocket, the high efflux velocity from the rocket is generated without any chemical reaction taking place. For example, a gas could be heated to high pressure and temperature by passing it through a nuclear reactor, and it could then be expanded through a nozzle to give a high efflux velocity.

14.27 NON-CONSERVATIVE SYSTEM

A non-conservative system is the one in which work done by a force is dependent on the path. It is not equal to the difference between the final and initial values of an energy function.

14.28 NON-NEWTONIAN FLUID

Fluids for which stress is not proportional to the time rate of strain are called non-Newtonian fluids. Non-Newtonian fluids do not obey Newton's law of viscosity. They show a nonlinear dependence of shearing stress on velocity gradient. Fluids such as honey, paste, printer's ink, tar, etc. are non-Newtonian fluids.

14.29 NON-SIMPLE REGION

The non-simple region in a supersonic flow field is that with curved left- and right-running characteristic lines, as shown in Figure 14.2.

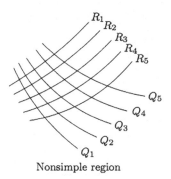

Figure 14.2 Non-simple region.

14.30 NORMAL BP RANGE

A normal blood pressure level is <120/80 mm Hg. No matter your age, you can take steps each day to keep your blood pressure in a healthy range.

14.31 NORMAL PULSE RATE

A normal resting heart rate should be 60–100 beats per minute, but it can vary from minute to minute. It can go up to 130–150 beats or higher per minute when you're exercising which is normal because the body needs to pump more oxygen-rich blood around the body.

14.32 NORMAL SHOCK

Normal shock may be described as a compression front in a supersonic flow field across which the flow properties jump. The thickness of the shock is comparable to the mean free path of the gas molecules in the flow field.

There is no heat added to or taken away from the flow as it traverses a shock wave; that is, the flow process across the shock wave is adiabatic. Therefore, the total temperature remains the same ahead of and behind the wave, $T_{02} = T_{01}$.

For a stationary normal shock, the total enthalpy is always constant across the wave, which, for calorically or thermally perfect gases, translates into a constant total temperature across the shock. However, for a chemically reacting gas, the total temperature is not constant across the shock. Also, if the shock wave is not stationary (that is, for a moving shock), neither the total enthalpy nor the total temperature is constant across the shock wave.

14.33 NORMAL SHOCK RELATIONS

Considering the normal shock shown in Figure 14.3, the following normal shock relations, assuming the flow to be one-dimensional, can be obtained:

$$M_2^2 = \frac{2 + (\gamma - 1) M_1^2}{2\gamma M_1^2 - (\gamma - 1)}$$

$$\frac{p_2}{p_1} = 1 + \frac{2\gamma}{\gamma + 1} \left(M_1^2 - 1 \right)$$

$$\frac{\rho_2}{\rho_1} = \frac{V_1}{V_2} = \frac{(\gamma + 1) M_1^2}{(\gamma - 1) M_1^2 + 2}$$

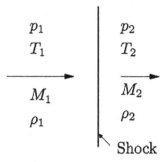

Figure 14.3 Flow through a normal shock.

$$\frac{T_2}{T_1} = \frac{h_2}{h_1} = \frac{a_2^2}{a_1^2} = 1 + \frac{2(\gamma-1)}{(\gamma-1)^2}\frac{(\gamma M_1^2 + 1)}{M_1^2}(M_1^2 - 1)$$

where h_1 and h_2 are the static enthalpies upstream and downstream of the shock, respectively.

The stagnation pressure ratio across a normal shock, in terms of the upstream Mach number, is

$$\frac{p_{02}}{p_{01}} = \left[1 + \frac{2\gamma}{\gamma+1}(M_1^2 - 1)\right]^{\frac{-1}{\gamma-1}}\left[\frac{(\gamma+1)M_1^2}{(\gamma-1)M_1^2 + 2}\right]^{\frac{\gamma}{\gamma-1}}$$

The change in entropy across the normal shock is given by

$$s_2 - s_1 = R\ln\frac{p_{01}}{p_{02}}$$

14.34 NORMAL SHOCK TABLES

In aerodynamics, the normal shock tables are a series of tabulated data listing the various properties before and after the occurrence of a normal shock wave. With a given upstream Mach number, the post-shock Mach number can be calculated along with the pressure, density, temperature, and stagnation pressure ratios. Such tables are used to calculate the properties after a normal shock.

14.35 NOSEDIVE

Nosedive is a form of gliding with a gliding angle close to $90°$, as shown in Figure 14.4.

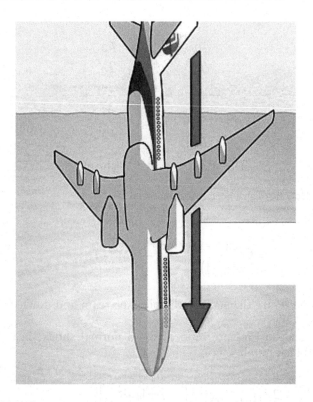

Figure 14.4 Nosedive.

14.36 NOZZLE

A nozzle is a passage used to transform pressure energy into kinetic energy.

A nozzle is a passage in which the flow accelerates. A convergent-divergent nozzle is essential for generating a supersonic flow of desired properties. Also, a convergent nozzle can accelerate a flow to a maximum of Mach 1.0 only. For accelerating the flow choked at the exit of a convergent nozzle to a supersonic level, a divergent portion must be added to the convergent duct with the convergent nozzle exit with sonic condition becoming the throat of the convergent-divergent nozzle. The convergent-divergent nozzles are of two types: straight convergent-divergent nozzle and contoured or de Laval nozzle. For generating uniform as well as unidirectional supersonic flow, a de Laval nozzle should be used and if only the uniform Mach number is of interest and unidirectional quality is not mandatory, straight convergent-divergent nozzles can be employed.

14.37 NOZZLE DISCHARGE COEFFICIENT

The nozzle discharge coefficient C_d is the ratio of the actual mass flow rate through the nozzle to the mass flow calculated from the isentropic relations for the initial and final pressures of the actual nozzle, that is,

$$C_d = \frac{\dot{m}_{actual}}{\dot{m}_{isentropic}}$$

14.38 NOZZLE EFFICIENCY

Nozzle efficiency is defined as the ratio of the actual kinetic energy at the nozzle exit to the kinetic energy that would be obtained in a frictionless nozzle expanding the gas to the same final pressure. In other words, nozzle efficiency is the ratio of the actual to ideal kinetic energy at the nozzle exit. The nozzle efficiency plays a dominant role in turbine design where it is important to estimate accurately the average velocity of the flow leaving the nozzle.

14.39 NUSSELT NUMBER

In fluid dynamics, the Nusselt number (Nu) is the ratio of convective to conductive heat transfer at a boundary in a fluid. Convection includes both advection (fluid motion) and diffusion (conduction). The conductive component is measured under the same conditions as the convective but for a hypothetically motionless fluid. It is a dimensionless number, closely related to the fluid's Rayleigh number.

Chapter 15

OASPL to Ozone Hole

15.1 OASPL

The overall sound pressure level, OASPL, is the jet noise. Sound pressure levels are usually measured on a logarithmic scale but the unit is the *decibel*. There is an advantage in using the decibel scale. Because the ear is sensitive to noise in a logarithmic fashion, the decibel scale more nearly represents how we respond to noise.

15.2 OBLIQUE SHOCK

It is a compression wave at an angle to the flow. The inclination angle of the wave to the upstream flow direction is called the *shock angle*. A flow through an oblique shock also is decelerated to a lower Mach number. The reduced Mach number behind the wave can be either supersonic or subsonic. Shock, which decelerates the flow to become subsonic, is called a strong *shock* and that for which the downstream flow, even though is decelerated to a lower Mach number than the upstream Mach number, continues to be supersonic is called a *weak shock*. In practice, all oblique shocks encountered are weak. To generate a strong oblique shock, one must make special arrangements in the device. One such arrangement is increasing the backpressure for an engine intake through means such as blocking the exit area with a solid object.

A flow through an oblique shock is represented in Figure 15.1.

The oblique shock can be visualised as a normal shock with an upstream Mach number $M_1 \sin \beta$. Thus, the replacement of M_1 in normal shock relations by $M_1 \sin \beta$ results in the corresponding relations for the oblique shock.

The main governing equation for oblique shocks is the relation between the flow turning angle, shock angle, and Mach number, popularly known as the $(\theta - \beta - M)$ relation,

$$\tan \theta = 2 \cot \beta \left(\frac{M_1^2 \sin^2 \beta - 1}{M_1^2 [\gamma + \cos 2\beta] + 2} \right)$$

DOI: 10.1201/9781003348405-15

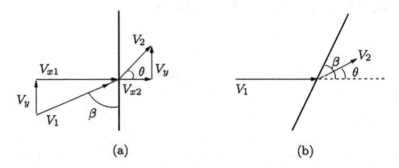

Figure 15.1 (a) Flow through a normal shock with superposition of velocity V_y, (b) oblique shock with shock angle beta β.

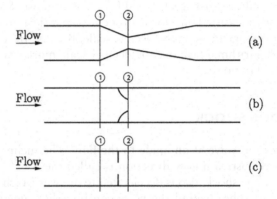

Figure 15.2 Schematic of (a) venturi, (b) flow nozzle, and (c) orifice.

15.3 OBSTRUCTION OR CONSTRICTION METERS

The three types of obstruction or constriction meters that are widely used for flow measurement are the venturi, flow nozzles, and orifice. Typical shapes of these constriction meters are shown schematically in Figure 15.2.

15.4 OCEAN

The ocean (also the sea or the world ocean) is a body of salt water that covers approximately 71% of the Earth's surface.

15.5 OCEAN AND SEA DIFFERENCE

In terms of geography, seas are smaller than oceans and are usually located where the land and ocean meet. Typically, seas are partially enclosed by land.

Figure 15.3 Ocean eddy.

15.6 OCEAN DEPTH

The average depth of the ocean is about 12,100 feet. The deepest part of the ocean is called the Challenger Deep and is located beneath the western Pacific Ocean at the southern end of the Mariana Trench, which runs several 100 km southwest of the US territorial island of Guam.

15.7 OCEAN EDDY

Ocean eddies are like giant whirlpools. The swirling motion of eddies in the ocean cause nutrients, that are normally found in colder, deeper waters, to come to the surface as shown in Figure 15.3.

15.8 OCEANOGRAPHY

Oceanography also known as oceanology is the scientific study of the ocean. It is an important Earth science, which covers a wide range of topics, including ecosystem dynamics; ocean currents, waves, and geophysical fluid dynamics; plate tectonics and the geology of the sea floor; and fluxes of various chemical substances and physical properties within the ocean and across its boundaries. These diverse topics reflect multiple disciplines that oceanographers utilise to glean further knowledge of the world ocean, including astronomy, biology, chemistry, climatology, geography, geology, hydrology, meteorology, and physics. Paleoceanography studies the history of the oceans in the geologic past. An oceanographer is a person who studies many things concerned with oceans including marine geology, physics, chemistry, and biology.

15.9 OLEO STRUT

An oleo strut is a pneumatic air–oil hydraulic shock absorber used on the landing gear of most large aircraft and many smaller ones. This design cushions the impacts of landing and damps out vertical oscillations.

15.10 ONE-DIMENSIONAL FLOW

One-dimensional flow is that in which the radius of curvature of the streamlines is very large and the cross-sectional area of the passage (streamtube) does not change abruptly.

15.11 OOZING

To flow slowly out or to allow something to flow slowly out.

15.12 OPEN-CHANNEL FLOW

When the liquid is bounded by sidewalls – such as the banks of a river or canal – the flow is said to take place in an open channel.

Open-channel flow, a branch of hydraulics and fluid mechanics, is a type of liquid flow within a conduit or in a channel with a free surface, known as a channel. The other type of flow within a conduit is pipe flow. These two types of flow are similar in many ways but differ in one important respect: the free surface. Open-channel flow has a free surface, whereas pipe flow does not.

15.13 OPEN-CIRCUIT TUNNELS

Open-circuit tunnels are tunnels without guided return of air. If the tunnel draws air directly from the atmosphere, fresh air flows through the tunnel all the time.

15.14 OPEN-JET TUNNEL

If the test section is bounded by air at different velocities (usually at rest), the tunnel is called an open-jet tunnel.

15.15 OPEN SYSTEM

An open system is a properly selected region in space. It usually encloses a device, which involves mass flow, such as a nozzle, diffuser, compressor, or turbine.

Both energy and mass can cross the boundary of a control volume, which is called a control surface, but the shape of the control volume will remain unchanged.

Open systems are capable of delivering work continuously because in the system, the medium that transforms energy is continuously replaced. This useful work, which a machine continuously delivers, is called the shaft work.

15.16 OPPOSING JETS

If a jet is opposed by another jet, the combination is termed opposing jets.

15.17 OPTICS

Optics is the branch of physics that studies the behaviour and properties of light, including its interactions with matter and the construction of instruments that use or detect it. Optics usually describes the behaviour of visible, ultraviolet, and infrared light. Because light is an electromagnetic wave, other forms of electromagnetic radiation such as X-rays, microwaves, and radio waves exhibit similar properties.

15.18 OPTICAL FIBRE

An optical fibre is a flexible, transparent fibre made by drawing glass (silica) or plastic to a diameter slightly thicker than that of a human hair. Optical fibres are used most often as a means to transmit light between the two ends of the fibre and find wide usage in fibre-optic communications, where they permit transmission over longer distances and at higher bandwidths (data transfer rates) than electrical cables. Fibres are used instead of metal wires because signals travel along them with less loss; in addition, fibres are immune to electromagnetic interference – a problem from which metal wires suffer. Fibres are also used for illumination and imaging and are often wrapped in bundles so they may be used to carry light into, or images out of confined spaces, as in the case of a fiberscope. Specially designed fibres are also used for a variety of other applications, some of them being fibre-optic sensors and fibre lasers.

15.19 OPTICAL PYROMETER

Some solids, such as metals, when they become hot, begin to emit light with a very dull red colour, and as the temperature increases, the colour becomes a brighter red, orange, yellow, and so on. The particular colour emission occurs at a particular temperature, and hence, can be used as a means of measuring the temperature if suitable calibration is adopted. It is this principle that is used in the optical pyrometer.

15.20 OPTICAL-TYPE PRESSURE TRANSDUCER

Schematic sketch of an optical-type pressure transducer is shown in Figure 15.4. The pressure-sensing element is a thin diaphragm that is highly polished on both sides. When the pressures p_1 and p_2 are equal, the diaphragm remains flat and the width of the reflected light beams 'a' and 'b' is the same. Under this condition, the differential output from the photocells is zero. But when the diaphragm gets deflected due to pressure difference, the width of the reflected beams is different. The amount of light falling on the photocells through the slits is different. This causes a differential output proportional to the difference between the pressures p_1 and p_2.

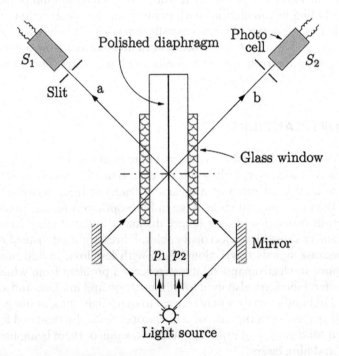

Figure 15.4 Optical-type pressure transducer.

Figure 15.5 Ornithopter.

15.21 ORIFICE PLATE

An orifice plate is a device used for measuring flow rate, reducing pressure, or restricting flow (in the latter two cases, it is often called a restriction plate).

15.22 ORNITHOPTER

The ornithopter (shown in Figure 15.5) obtains thrust by flapping its wings. It has found practical use in a model hawk to freeze prey animals into stillness so that they can be captured and also in toy birds.

15.23 ORSAT GAS ANALYSER

The Orsat gas analyser is a commonly used device to analyse the composition of combustion gases.

15.24 OSCILLATORY MOTION

Oscillatory motion can be termed as the repeated motion in which an object repeats the same movement over and over. In the absence of friction, the oscillatory motion would continue forever; but in the real world, the system eventually settles into equilibrium.

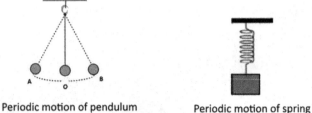

Periodic motion of pendulum   Periodic motion of spring

Figure 15.6 Oscillatory motion types.

15.25 OSCILLATORY MOTION TYPES

There are two types of oscillatory motions, namely, Linear Oscillatory Motion and Circular Oscillatory Motion, as illustrated in Figure 15.6. In linear motion, the object moves left and right or up and down. In the circular motion, though the object moves left to right but in circular form.

15.26 OSMOSIS

Osmosis is the movement of a solvent across a semipermeable membrane towards a higher concentration of solute (lower concentration of solvent). In biological systems, the solvent is typically water, but osmosis can occur in other liquids, supercritical liquids, and even gases.

15.27 OSMOTIC PRESSURE

Osmotic pressure is defined as the pressure that must be applied to the solution side to stop fluid movement when a semipermeable membrane separates a solution from pure water.

15.28 OSWALD EFFICIENCY NUMBER

The Oswald efficiency, similar to the span efficiency, is a correction factor that represents the change in drag with the lift of a three-dimensional wing or aeroplane as compared with an ideal wing having the same aspect ratio and an elliptical lift distribution.

15.29 OSWALD WING EFFICIENCY

The induced drag coefficient can be expressed as

$$C_{D_v} = \frac{C_L^2}{\pi e A\mathbb{R}}$$

where e, which is the inverse of induced drag factor k, is called the Oswald wing efficiency.

15.30 OTTO CYCLE

It is the ideal cycle for spark ignition (SI) engines, named after Nikolaus A. Otto.

15.31 OUTER SPACE

Outer space is the expanse that exists beyond the Earth and between celestial bodies.

15.32 OVERALL PRESSURE RATIO

In aeronautical engineering, the overall pressure ratio, or overall compression ratio, is the ratio of the stagnation pressure as measured at the front and rear of the compressor of a gas turbine engine. The terms compression ratio and pressure ratio are used interchangeably. The overall compression ratio also means the overall cycle pressure ratio that includes intake ram.

15.33 OVERALL PROPULSIVE EFFICIENCY

The overall propulsive efficiency η is the efficiency with which the energy contained in a vehicle's propellant is converted into useful energy to replace losses due to air drag, gravity, and acceleration.

15.34 OVEREXPANDED NOZZLE

A nozzle is said to be overexpanded when the pressure at the nozzle exit p_e is less than the backpressure p_b.

15.35 OXALIC ACID

Oxalic acid is an organic compound with the IUPAC name ethanedioic acid and the formula $HO_2C\text{-}CO_2 H$. It is the simplest dicarboxylic acid. It is a white crystalline solid that forms a colourless solution in water.

15.36 OXIDISER

An oxidiser is a type of chemical that fuel requires to burn. Most types of burning on Earth use oxygen, which is prevalent in the atmosphere. However, in space, there is no atmosphere to provide oxygen or other oxidisers so rockets need to carry up their oxidisers.

15.37 OXYGEN

Oxygen is a colourless, odourless, and tasteless gas essential to living organisms. Oxygen is an element with atomic symbol O, atomic number 8, and atomic weight 16.

15.38 OZONE

Ozone is a colourless or pale blue gas, slightly soluble in water and much more soluble in inert non-polar solvents such as carbon tetrachloride or fluorocarbons. Ozone is a gas made up of three oxygen atoms (O_3). It occurs naturally in small (trace) amounts in the upper atmosphere (the stratosphere).

15.39 OZONE HOLE

The term 'ozone hole' refers to the depletion of the protective ozone layer in the upper atmosphere (stratosphere) over Earth's polar regions.

Chapter 16

Pacific Ocean to Pyrometers

16.1 PACIFIC OCEAN

The Pacific Ocean is the largest and deepest of Earth's oceanic divisions. It extends from the Arctic Ocean in the north to the Southern Ocean in the south and is bounded by the continents of Asia and Australia in the west and the Americas in the east.

16.2 PANEL METHOD

The panel method is a numerical technique to solve flow past bodies by replacing the bodies with mathematical models consisting of source or vortex panels.

Essentially, the surface of the body to be studied will be represented by panels consisting of sources and free vortices. These are referred to as source panel and vortex panel methods, respectively. If the body is a lift-generating geometry, such as an aircraft wing, the vortex panel method will be appropriate for solving the flow past, since the lift generated is a function of the circulation or the vorticity around the wing. If the body is a non-lifting structure, such as a pillar of a river bridge, the source panel method might be employed for solving the flow past that.

16.3 PAPER

Paper is a material made of wood pulp, rags, straw, or other fibrous material in the form of thin sheets and used for writing, printing, wrapping things, etc.

16.4 PARACHUTE

A parachute is a device used to slow the motion of an object through an atmosphere by creating drag.

DOI: 10.1201/9781003348405-16

16.5 PARAFFIN

Paraffin is a waxy, white, or colourless solid mixture of hydrocarbons made from petroleum and used to make candles, wax paper, lubricants, and waterproof coatings. It also called paraffin wax.

16.6 PARAGLIDING

Paragliding is a recreational and competitive adventure sport of flying a paraglider which is lightweight, free-flying, foot-launched glider aircraft with no rigid primary structure. The pilot sits in a harness or lies supine in a cocoon-like 'speed bag' suspended below a fabric wing.

16.7 PARASITE DRAG

In aviation, parasite drag is defined as a drag that is not associated with the production of lift.

Parasitic drag acts on any object when the object is moving through a fluid. It is a combination of form drag and skin friction drag. It affects all objects regardless of whether they are capable of generating lift.

16.8 PARASOL-WING

In a parasol-wing, the wing is located above the fuselage and is not directly connected to it. The structural support is typically provided by a system of struts, especially in the case of older aircraft, wire bracing, as seen in Figure 16.1.

Figure 16.1 Parasol-wing.

16.9 PARTIALLY FILLED FLOW

Partially filled flow through closed conduits is encountered in civil engineering practice. Flow in sewerage and drainage passages are typical examples of partially filled flow.

16.10 PARTIAL PRESSURE

In a mixture of gases, each constituent gas has a partial pressure, which is the notional pressure of that constituent gas if it alone occupies the entire volume of the original mixture at the same temperature. The total pressure of an ideal gas mixture is the sum of the partial pressures of the gases in the mixture.

16.11 PARTICLE

Particle is the smallest entity of a fluid element.

16.12 PARTICLE IMAGE VELOCIMETRY (PIV)

Particle image velocimetry (PIV) is an optical method of flow visualisation. It can be used to obtain the instantaneous velocities and the related properties in fluid flows. The fluid is seeded with tracer particles, which are sufficiently small so that they can flow with the fluid at the same speed. The fluid with seeding particles is illuminated so that particles are visible. The motion of the seeding particles is used to calculate the speed and direction of the flow being studied.

The main difference between PIV and laser Doppler velocimetry is that PIV produces two-dimensional or even three-dimensional vector fields, while the laser Doppler velocimetry measures only the velocity at a point. In PIV, the particle concentration is such that it is possible to identify individual particles in an image. When the particle concentration is so low that it is possible to follow an individual particle, it is called particle-tracking velocimetry. Laser speckle velocimetry has to be used for cases where the particle concentration is so high that it is difficult to observe individual particles in an image.

Some of the vital details about vortex formation due to the differential shear at the interface of the jet edge and the surrounding environment, which is stagnant, captured with PIV, are shown in Figure 16.2.

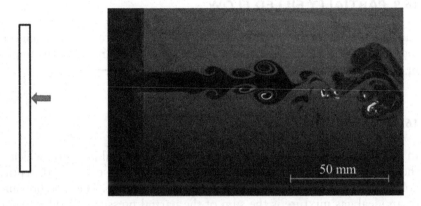

Figure 16.2 Vortices at the edge of rectangular jet at Re = 1500.

16.13 PARTITION FUNCTION

A partition function describes the statistical properties of a system in thermodynamic equilibrium. They are functions of the thermodynamic state variables, such as temperature and volume. The partition function is dimensionless. It is a pure number.

16.14 PASCAL

The SI unit of pressure is N/m^2 ($1\,N/m^2$ is ~0.000145 pounds per square inch or 9.9×10^{-6} atmospheres).

16.15 PASCAL'S LAW

The pressure acting at a point in a fluid at rest is the same in all directions. This is known as *Pascal's* law.

16.16 PASSIVE CONTROL

Passive control usually uses geometrical modifications of the element from which flow separation occurs to change the shear layer stability characteristics. Some examples of these modifications are trip wires in plane shear layers, convoluted splitter plates, and non-circular jets such as square jets and elliptic jets.

Passive control techniques range from alterations in the exit shape of the nozzle to the implementation of tooth-like tabs and vortex generators in

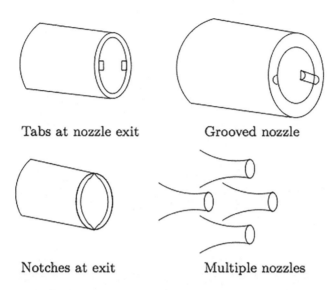

Tabs at nozzle exit   Grooved nozzle

Notches at exit   Multiple nozzles

Figure 16.3 Schematic of a few passive controls.

the jet. Some of the commonly used passive control methods are shown in Figure 16.3.

16.17 PATHLINE

Pathline may be defined as a line in the flow field describing the trajectory of a given fluid particle. From the Lagrangian viewpoint, namely, a closed system with a fixed identifiable quantity of mass, the independent variables are the initial position, with which each particle is identified, and the time. Hence, the locus of the same particle from time t_0 to t_n is called the pathline.

16.18 PATH FUNCTIONS

Path functions have inexact differentials designated by the symbol ∂. Therefore, a differential amount of heat or work is represented by ∂Q or ∂W, instead of dQ or dW.

16.19 PECLET NUMBER

Peclet number is a dimensionless group representing the ratio of heat transfer by the motion of fluid to heat transfer by thermal conduction. It is a

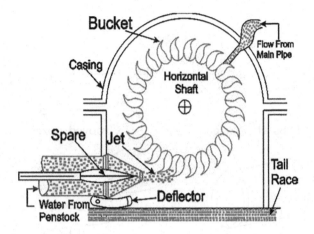

Figure 16.4 Pelton wheel turbine.

measure of the relative importance of advection versus diffusion, where a large number indicates an advectively dominated distribution, and a small number indicates a diffuse flow.

16.20 PELTON WHEEL TURBINE

The impulse turbine is also called the Pelton wheel in honour of Lester A. Pelton (1829–1908), who contributed to much of its development in the early gold-mining days in California.

   Pelton turbine (shown in Figure 16.4) is a tangential flow impulse turbine in which the pressure energy of water is converted into kinetic energy to form a high-speed water jet and this jet strikes the wheel tangentially to make it rotate. It is also called the Pelton wheel.

16.21 PENDULUM

A pendulum is a weight suspended from a pivot so that it can swing freely. When a pendulum is displaced sideways from its resting, equilibrium position, it is subjected to a restoring force due to gravity that will accelerate it back towards the equilibrium position.

16.22 PERCOLATION

Percolation is the process of a liquid slowly passing through a filter. It's how coffee is usually made. Percolation comes from the Latin word *percolare*, which means 'to strain through'. Percolation happens when liquid is strained through a filter like when someone makes coffee.

16.23 PERFECT GAS

A perfect gas is an imaginary substance that obeys the relation $p = \rho R T$. In perfect gas, the intermolecular forces are negligible. A perfect gas obeys the thermal and caloric equations of state. This is even a greater specialisation than thermally perfect gas. For a perfect gas, both c_p and c_v are constants and are independent of temperature. Such a gas with constant c_p and c_v is called a calorically perfect gas. Therefore, a perfect gas should be thermally as well as calorically perfect. For a thermally perfect gas, u, h, c_p and c_v are functions of temperature alone.

16.24 PERFECT GAS EQUATION OF STATE

For a perfect gas at equilibrium, this relation has the following form:

$$pv = RT$$

where p is the pressure, v is the specific volume, R is the gas constant, and T is the temperature. This is known as the perfect gas state equation. This equation is also called an ideal gas equation of state or simply the ideal gas relation, and a gas that obeys this relation is called an ideal gas. But it is essential to understand the difference between perfect and ideal gases. A perfect gas is calorically perfect. That is, its specific heats at constant pressure (c_p) and constant volume (c_v) are constants and independent of temperature.

A perfect gas can be viscous or inviscid. Also, a perfect gas flow may be incompressible or compressible. But an ideal gas is assumed to be inviscid and incompressible.

16.25 PERFORMANCE

Performance of an aircraft consists of take-off, climb, cruise or level flight, descend or gliding, and landing. Level flight is the standard condition of flight with which all other manoeuvres are compared. Gliding involves simple fundamental principles that are more elementary than those of level flight. Landing is something special involving principles of flight at low speeds.

16.26 PERFORMANCE OF ACTUAL NOZZLES

Because of frictional effects, the performance of real nozzles differs slightly from that determined from the isentropic flow relations. However, the departure of the actual performance from that computed using the isentropic assumption is usually small. Therefore, the usual design procedure is based on the use of isentropic relations modified by two types of empirically determined coefficients, namely the nozzle efficiency and the coefficient of discharge.

16.27 PERIODIC MOTION

Periodic motion, in physics, is a motion repeated in equal intervals of time.

16.28 PERIPHERAL-VELOCITY FACTOR

The peripheral-velocity factor for a pump impeller is the ratio of the peripheral velocity to $\sqrt{2gh}$.

16.29 PERPETUAL MOTION

Perpetual motion is the action of a device that once set in motion would continue in motion forever with no additional energy required to maintain it. Such devices are impossible on grounds stated by the first and second laws of thermodynamics.

A perpetual-motion machine is a hypothetical machine that can do work infinitely without an external energy source. This kind of machine is impossible, as it would violate either the first or second law of thermodynamics or both.

16.30 PERPETUAL-MOTION MACHINE

Any device that violates either the first or second law of thermodynamics is called a perpetual-motion machine. It is only a hypothetical one, and despite numerous attempts, no perpetual-motion machine is known to have worked.

16.31 PERPETUAL-MOTION MACHINE OF THE FIRST KIND (PMM1)

A device that violates the first law of thermodynamics (creates energy) is called a perpetual-motion machine of the first kind (PMM1).

16.32 PERPETUAL-MOTION MACHINE OF THE SECOND KIND (PMM2)

A device that violates the second law of thermodynamics is called a perpetual-motion machine of the second kind (PMM2).

16.33 PETROLEUM

Petroleum is a naturally occurring liquid found beneath the Earth's surface that can be refined into fuel. Petroleum is used as fuel to power vehicles, heating units, and machines and can be converted into plastic and other materials.

16.34 PETROLEUM JELLY

Petroleum jelly (also called petrolatum) is a mixture of mineral oils and waxes.

16.35 PHASE

Phase is structurally homogeneous and physically distinct, at least microscopically. A phase is a substance that has a particular physical state, either a solid, liquid, or gas, and a particular chemical composition.

16.36 PHASE EQUILIBRIUM

If a system involves more than one phase, it is said to be in phase equilibrium when the mass of each phase reaches an equilibrium level and stays there.

16.37 PHASES OF A PURE SUBSTANCE

The three principal phases of a substance are solid, liquid, and gas. However, a substance may have several phases within a principal phase, each with a different molecular structure.

16.38 PHENOL

Phenol is an aromatic organic compound with the molecular formula C_6H_5OH. It is a white crystalline solid that is volatile.

16.39 PHOTOCHEMICAL SMOG

Photochemical smog is a mixture of pollutants that are formed when nitrogen oxides and volatile organic compounds (VOCs) react to sunlight, creating a brown haze above cities. It tends to occur more often in summer because that is when we have the most sunlight.

16.40 PHOTOMULTIPLIER

The Doppler shift is detected by a device called a photomultiplier, which gives an electrical output whose frequency is proportional to the velocity of the scattering particle.

16.41 PHUGOID

A phugoid, shown in Figure 16.5, is the path of a particle that moves under gravity in a vertical plane and which is acted upon by a force L normal to the path and proportional to V^2.

16.42 PHYSICAL PROPERTIES OF FLUIDS

Density, viscosity, and surface tension are the three physical properties of fluids that are particularly important.

16.43 PHYSICS

Physics is the natural science that studies matter, its motion, and behaviour through space and time, and the related entities of energy and force. Physics

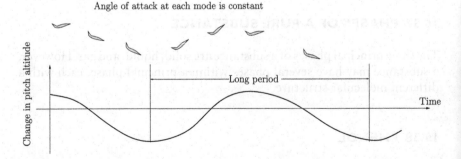

Figure 16.5 Phugoid.

is one of the most fundamental scientific disciplines, and its main goal is to understand how the universe behaves.

16.44 PIGMENT

A pigment is a coloured material that is completely or nearly insoluble in water. In contrast, a dye is typically soluble, at least at some stage in its use. Generally, dyes are often organic compounds whereas pigments are often inorganic compounds.

16.45 PIPE FLOW

Pipe flow, a branch of hydraulics and fluid mechanics, is a type of liquid flow within a closed conduit.

16.46 PIRANI GAUGE

Pirani gauge is a device that measures pressure through the change in thermal conductance of the gas. The fact that the effective thermal conductivity of a gas decreases at low pressures is made use of in the measurement of pressures by the Pirani gauge. It consists of an electrically heated filament placed inside a vacuum space.

16.47 PITCH OF A PROPELLER

The forward distance travelled by a propeller per revolution, x, is called the geometric pitch since it depends only on the geometric dimensions and not on the performance of the propeller. The value of the geometric pitch of a fixed-pitch propeller may vary from about 1 m for a slow type of aircraft to 5 or 6 m for a faster aircraft.

16.48 PITCHING

Rotary motion of the aircraft about the lateral axis, as seen in Figure 16.6, is called pitching. A pitch motion is an up or down movement of the nose of the aircraft. The pitching motion is caused by the deflection of the elevator of the aircraft.

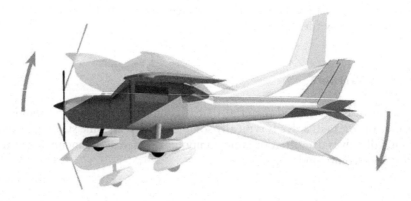

Figure 16.6 Pitching motion of aircraft.

16.49 PITCHING MOMENT

Pitching moment is the moment acting about the lateral axis (y-axis). It is the moment due to the lift and drag acting on the aircraft. Pitching moment causing nose-up is regarded as positive. In aerodynamics, the pitching moment on an aerofoil is the moment (or torque) produced by the aerodynamic force on the aerofoil if that aerodynamic force is considered to be applied, not at the centre of pressure, but at the aerodynamic centre of the aerofoil. The pitching moment on the wing of an airplane is part of the total moment that must be balanced using the lift on the horizontal stabiliser. More generally, a pitching moment is any moment acting on the pitch axis of a moving body.

16.50 PITOT PRESSURE

Pitot pressure is the sum of dynamic and static pressure. It is used to measure airspeed.

16.51 PITOT PROBE

A pitot probe is a tube with a blunt end facing the air stream. The tube will normally have an inside to outside diameter ratio of 0.5 to 0.75, and a length aligned with the air stream of 15–20 times the tube diameter. The inside diameter of the tube forms the pressure orifice. It is widely used to determine the airspeed of an aircraft, water speed of a boat, and to measure liquid, air, and gas flow velocities in certain industrial applications.

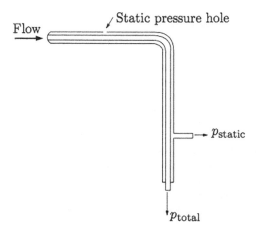

Figure 16.7 Pitot-static probe.

16.52 PITOT-STATIC PROBE

A pitot-static tube is used for measuring the total and static pressures simultaneously. It is usually a blunt-nosed tube with an opening at its nose to sense the total pressure and a set of holes on the surface of the tube to sense the static pressure.

The dynamic pressure of a flow can be measured directly by a special probe called a pitot-static probe, which is a combination of the pitot and static probes. A typical pitot-static probe is shown in Figure 16.7.

16.53 PITOT-STATIC TUBE CHARACTERISTICS

The pressure distribution along the surface of the horizontal stem of a pitot-static tube, in the absence of a vertical stem, is something similar to the pressure distribution on a streamlined body of revolution.

16.54 PITOT-STATIC TUBE LIMITATIONS

Like any instrument, the pitot-static tube also has its limitations, especially in the measurement of small differential pressures experienced at low speeds. For measuring such small pressure differences, the requirement of ultra-high sensitive manometers poses severe limitations on the use of pitot-static probes for such measurements. For instance, to measure the velocity head for an air speed of 0.6 m/s within ±1% accuracy, the manometer has to be sensitive to about 0.02 mm of the water column. An instrument of this

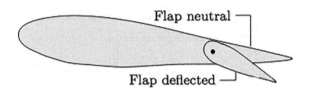

Figure 16.8 An aircraft wing with flap.

sensitivity has to be designed with special care. Such an instrument can only be used in laboratories. If we take 1 mm of water as a limiting sensitivity for a manometer, then the lowest airspeed that can be measured with an accuracy of ±1% with a pitot-static tube is about 4 m/s.

16.55 PIV SEEDING PARTICLES

Due to its almost perfectly spherical particle shape, the polyamide particles offered by LaVision are an ideal seeding material for PIV applications in liquids.

16.56 PLAIN FLAP

A plain flap is illustrated in Figure 16.8.

The main effect caused by flap deflection is an increase in the effective camber of the wing. This increase of camber reduces the zero-lift incidence without affecting the lift-curve slope. Thus, at incidences well below the stall, there is a constant increment in the lift coefficient. This effect may be slightly enhanced because flap deflection additionally produces a slight increase in the effective incidence. Another effect of the flap is that its deflection delays the onset of the stall.

16.57 PLAIN GLASS

Plain glass is colourless and transparent. Sheet glass is either drawn or float glass – the type usually used in household windows.

16.58 PLANE GLASS PLATE

Plane glass plate is a surface through which when a parallel beam of rays is passed, it will emerge in the form of a parallel beam. In other words, the rays do not converge at all after getting refracted from the plate itself, hence, it is said that they converge at infinity.

16.59 PLANFORM AREA

Planform area is the area of the body as seen from above. This is suitable for wide flat bodies such as aircraft wings and hydrofoils. For example the planform area is the area of the wing as viewed from above the wing.

16.60 PLASMA

Plasma is an ionised gas. A gas at normal and moderately high temperature is a nonconductor. But at very high temperatures of the order of 10,000 K and above, thermal excitation sets in. This leads to dissociation and ionisation. Ionised gas is called plasma, which is an electrically conducting medium. Electrically conducting fluids are encountered in engineering problems, such as re-entry of spacecraft and missiles, plasma jet, controlled fusion, and magnetohydrodynamic generators.

16.61 PLASMA ARC TUNNELS

Plasma arc tunnels are devices capable of generating high-speed flows at very high temperatures. The test gas can reach temperatures above 13,000 °C.

16.62 PLASMA ENGINE TYPES

The following are the types of plasma engines developed.

Helicon double-layer thruster: A helicon double-layer thruster uses radio waves to create a plasma and a magnetic nozzle to focus and accelerate the plasma away from the rocket engine. A mini-helicon plasma thruster is ideal for space manoeuvres, runs off of nitrogen, and the fuel has an exhaust velocity (specific impulse) ten times that of chemical rockets.

Magnetoplasmadynamic thruster: This thruster (MPD) uses the Lorentz force (a force resulting from the interaction between a magnetic field and an electric current) to generate thrust – the electric charge flowing through the plasma in the presence of a magnetic field causing the plasma to accelerate due to the generated magnetic force.

Hall effect thruster: This thruster combines a strong localised static magnetic field perpendicular to the electric field created between an upstream anode and a downstream cathode called a neutraliser to create a 'virtual cathode' (area of high electron density) at the exit of the device. This virtual cathode then attracts the ions formed inside the thruster closer to the anode. Finally, the accelerated ion beam is neutralised by some of the electrons emitted by the neutraliser.

Electrodeless plasma thruster: This thruster uses the ponderomotive force1 that acts on any plasma or charged particle when under the influence of a strong electromagnetic energy gradient to accelerate the plasma.

16.63 PLASMA PROPULSION ENGINE

A plasma propulsion engine is a type of electric propulsion that generates thrust from quasi-neutral plasma.

16.64 PLATINUM

Platinum is a chemical element with the symbol Pt and atomic number 78. It is a dense, malleable, ductile, highly unreactive, precious, and silverish-white transition metal. Its name is derived from the Spanish term *platino* meaning 'little silver'. Platinum is one of the least reactive metals.

16.65 PLATINUM RHODIUM THERMOCOUPLE

For measuring high temperatures of up to 1,700°C, thermocouple wires made of high-purity platinum or platinum/rhodium alloys are therefore the solution.

16.66 PLENUM CHAMBER

A plenum chamber is a pressurised housing containing a fluid (typically air) at positive pressure.

16.67 PLENUM SPACE

A plenum space is a part of a building that can facilitate air circulation for heating and air conditioning systems by providing pathways for either heated/conditioned or return airflows, usually at greater than atmospheric pressure.

16.68 PLUG NOZZLE

The plug nozzle, shown in Figure 16.9, is a type of nozzle that includes a centre body or plug around which the working fluid flows. Plug nozzles have applications in aircrafts, rockets, and numerous other fluid flow devices.

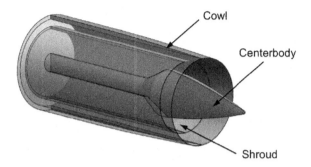

Cowl

Centerbody

Shroud

Figure 16.9 Plug nozzle.

Figure 16.10 Plumb bob.

16.69 PLUMB BOB

An indication of the nature of the bank would be a plumb bob (shown in Figure 16.10) hung in the cockpit out of contact with the wind. In normal flight, this would hang vertically. During a correct bank, it would not hang

Figure 16.11 Plume.

vertically but in exactly the same position relative to the aircraft as it would in a normal fight, that is, it would bank with the aircraft. If over-banked, the plumb line would be inclined inwards and, if underbanked, outwards from the above position. This plumb bob idea, in the form of a pendulum, forms the basis of the sideslip indicator that is provided by the top pointer of the so-called turn and bank indicator. The pointer is geared to move in such a way that the pilot must move the control column away from the direction of the pointer – this being the instinctive reaction.

16.70 PLUMES

Plumes are shear flows produced by buoyancy forces, as illustrated in Figure 16.11. This is caused by thermal potential, unlike jets, which are caused by momentum flux.

16.71 PLUNGE

To jump, drop, or fall suddenly and with force.

16.72 PNEUMATICS

Pneumatics (from the Greek word *pneuma* 'wind, breath') is a branch of engineering that makes use of gas or pressurised air. The pneumatic systems used in industries are commonly powered by compressed air or compressed inert gases.

16.73 PNEUMATIC ACTUATORS

Pneumatic actuators are mechanical devices that use compressed air acting on a piston inside a cylinder to move a load along a linear path.

16.74 PNEUMATIC MOTORS

Pneumatic motors generally convert the compressed air energy to mechanical work through either linear or rotary motion.

16.75 PNEUMATIC SYSTEM

A pneumatic system is a collection of interconnected components using compressed air to do work for automated equipment. The examples can be found in industrial manufacturing, a home garage, or a dentist's office. This work is produced in the form of linear or rotary motion. The examples of pneumatic systems and components are air brakes on buses and trucks, air brakes on trains, air compressors, air engines for pneumatically powered vehicles, barostat systems used in neurogastroenterology and for researching electricity, and cable jetting – a way to install cables in ducts and dental drill.

16.76 POHLHAUSEN'S METHOD

One of the earliest and, until recently, most widely used approximate methods for the solution of the boundary layer equation is that developed by Pohlhausen. This method is based on the momentum equation of Karman, which is obtained by integrating the boundary layer equation across the layer.

16.77 POINT FUNCTIONS

Properties are point functions. That is, they depend on the state only and not on how a system reaches that state. Property changes are designated by the symbol d. For example, a differential amount of pressure, p, or temperature, T, is represented by dp or dT.

16.78 POINT RECTILINEAR VORTEX

It is the limiting case of a circular vortex of constant strength γ with radius a tending towards zero.

16.79 POISSON'S EQUATION

Poisson's equation is the relation connecting the pressure and density of an isentropic process as

$$\frac{p_2}{p_1} = \left(\frac{\rho_2}{\rho_1}\right)^\gamma$$

where subscripts 1 and 2 refer to states before and after the process and γ is the ratio of specific heats.

16.80 POISEUILLE'S LAW

Poiseuille's law states that 'the velocity of a liquid flowing through a capillary is directly proportional to the pressure of the liquid and the fourth power of the radius of the capillary and is inversely proportional to the viscosity of the liquid and the length of the capillary'.

$$Q = \frac{\pi}{8} \frac{\Delta p}{L} \frac{R^4}{\mu}$$

Flow rate resulting from a pressure drop through a circular tube of radius R and length L. Valid for steady, well-developed laminar flow (Poiseuille flow).

16.81 POLAR CURVE

The variation of lift coefficient, C_L, with induced drag coefficient, C_{D_v}, is called the polar curve of the aerofoil.

16.82 POOL BOILING

Pool boiling is boiling in the absence of bulk fluid motion and forced convection.

16.83 POTENTIAL CORE REGION

This region consists of a core zone of constant axial velocity equal to the jet (nozzle) exit velocity surrounded by a rapidly growing and predominantly shear-dominated annulus of mixing layer or shear layer with intense turbulence. The potential core of a subsonic jet typically extends to about six times the nozzle exit diameter (D_e) downstream from the nozzle exit. This is because the mixing initiated at the jet boundary (periphery) has not yet permeated into the entire flow field, thus leaving a region that is characterised by a constant axial velocity.

16.84 POTENTIAL ENERGY

The potential energy is the energy that a system possesses as a result of its elevation in a gravitational field.

16.85 POTENTIAL FLOW

Potential flow is an imaginary flow of zero viscosity. It is based on the concept that the flow field can be represented by a potential function ϕ such that

$$\nabla^2 \phi = 0$$

This linear partial differential equation is popularly known as the Laplace *equation*. Inviscid flow is also called potential flow, and in this case, the Navier–Stokes equation can be simplified to become linear.

16.86 POTENTIAL FLOW PAST A TWO-DIMENSIONAL CIRCULAR CYLINDER

The streamlines around the circular disc, shown in Figure 16.12, exhibit a pattern identical to potential flow past a two-dimensional circular cylinder.

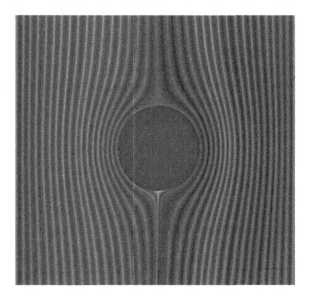

Figure 16.12 Flow past a circular disc in a Hele-Shaw apparatus.

It is seen that the flow is symmetrical about both the horizontal and vertical axes of the cylinder. The forward and rear stagnation points are clearly seen. This flow pattern clearly demonstrates that the flow in the Hele-Shaw device is analogous to potential flow.

16.87 POTENTIAL FUNCTION

Potential function is a mathematical function representing irrotational flows.

16.88 PRANDTL–GLAUERT TRANSFORMATION

The **Prandtl–Glauert transformation** is a mathematical technique that allows solving certain compressible flow problems by incompressible-flow calculation methods.

16.89 PRANDTL–MEYER EXPANSION

A supersonic expansion fan is technically known as the Prandtl–Meyer expansion fan. It is a two-dimensional simple wave and is a centred expansion process that occurs when a supersonic flow turns around a convex corner. The fan consists of an infinite number of Mach waves diverging from a sharp corner. When a flow turns around a smooth and circular corner, these waves can be extended backwards to meet at a point.

Supersonic flow passing through a Prandtl–Meyer expansion would experience a smooth and gradual change of flow properties. The Prandtl–Meyer fan consists of an infinite number of Mach (expansion) lines centred at the convex corner. The expansion fan has a wedge-like shape with the corner as the apex.

As the streamlines turn smoothly across the expansion fan, the flow velocity increases continuously and the pressure, density, and temperature decrease continuously. This type of flow was first studied by Meyer, a student of Prandtl in 1907. This is a turning problem in which the streamlines are continuous and the flow is isentropic everywhere except at the vertex of the expansion fan, where the flow experiences a sudden change of flow properties and the process is nonisentropic.

16.90 PRANDTL–MEYER FUNCTION

It is known from basic studies on fluid flows that a flow that preserves its geometry in space or time or both is called a self-similar flow. In the simplest cases of flows, such motions are described by a single independent variable, referred to as the similarity variable. The Prandtl–Meyer function is such a similarity variable.

The Prandtl–Meyer function in terms of the Mach number M_1 just upstream of the expansion fan can be written as

$$v = \frac{\gamma+1}{\gamma+1} \arctan\sqrt{\frac{\gamma-1}{\gamma+1}(M^2-1)} - \arctan\sqrt{(M^2-1)}$$

From this, it is seen that, for a given M, v is fixed.

16.91 PRANDTL'S MIXING LENGTH THEORY

Prandtl's mixing length theory is a two-dimensional model attempting to describe the momentum transfer within a turbulent fluid flow. It can also be defined as the average distance that a small mass of fluid will travel before it exchanges its momentum with another mass of fluid.

16.92 PRANDTL NUMBER

Prandtl number (Pr) is defined as the ratio of momentum diffusivity (kinematic viscosity) to thermal diffusivity. The Prandtl number, which is the ratio of kinematic viscosity and thermal diffusivity, is a measure of the relative importance of velocity and heat conduction.

For air at room temperature, Pr is 0.71 and most common gases have similar values. The Prandtl number of water at 17°C is 7.56.

Prandtl number, $Pr = v/\alpha$, may be interpreted as the ratio of momentum diffusivity v to thermal diffusivity α. The Prandtl number may be viewed as a measure of the relative effectiveness of momentum transport and energy transport by diffusion in the velocity (momentum) and thermal boundary layers, respectively.

16.93 PRANDTL RELATION

Prandtl relation is an expression relating the velocity, V_1, ahead of and velocity, V_2, behind a normal shock with the critical speed of sound, a^{*2},

$$V_1 V_2 = a^{*2}$$

This equation implies that the velocity change across a normal shock must be from supersonic to subsonic and vice versa. But it can be shown that only the former is possible. Hence, the Mach number behind a normal shock is always subsonic. This is a general result, not limited just to a calorically perfect gas.

16.94 PRECIPITATION

Precipitation is water released from clouds in the form of rain, freezing rain, sleet, snow, or hail.

It is the primary connection in a water cycle that provides for the delivery of atmospheric water to the Earth. Most precipitation falls as rain.

16.95 PRECISION

Precision is the quality or state of being precise – exactness.

16.96 PREMIXED FLAME

A premixed flame is formed under certain conditions during the combustion of a premixed charge (also called pre-mixture) of fuel and oxidiser. Since the fuel and oxidiser are the key chemical reactants of combustion that are available throughout a homogeneous stoichiometric premixed charge, the combustion process once initiated sustains itself by way of its own heat release. The majority of the chemical transformation in such a combustion process occurs primarily in a thin interfacial region that separates the unburned and the burned gases. The premixed flame interface propagates through the mixture until the entire charge is depleted. The propagation speed of a premixed flame is known as the flame speed (or burning velocity), which depends on the convection-diffusion-reaction balance within the flame, i.e., on its inner chemical structure. The premixed flame is characterised as laminar or turbulent depending on the velocity distribution in the unburned pre-mixture (which provides the medium of propagation for the flame).

16.97 PRESERVATIVE

A preservative is a substance or a chemical that is added to products such as food products, beverages, pharmaceutical drugs, paints, biological samples, cosmetics, wood, and many other products to prevent decomposition by microbial growth or by undesirable chemical changes.

Natural preservatives include rosemary and oregano extract, hops, salt, sugar, vinegar, alcohol, diatomaceous earth, and castor oil.

16.98 PRESSURE

Pressure is force per unit area that acts normal to the surface of any object, which is immersed in a fluid. Since pressure is the intensity of force, it has the dimensions

$$\frac{[\text{Force}]}{[\text{Area}]} = \frac{\left[MLT^{-2}\right]}{L^2} = \left[ML^{-1}T^{-2}\right]$$

and is expressed in the units of Newton per square metre (N/m²) or simply pascal (Pa). At standard sea level conditions, the atmospheric pressure is 101,325 Pa, which corresponds to 760 mm of mercury column height.

For a fluid at rest, at any point, the pressure is the same in all directions. The pressure in a stationary fluid varies only in the vertical direction and is constant in any horizontal plane. That is, in stationary fluids, the pressure increases linearly with depth.

16.99 PRESSURE ALTITUDE

To calculate the standard pressure p at a given altitude, the temperature is assumed standard, and the air is assumed as a perfect gas. The altitude obtained from the measurement of the pressure is called pressure altitude (PA).

16.100 PRESSURE COEFFICIENT

Pressure coefficient, C_p, is the non-dimensional difference between a local pressure and the freestream pressure. Thus,

$$C_p = \frac{p - p_\infty}{\frac{1}{2}\rho_\infty V_\infty^2}$$

16.101 PRESSURE DRAG

The drag caused by pressure loss is called pressure drag. This is also referred to as form drag because the form or shape of the moving object dictates the separation and the expanse of the separated zone.

The pressure drag arises due to the separation of the boundary layer caused by the adverse pressure gradient.

16.102 PRESSURE DROP

Pressure drop is the difference in the total pressure between two points in a fluid-carrying network.

16.103 PRESSURE GRADIENT

A pressure gradient exists when there is a difference in pressure between two points. The flow will occur from the area of high pressure to the area of low pressure.

In atmospheric science, the pressure gradient is a physical quantity that describes in which direction and at what rate the pressure increases most rapidly around a particular location. The pressure gradient is a dimensional quantity expressed in units of pascals per metre (Pa/m). Mathematically, it is obtained by applying the del operator to a pressure function of position. The negative gradient of pressure is known as force density.

16.104 PRESSURE-GRADIENT FORCE

The pressure-gradient force is the force that results when there is a difference in pressure across a surface. In general, pressure is force per unit area across a surface. The resulting force is always directed from the region of higher pressure to the region of lower pressure.

16.105 PRESSURE HEAD

In fluid mechanics, pressure head is the height of a liquid column that corresponds to a particular pressure exerted by the liquid column on the base of its container.

16.106 PRESSURE-HILL

It is the positive pressure or compression zone near the nose of the body (in a flow) and the pressure becomes negative or suction, downstream of this positive pressure zone.

16.107 PRESSURE IN LPG CYLINDER

The regular LPG (liquified petroleum gas) cylinders have pressure from 5.5 to 6.5 kg/cm^2 depending on the surrounding temperature. A new sealed LPG cylinder has ~85% liquid and the rest is vapour.

16.108 PRESSURES IN THE BODY

Body system pressure in mm Hg (gauge): Maximum (systolic) 100–140, minimum (diastolic) 60–90; Blood pressure in large veins 4–15, eyes 12–24, brain and spinal fluid (lying down) 5–12, bladder while filling 0–25, when full 100–150; chest cavity between lungs and ribs –8 to –4, inside lungs –2 to +3, digestive tract oesophagus –2, stomach 0–20, intestines 10–20, and middle ear <1.

16.109 PRESSURE MEASUREMENT

Pressure measurement is the analysis of an applied force by a fluid (liquid or gas) on a surface. Pressure is typically measured in units of force per unit of surface area. Many techniques have been developed for the measurement of pressure and vacuum. Instruments used to measure and display pressure in an integral unit are called pressure meters or pressure gauges or vacuum gauges. A manometer is a good example as it uses the surface area and weight of a column of liquid to both measure and indicate pressure. Likewise, the widely used Bourdon gauge is a mechanical device, that both measures and indicates and is probably the best-known type of gauge.

16.110 PRESSURE MEASURING DEVICES

The pressure measuring devices meant for measurements in fluid flow may broadly be grouped into manometers and pressure transducers. Various types of liquid manometers are employed depending upon the range of pressures to be measured and the degree of precision required. U-tube manometers, multi-tube manometers, and Betz-type manometers are some of the popular liquid manometers used for pressure measurements in fluid flow. The pressure transducers used for fluid flow experimentation may be classified as electrical-type transducers, mechanical-type transducers, and optical-type transducers based on the functioning principle of the sensor in the transducer.

16.111 PRESSURE RECOVERY FACTOR

It represents the actual stagnation pressure of a fluid stream at the diffuser exit relative to the maximum possible stagnation pressure.

16.112 PRESSURE-REGULATING VALVE

A pressure-regulating valve (PRV) controls the pressure of a fluid or gas to the desired value. Regulators are used for gases and liquids and can be an integral device with a pressure setting, a restrictor, and a sensor all in one body, or consist of a separate pressure sensor, controller, and flow valve.

16.113 PRESSURE RELIEF VALVE

A pressure relief valve is a safety device designed to protect a pressurised vessel or system during an overpressure event.

16.114 PRESSURE TRANSDUCERS

Pressure transducers are electromechanical devices that convert pressure to an electrical signal, which can be recorded with a data acquisition system such as that used for recording strain gauge signals. These transducers are generally classified as mechanical, electrical, and optical types. Commonly used transducers employ elastic diaphragms (of various shapes), which are subjected to a displacement whenever pressure is applied. This movement is generally small and kept within the linear range, and is amplified using mechanical, electrical, electronic, or optical systems.

16.115 PRESTON/STANTON TUBES

The Preston/Stanton tubes are essentially small-impact (pitot) tubes meant for the indirect measurement of skin friction, using some well-established mean velocity correlations to infer mean shear stress at the wall. Preston tube used with success is a hypodermic needle tube to determine the local wall shear stress from the dynamic pressure indicated by the tube. To this end, a circular tube with 1-mm outer diameter and inner to outer diameter ratio of 0.6 was placed parallel to the mainstream and against the wall, as shown in Figure 16.13.

The diameter of the tube is still large enough to ensure that the main part of the tube diameter is in the fully turbulent zone of the flow along the wall. The dynamic pressure indicated by the tube depends on the local velocity distribution close to the wall.

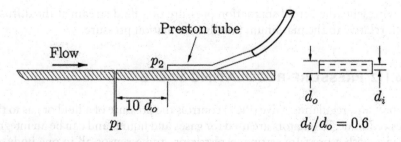

Figure 16.13 Preston tube.

16.116 PRISMATIC BODY

A floating object of a constant cross-section is called a prismatic body. A body with a variable cross-section is called a *non-prismatic body*.

16.117 PROBABILITY

The probability of an event may be defined as the ratio of the number of ways the event can occur to the total number of all possible events. A probability of unity means certainty of occurrence, while a probability of zero corresponds to non-occurrence.

16.118 PROCESS

Process may be defined as a change of a system from one equilibrium state to another. The states through which a system passes through during a process is called the path of the process.

16.119 PROFESSOR

A professor is a university teacher of the highest level.

16.120 PROFILE

The section of a wing by a plane parallel to the plane of symmetry is called a profile.

16.121 PROFILE DRAG

The drag of a two-dimensional aerofoil is called profile drag. It is the sum of pressure (or form) drag and skin friction drag caused by viscosity. For a well-designed aerofoil at a low angle of attack, α, the wake is thin, and the form drag is significantly smaller than the skin friction drag.

Profile drag or sometimes called form drag is the drag caused by the separation of the boundary layer from a surface and the wake created by that separation. It is primarily dependent upon the shape of the object. The air that is flowing over the aircraft or aerofoil causes form or pressure drag.

Figure 16.14 Propeller.

16.122 PROPELLENT

A propellent is a chemical substance used in the production of energy or pressurised gas that is subsequently used to create a movement of a fluid or to generate propulsion of a vehicle, projectile, or another object.

16.123 PROPELLER

The propeller or airscrew, shown in Figure 16.14, is the most used of the various systems of propulsion in the past. However, at present, mostly gas turbine, rather than reciprocating engine, is used for driving propellers. The objective of the propeller is to convert the torque, or turning effect, given by the power of the engine into a straightforward pull or push called thrust. If a propeller is in front of the engine, it will cause tension in the shaft and so will pull the aircraft – such an airscrew is called a tractor. If the propeller is behind the engine, it will push the aircraft forward, and this is called a pusher.

16.124 PROPELLER EFFICIENCY

The efficiency of a propeller is the ratio of the useful work delivered by the propeller to the work into it by the engine. We know that the force multiplied by the distance moved gives the mechanical work done; therefore, when either the force or the distance is zero, the useful work done becomes zero and the efficiency is zero. Thus, when the propeller moves forward in each revolution at a distance equal to the experimental pitch, the fact that there is no thrust means that the efficiency is nil. Also, when there is no forward speed, the distance moved is zero, and hence, there is no work done. Therefore, the efficiency is zero. Between these two extremes are the normal conditions of flight.

16.125 PROPELLER ENGINES

Propeller engines are essentially internal combustion engines. The propeller in the engine driven by the internal combustion engine takes in a large air mass and pushes it with a marginally increased speed. Because of the large amount of air mass, the thrust generated becomes significant, even though the velocity increase induced by the airflow is small.

In a propeller engine, the engine that drives the propeller may itself be a gas turbine, and in this case, the designer has the choice to allot the proportion of the power delivered by the propeller and jet as per the design considering the turboprop system.

16.126 PSI

The pound per square inch (symbol: lbf/in^2; abbreviation: psi) is a unit of pressure or stress based on avoirdupois units. It is the pressure resulting from a force of one pound force applied to an area of one square inch.

16.127 PUDDLE

A puddle is a small accumulation of liquid, usually water, on a surface. It can form either by pooling in a depression on the surface or by surface tension upon a flat surface. A puddle is generally shallow enough to walk through and too small to traverse with a boat or raft.

16.128 PULL-UP PHASE

This is the phase after landing. For a quick pull-up, drag is the essential component required. The drag provided by wheel brakes and air brakes is a better-suited drag, provided the aircraft can stand it and does not tip on its nose. In addition to air brakes, a type of flap, when fully lowered, gives a good braking effect, and so do the wings at an angle of 16° or so with a tail wheel-type of undercarriage.

16.129 PULSED LASERS

Pulsed lasers are lasers which emit light not in a continuous mode, but rather in the form of optical pulses (light flashes). Depending on the pulse duration, pulse energy, pulse repetition rate, and wavelength required, very different methods for pulse generation and very different types of pulsed lasers are used.

16.130 PULSED PLASMA THRUSTER

A pulsed plasma thruster (PPT), also known as a plasma jet engine, is a form of electric spacecraft propulsion.

16.131 PULSEJET

A pulsejet engine is a type of jet engine in which combustion occurs in pulses. A pulsejet engine can be made with a few or no moving parts and is capable of running statically. Pulsejet engines are a lightweight form of jet propulsion, but usually have a poor compression ratio, and hence, give a low specific impulse.

16.132 PUMPS

The device which converts mechanical energy to fluid energy is called a pump.
   or
   A pump is a device that moves fluids (liquids or gases), or sometimes slurries, by mechanical action, typically converted from electrical energy into hydraulic energy.

16.133 PUMP-TURBINE HYDRAULIC MACHINES

Pump-turbine hydraulic machines are similar in design and construction to the Francis turbine.
   The pump turbine is used at pumped-storage hydroelectric plants, which pump water from a lower reservoir to an upper reservoir during off-peak load periods so that water is available to drive the machine as a turbine when peak power generation is required.

16.134 PURE ROTATION

In pure rotation, the fluid elements rotate about their axes, which remain fixed in space.

16.135 PURE SUBSTANCE

A pure substance is a substance with fixed chemical composition throughout its mass. For example, water, helium, nitrogen, and carbon dioxide are all pure substances. A pure substance need not be a single chemical element or compound. For example, air, which is a mixture of several gases, is a pure substance since it has a uniform chemical composition. A mixture of two or more phases of a pure substance is also a pure substance as long as the chemical composition of all phases is the same. A mixture of liquid water and ice is a pure substance.

16.136 PURE TONE

When a sound has only one frequency, it is referred to as a pure tone. However, in practice, pure tone never exists. Most sounds are made up of different frequencies.

16.137 PURE TRANSLATION

The fluid elements are free to move anywhere in space but continue to keep their axes parallel to the reference axes fixed in space. The flow in the potential flow zone, outside the boundary layer over an aerofoil, is substantially this type of flow.

16.138 PURE WATER

Due to its capacity to dissolve numerous substances in great amounts, pure water almost does not exist in nature. Water, the only substance on the Earth that can exist as vapour, liquid, and solid, is the most abundant substance in the human body.

16.139 PUSHER

If the propeller is behind the engine, it will push the aircraft forward, and this is called a pusher.

16.140 PYROCLASTIC FLOW

A pyroclastic flow is a dense, fast-moving flow of solidified lava pieces, volcanic ash, and hot gases. It occurs as part of certain volcanic eruptions.

A pyroclastic flow is extremely hot burning anything in its path. It may move at speeds as high as 200 m/s. Pyroclastic flows form in various ways.

16.141 PYROMETERS

Some solids, such as metals, when they become hot, begin to emit light with a very dull red colour, and as the temperature increases, the colour becomes a brighter red, orange, yellow, and so on. The particular colour emission occurs at a particular temperature, and hence, can be used as a means of measuring the temperature if suitable calibration is adopted. It is this principle that is used in the optical pyrometer.

Chapter 17

Quality of Energy to Quasi-One-Dimensional Flow

17.1 QUALITY OF ENERGY

The quality of energy may be defined as its potential ability to do work, in a broad sense.

17.2 QUANTISED LEVELS OF DIFFERENT MODES OF ENERGY

Translational, rotational, vibrational, and electronic energies are the quantised levels of different modes of energy.

17.3 QUANTUM MECHANICS

The branch of mechanics that deals with the mathematical description of the motion and interaction of subatomic particles, incorporating the concepts of quantisation of energy, wave-particle duality, the uncertainty principle, and the correspondence principle.

Quantum mechanics is a fundamental theory in physics that describes the physical properties of nature at the scale of atoms and subatomic particles. It is the foundation of all quantum physics including quantum chemistry, quantum field theory, quantum technology, and quantum information science.

17.4 QUANTUM STATE

In quantum physics, a quantum state is a mathematical entity that provides a probability distribution for the outcomes of each possible measurement on a system. Knowledge of the quantum state together with the rules for the system's evolution in time exhausts all that can be predicted about the system's behaviour. A mixture of quantum states is again a quantum state.

DOI: 10.1201/9781003348405-17

Quantum states that cannot be written as a mixture of other states are called pure quantum states, while all other states are called mixed quantum states.

17.5 QUASI-EQUILIBRIUM PROCESS

A quasi-equilibrium process is a slow process, which allows the system to adjust itself internally so that all properties in one part of the system do not change any faster than those in other parts.

17.6 QUASI-ONE-DIMENSIONAL FLOW

One-dimensional flow is that in which the radius of curvature of the streamlines is very large and the cross-sectional area of the passage (streamtube) does not change abruptly.

A quasi-one-dimensional flow is a steady flow and all flow properties are uniform across any given cross-section of the streamtube, and hence, are functions of x only. That is, steady flow with $A = A(x), p = p(x), r = r(x)$ and $V = V(x)$ is defined as quasi-one-dimensional flow.

Chapter 18

Radar to Runway

18.1 RADAR

Radar is a detection system that uses radio waves to determine the distance (range), angle, or velocity of objects.

18.2 RADIAL FLOW

Flow in which the fluid flows radially inwards, or outwards, from a centre is called a radial flow.

18.3 RADIAN

A unit of measure for angles. One radian is the angle made at the centre of a circle by an arc whose length is equal to the radius of the circle.

18.4 RADIATION

In physics, radiation is the emission or transmission of energy in the form of waves or particles through space or through a material medium.

18.5 RADIATION PYROMETER

In this type of pyrometer, an arrangement is made to focus the radiant energy on the hot junction of a thermocouple. The cold junctions of the thermocouple are shielded from the radiation and are coupled externally to a galvanometer.

DOI: 10.1201/9781003348405-18

18.6 RAIN

Rain is liquid water in the form of droplets that have condensed from atmospheric water vapour and have become heavy enough to fall under gravity.

18.7 RAIN GAUGE

A rain gauge is an instrument to measure rainfall and is measured in mm.

18.8 RAINBOW

A rainbow, shown in Figure 18.1, is a meteorological phenomenon that is caused by reflection, refraction, and dispersion of light in water droplets resulting in a spectrum of light appearing in the sky.

18.9 RAKES

For the simultaneous measurement of many total-head readings, a bank of total-head tubes (shown in Figure 18.2), generally termed rake, is commonly employed.

18.10 RAMJET

Ramjet is a simple air-breathing engine. It consists of a diffuser, a combustion chamber, and an exhaust nozzle, as shown in Figure 18.3.

Figure 18.1 Rainbow.

Figure 18.2 Total pressure rake.

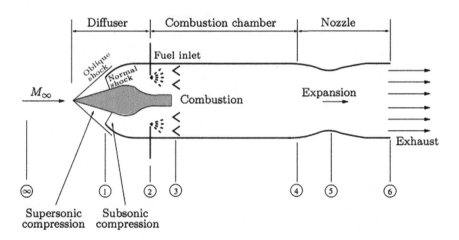

Figure 18.3 Schematic of a ramjet engine. (∞) Freestream; (1) oblique shock; (2) fuel spray; (3) flame holder, (4) nozzle entry; (5) nozzle throat; (6) nozzle exit.

Air entering the diffuser is compressed to attain the low-speed level required for combustion. Fuel is mixed with air in the combustion chamber and burned. The hot gas mixture after the combustion is expanded through the nozzle, which converts the high-pressure energy of the combustion gases to kinetic energy, resulting in the generation of thrust. It is important to note that in the diffuser, the incoming air is decelerated from the high flight speed, M_∞, which is usually supersonic, to a relatively low velocity through the compression caused by the shock system at the intake and the ramming action inside the duct from just downstream of the inlet to the combustion chamber entrance. Therefore, although ramjets can operate at subsonic flight speeds, the increasing pressure rise accompanying higher flight speeds renders ramjets more suitable for supersonic flights.

The ramjet engine shown in Figure 18.3 is typical of supersonic ramjets that employ supersonic diffusion through a system of shocks. Since the supersonic flow stream entering the intake must be decelerated to about Mach 0.2–0.3 before entry to the combustion chamber, the pressure rise can be substantial. For example, for isentropic deceleration from Mach 4 to Mach 0.3, the static pressure ratio between the ambient and the combustion chamber should be around 145. But only a fraction of the isentropic pressure ratio is achieved. This is because, at high Mach numbers, the total pressure losses associated with shocks are substantial.

18.11 RANDOM ERROR

Random error (non-repeatability) is different for every reading, and hence, cannot be removed. The factors that introduce random error are uncertain by their nature.

18.12 RANDOM MOTION

Random motion is a motion in which the particle moves in a zigzag manner and not in a straight line. It is a type of motion that is unpredictable. In this type of motion, the object moves in any direction, and the direction continually changes.

18.13 RANGE OF AN AIRCRAFT

Range of an aircraft is the distance it can cover in flight for a given quantity of fuel. Therefore, for an aircraft flying for maximum range means flying with minimum drag. In other words, an aircraft flies with maximum efficiency no matter if it uses a propeller engine or a jet engine.

The maximal total range is the maximum distance an aircraft can fly between take-off and landing, as limited by fuel capacity in powered aircraft, or cross-country speed and environmental conditions in unpowered aircraft. The range can be seen as the cross-country ground speed multiplied by the maximum time in the air. The fuel time limit for a powered aircraft is fixed by the fuel load and rate of consumption. When all fuel is consumed, the engines stop and the aircraft will lose its propulsion.

18.14 RANKINE CYCLE

Rankine cycle is the ideal case for vapour power cycles. Many of the impracticalities associated with the Carnot cycle can be eliminated by superheating the steam in the boiler and condensing it completely in the condenser. The cycle that results from these processes is the Rankine cycle. The processes involved are isentropic compression in a pump, constant pressure heat addition in a boiler, isentropic expansion in a turbine, and constant pressure heat rejection in a condenser.

18.15 RANKINE'S HALF-BODY

An interesting pattern of flow past a half-body, known as Rankine's half-body, shown in Figure 18.4, can be obtained by combining a source and a uniform flow parallel to the x-axis.

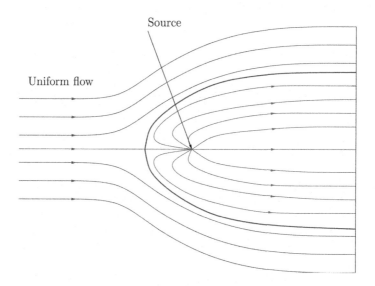

Figure 18.4 Uniform flow and a source.

The half-body is obtained by the linear combination of the individual stream functions of a source and a uniform flow, as per Rankine's theorem, which states that 'the resulting stream function of n potential flows can be obtained by combining the stream functions of the individual flows'.

18.16 RANKINE SCALE

In the English system, the absolute temperature scale is the Rankine scale. It is related to the Fahrenheit scale by $T(R) = T(°F) + 459.67$.

18.17 RANKINE'S THEOREM

Rankine's theorem states that 'the resulting stream function of n potential flows can be obtained by combining the stream functions of the individual flows'.

18.18 RANQUE–HILSCH VORTEX TUBE

It is a mechanical device that separates a compressed gas into hot and cold streams.

18.19 RAREFIED FLOW

The flow in which the mean free path of the molecules is of the same order or more than the characteristic dimension of the problem is termed a rarefied flow.

To deal with highly rarefied gases, we should resort to the microscopic approach of kinetic theory, since the continuum approach of classical fluid mechanics and thermodynamics is not valid here.

When the Reynolds number is low because of low density, the flow is termed rarefied flow.

Flow in space and very high altitudes, in the Earth's atmosphere, are rarefied flows.

The concept of continuum fails when the mean free path of fluid molecules is comparable to some characteristic geometrical parameter in the flow field. A dimensionless parameter, Knudsen number, Kn, defined as the ratio of the mean free path to a characteristic length, aptly describes the degree of departure from continuum flow. Based on the Knudsen number, the flow regimes are grouped as continuum, slip flow, transition flow, free molecular flow, and magnetofluid mechanics.

18.20 RAREFIED GAS DYNAMICS

Rarefied gas dynamics is concerned with flows at such low density that the molecular mean free path is not negligible.

18.21 RATHAKRISHNAN LIMIT

The jet control in the form of a cross-wire is known as the *Rathakrishnan limit*.

18.22 RATIO OF SPECIFIC HEATS

The ratio of specific heats $\gamma = c_p / c_v$ is a measure of the relative internal complexity of the molecules of the gas. It has been determined from the kinetic theory of gases that the ratio of specific heats can be related to the number of degrees of freedom n of the gas molecules by the relation $\gamma = (n+2)/n$.

The ratio of specific heats varies from 1 to 1.67. For monatomic gases like argon, $\gamma = 1.67$. Diatomic gases such as oxygen and nitrogen have $\gamma = 1.4$; for gases with extremely complex molecular structures, γ is slightly more than unity.

18.23 RAYLEIGH FLOW

Rayleigh flow is a flow with simple T_0 change. In other words, Rayleigh flow is that in which a change of state from one to another is caused solely by a change of stagnation temperature.

In a flow with area change, the process is considered to be isentropic with the assumption that the frictional and energy effects are absent. In a Fanno line flow, only the effect of wall friction is taken into account in the absence of area change and energy effects. In the Rayleigh flow, the processes involving a change in the stagnation temperature or the stagnation enthalpy of a gas stream are the main cause for a state change. From one-dimensional point of view, this is yet another effect producing continuous changes in the state of a flowing stream and this factor is called the energy effect, such as external heat exchange, combustion, or moisture condensation. Though a process involving simple stagnation temperature (T_0) change is difficult to achieve in practice, many useful conclusions of practical significance may be drawn by analysing the process of simple T_0 change. That is, the Rayleigh flow refers to frictionless, non-adiabatic flow through a constant area duct where the effect of heat addition or rejection is considered. Compressibility effects often come into consideration, although the Rayleigh flow model certainly also applies to incompressible flow. For this model, the duct area remains constant and no mass is added within the duct. Therefore, unlike the Fanno flow, the stagnation temperature is a variable. The heat addition causes a decrease in stagnation pressure, which is known as the Rayleigh effect and is critical in the design of combustion systems. Heat addition will cause both supersonic and subsonic Mach numbers to approach Mach 1, resulting in a choked flow. Conversely, heat rejection decreases a subsonic Mach number and increases a supersonic Mach number along the duct. It can be shown that for calorically perfect flows, the maximum entropy occurs at $M = 1$. The Rayleigh flow is named after John Strutt, 3rd Baron Rayleigh.

18.24 RAYLEIGH NUMBER

In fluid mechanics, the Rayleigh number (Ra) for a fluid is a dimensionless number associated with buoyancy-driven flow also known as free or natural convection. It characterises the fluid's flow regime: a value in a certain lower range denotes laminar flow; a value in a higher range, turbulent flow. Below a certain critical value, there is no fluid motion and heat transfer is by conduction rather than convection.

18.25 RAYLEIGH SUPERSONIC PITOT FORMULA

Rayleigh supersonic pitot formula is the ratio of static pressure to the total pressure behind the normal shock in terms of the Mach number,

$$\frac{p_1}{p_{02}} = \frac{\left(\dfrac{2\gamma}{\gamma+1} M_1^2 - \dfrac{\gamma-1}{\gamma+1} \right)^{1/(\gamma-1)}}{\left(\dfrac{\gamma+1}{2} M_1^2 \right)^{\gamma/(\gamma-1)}}$$

18.26 REACTANTS

Reactants are the components that exist before a reaction during a combustion process.

18.27 REACTION TURBINE

A reaction turbine is that in which flow takes place in a closed chamber under pressure. The flow through a reaction turbine may be radially inward, axial, or mixed (that is partially radial and partially axial).

18.28 REAGENT

A substance used to cause a chemical reaction, especially to find out if another substance is present.

18.29 REAL GAS

Real gas is that in which intermolecular forces are important and must be accounted for. For all real gases c_p, c_v, and γ vary with temperature, but only moderately. For example, c_p of air increases about 30% as temperature increases from 0 to 3,000°C.

18.30 REAL-GAS EQUATION OF STATE

The most well known of these equations is the Van der Waal's equation,

$$p = \frac{RT}{V-b} - \frac{a}{V^2}$$

where p is pressure, R is universal gas constant, T is absolute temperature, V is molar volume, a and b are constants. This equation expresses the relationship between the thermodynamic variables in the same way as the perfect gas equation, but with an important modification to account for the attractive forces of real-gas behaviour at high density in the second term.

18.31 RECIPROCATING ENGINE

A reciprocating engine is basically a piston-cylinder device.

18.32 RECIPROCATING PUMP

A reciprocating pump is a class of positive-displacement pumps that includes the piston pump, plunger pump, and diaphragm pump.

18.33 RECTANGULAR AEROFOIL

This is an aerofoil whose plan form is a rectangle. An aerofoil whose shape is that of a cylinder erected on an aerofoil profile satisfies this requirement.

18.34 RECTILINEAR VORTEX

A rectilinear vortex is a vortex tube whose generators are perpendicular to the plane of motion.

18.35 REDUCED-GRAVITY AIRCRAFT

A reduced-gravity aircraft is a type of fixed-wing aircraft that provides brief near-weightless environments for training astronauts, conducting research and making gravity-free movie shots.

18.36 REDUCED PRESSURE

The normalised pressure $p_R (= p/p_{cri})$ is called reduced pressure.

18.37 REDUCED TEMPERATURE

The normalised temperature $T_R (= T/T_{cri})$ is called reduced temperature.

18.38 REFLEXED CAMBER AEROFOIL

An aerofoil where the camber line curves back up near the trailing edge is called a reflexed camber aerofoil.

18.39 REFRACTION

Refraction is the bending of light (it also happens with sound, water, and other waves) as it passes from one transparent substance into another. This bending by refraction makes it possible for us to have lenses, magnifying glasses, prisms, and rainbows. Even our eyes depend upon this bending of light.

18.40 REFRACTIVE INDEX

Refractive index (index of refraction) is a value calculated from the ratio of the speed of light in a vacuum to that in a second medium of greater density.

In optics, the refractive index (also known as refraction index or index of refraction) of a material is a dimensionless number that describes how fast light travels through the material. It is defined as $n = c/v$, where c is the speed of light in a vacuum and v is the phase velocity of light in the medium.

For example, the refractive index of water is 1.333, meaning that light travels 1.333 times slower in water than in a vacuum.

18.41 REFRIGERANT

A refrigerant is a substance used in a heat cycle to transfer heat from one area and remove it to another.

18.42 REFRIGERATION

Refrigeration is the transfer of heat from a lower temperature region to a higher temperature one.

18.43 REFRIGERATOR

Refrigerator is a special device meant for the transfer of heat from a low-temperature medium to a high-temperature medium. Like heat engines, refrigerators are also cyclic devices. The working fluid used in the refrigeration cycle is called a refrigerant. The most popular refrigeration cycle is the vapour-compression cycle. The devices that produce refrigeration are called refrigerators.

18.44 REGENERATOR

Regenerator is the device where feed-water is heated. A regenerator is also called a feed-water heater. It is essentially a heat exchanger where heat is transferred from the steam to the feed-water either by mixing or without mixing the two fluid streams. The mixing one is called an open or direct-contact feed-water heater and the other is called a closed feed-water heater.

18.45 REGIMES OF FLUID MECHANICS

Based on the flow properties that characterise the physical situation, the flows are classified into ideal fluid flow, viscous incompressible flow, gas dynamics, rarefied gas dynamics, flow of multicomponent mixtures, and non-Newtonian fluid flow.

18.46 RELATIVE DENSITY

The relative density σ, defined as the ratio of density at the relevant altitude to the density at the sea level in the ISA, is an important quantity in aerodynamics. Relative density is the ratio of the density of a substance to the density of some standard substance at a specified temperature.

18.47 RELATIVE HUMIDITY

It is the ratio of the amount of moisture the air holds to the maximum amount of moisture the air can hold at the same temperature.

18.48 RELATIVE VELOCITY

The concept of relative motion or relative velocity is all about understanding the frame of reference. It can be thought of as the state of motion of the observer of some event. For example, if you were sitting on a lawn chair watching a train pass by from left to right at 50 m/s, you would consider yourself in a stationary frame of reference. From your perspective, you are at rest and the train is moving.

18.49 RELAXATION TIME

The time a fluid needs to regain its equilibrium structure after being stressed. Relaxation is usually the result of thermal (Brownian) processes.

18.50 RESEARCH TUNNELS

These tunnels are meant for studying and exploring basic phenomena and validating theoretical methods and codes. Therefore, a full Reynolds number simulation on large models of flight vehicles will not be necessary. The tunnels shall, however, have the capability to reproduce all-important local phenomena on a smaller scale.

18.51 RESISTANCE TEMPERATURE DETECTOR

Temperature can also be measured using resistance temperature detector (RTD) devices. They work on the principle that metals have marked temperature dependence. The RTD devices are more suitable for measurements of temperature of a block of metal than for measurements in gases like air. This is due to the fact that in a block of metal, the temperature communication between the metal and the RTD wire is good but the communication between the wire and air is poor.

18.52 RESONANCE

Resonance describes the phenomenon of increased amplitude that occurs when the frequency of a periodically applied force (or a Fourier component of it) is equal or close to a natural frequency of the system on which it acts. When an oscillating force is applied at a resonant frequency of a dynamic system, the system will oscillate at a higher amplitude than when the same force is applied at other, non-resonant frequencies.

18.53 RESPIRATION

Respiration is the movement of air or dissolved gases into and out of the lungs.

18.54 REVERSE PROCESS

A reverse process can be reversed without leaving any trace on the surroundings, that is, both the system and the surroundings are returned to their initial states at the end of the reverse process.

18.55 REVERSE TRANSITION

Reverse transition is a process in which a turbulent flow changes over to a laminar nature. This phenomenon is also known as relaminarisation.

18.56 REVERSIBLE ADIABATIC PROCESS

Reversible adiabatic process is also called an isentropic process. It is an idealised thermodynamic process that is adiabatic and in which the work transfers of the system are frictionless; there is no transfer of heat or of matter and the process is reversible.

18.57 REVERSIBLE PROCESS

A reversible process is a process that can be reversed to its initial state without leaving any trace on the surroundings. That is, both the system and the surroundings are returned to their initial states at the end of the reverse process. Processes that are not reversible are called irreversible processes.

18.58 REVERSIBLE WORK

Reversible work is defined as the maximum amount of useful work that can be obtained as a system undergoes a process between the specified initial and final states.

18.59 REYNOLDS ANALOGY

Reynolds analogy is the relation between the Stanton number, St, and the skin friction coefficient, c_f; $St = \dfrac{1}{2} c_f$.

18.60 REYNOLDS EXPERIMENT

Flow transition from laminar to turbulent was demonstrated by Osborne Reynolds in the early 1880s with a simple visualisation popularly known as the Reynolds experiment. The apparatus used by Reynolds is shown in Figure 18.5.

18.61 REYNOLDS NUMBER

The dimensionless group $\dfrac{\rho VL}{\mu}$ is called the Reynolds number Re. The Reynolds number is the ratio of inertial force to viscous force. The *Reynolds number* is a similarity parameter, expressed as the ratio of inertia force to viscous force.

$$Re_L = \frac{\rho VL}{\mu}$$

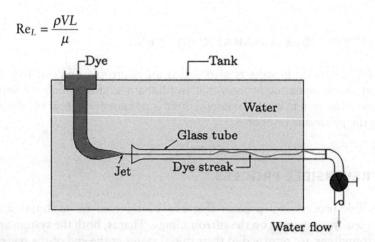

Figure 18.5 Reynolds apparatus.

where V and ρ are the velocity and density of the flow, respectively, μ is the dynamic viscosity coefficient of the fluid, and L is a characteristic dimension. For an inviscid fluid, $Re = \infty$.

The Reynolds number (Re) helps predict flow patterns in different fluid flow situations. At low Reynolds numbers, flows tend to be dominated by laminar (sheet-like) flow, while at high Reynolds numbers, flows tend to be turbulent. The turbulence results from the differences in the fluid's speed and direction, which may sometimes intersect or even move counter to the overall direction of the flow (eddy currents). These eddy currents begin to churn the flow, using up energy in the process, which for the liquids increases the chances of cavitation. The Reynolds number is an important dimensionless quantity in fluid mechanics.

18.62 REYNOLDS STRESSES

Reynolds stresses are additional stresses acting in a mean turbulent flow. The terms $-\rho u'^2$ and $-\rho u'v'$ are due to turbulence. They are popularly known as *Reynolds* or *turbulent stresses.* For a three-dimensional flow, the turbulent stress terms are $\rho u'^2, \rho v'^2, \rho w'^2, \overline{\rho u'v'}, \overline{\rho v'w'},$ and $\overline{\rho w'u'}$.

18.63 RHEOLOGY

Rheology is a branch of physics. It is the study of the relationship between stress and strain in fluids and the plastic flow of solids. Rheology studies the flow of matter primarily in a liquid or gaseous state and as 'soft solids' or solids under conditions in which they respond with a plastic flow rather than deforming elastically in response to an applied force. We can say that it is the science that deals with the deformation and flow of materials, both solids and liquids.

18.64 RHODIUM

Rhodium is a chemical element with the symbol Rh and atomic number 45. It is an extraordinarily rare, silvery-white, hard, corrosion-resistant, and chemically inert transition metal. It is a noble metal and a member of the platinum group.

18.65 RICH MIXTURE

In a rich mixture, the fuel will not burn properly and release lesser energy than can be obtained with an efficient burning when the mixture is weak.

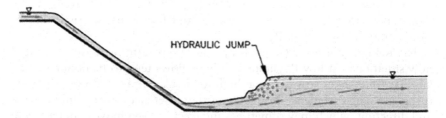

Figure 18.6 Right hydraulic jump.

18.66 RIDGE LIFT

Ridge lift is caused by rising air on the windward side of a slope. Ridge lift is used extensively by sea birds and by aircraft. In places where a steady wind blows, a ridge may allow virtually unlimited time aloft.

18.67 RIGHT HYDRAULIC JUMP

A shock (or hydraulic jump) in which the wave front is normal to the flow direction, as illustrated in Figure 18.6, is called a right hydraulic jump.

18.68 RIGHT-RUNNING WAVE

For an observer looking in the direction of flow towards the disturbance, the wave to the right is called the right-running wave.

18.69 RIVER

A large, natural flow of water that goes across land and into the sea is termed a river.

18.70 ROCKET ENGINE

A rocket engine is an engine that produces a force (a thrust) by creating a high-velocity output without using any of the constituents of the 'atmosphere' in which the rocket is operating. The thrust is produced because the exhaust from the rocket has a high velocity, and therefore, a high momentum.

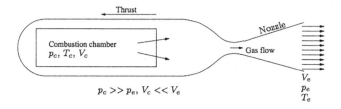

Figure 18.7 A simple rocket motor.

18.71 ROCKET ENGINE NOZZLE

A rocket engine nozzle is a propelling nozzle (usually of the de Laval type) used in a rocket engine to expand and accelerate the combustion gases produced by burning propellants so that the exhaust gases exit the nozzle at hypersonic velocities.

18.72 ROCKET PROPELLANT

Rocket propellant is the reaction mass of a rocket. This reaction mass is ejected at the highest achievable velocity from a rocket engine to produce thrust.

18.73 ROCKET PROPULSION

Rocket propulsion is straightforward in principle. As illustrated in Figure 18.7, in the simplest rocket motors, fuel is burned in a combustion chamber to create heat and a high-pressure gas. The hot gas then flows out through the specially shaped nozzle at high speed. The main difference between the rocket and other forms of propulsion is that air is not used as the oxidant in the burning process and the gases that are emitted from the outlet are all derived from the fuel. In other words, in a rocket motor, the fuel and oxidiser are stored in the motor itself. Thus, a rocket is essentially a non-air-breathing engine, unlike a turbine engine that takes in air from the atmosphere as the oxidiser. The rocket is mainly used for propelling missiles and spacecraft.

18.74 ROHSENOW CORRELATION

Rohsenow correlation is the most widely used correlation for the rate of heat transfer in the nucleate boiling regime.

18.75 ROLE OF LARGE-SCALE STRUCTURES IN SUBSONIC MIXING ENHANCEMENT

Existence of large-scale structures in the shear layer and their relation to the flow stability make it possible to control the development of the shear layer, and thus, affect its mixing characteristics. In general, large-scale structures are beneficial for the enhancement of bulk mixing, but they hinder fine-scale or molecular mixing, which is necessary, for example, in reacting flow applications. Enhancement of both large- and small-scale mixing can be achieved by exciting a combination of unstable modes.

18.76 ROLE OF SHEAR LAYER IN FLOW CONTROL

When a jet issuing from a nozzle propagates into the stagnant air, at the jet boundary, vortices are generated because of the shear between the fluid elements of the jet, which are in motion, and the stationary fluid elements of the atmosphere. These vortices entrain the stagnant air mass into the jet. Thus, an active shear zone is established at the jet boundary in the proximity of the nozzle exit. The shear action propagates towards the jet axis and reaches the axis at some downstream distance. From this location onwards, the viscous action dominates the entire jet field. A thorough understanding of the shear action is essential to control the mixing and acoustic characteristics of the jets.

18.77 ROLLING

Any rotary motion about the longitudinal axis is called rolling.

18.78 ROLLING MOMENT

Rolling moment is the moment acting about the longitudinal axis (x-axis) of the aircraft. It is generated by a differential lift generated by the ailerons, located closer to the wingtips. Rolling moment causing the right (starboard) wingtip to move downwards is regarded as positive.

18.79 ROOT CHORD

Root chord, c_r, is the chord at the wing centreline, that is, at the middle of the span, as shown in Figure 18.8.

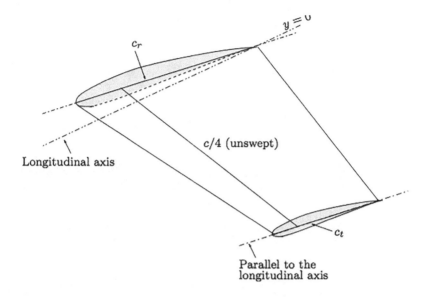

Figure 18.8 Unswept, tapered wing with geometric twist.

18.80 ROSSBY NUMBER

The Rossby number is a non-dimensional parameter used to describe the rotating flows.

18.81 ROTAMETER

A rotameter is a direct-reading meter, usually employed for flow rate measurements in passages with small diameters. Basically, it is a constant-pressure-drop variable-area meter. It consists of a tube with a tapered bore in which a float takes a vertical position corresponding to each flow rate through the tube, as shown in Figure 18.9.

For a given flow rate, the float remains stationary and the vertical component of the forces of differential pressure, viscosity, buoyancy, and gravity are balanced. This force balance is self-maintaining since the flow area of the rotameter, which is the annular area between the float and the tube, varies continuously with the vertical displacement of the float; thus, the device may be thought of as an orifice of adjustable area. The downward force, which is the gravity minus the buoyancy, is constant, therefore, the upward force, which is due to pressure drop, must also be constant. Since the float area is constant, the pressure drop should be constant. For a fixed flow area, the pressure drop Δp varies with the square of the flow rate, and so to keep the Δp constant for different flow rates, the area must vary. The

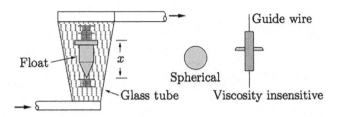

Figure 18.9 Rotameter.

tapered tube of the rotameter provides this variable area. The position of the float is a measure of the volume flow rate, and the position of the float can be made linear with the flow rate by making the tube area vary linearly with the vertical distance.

18.82 ROTARY VANE PUMP

A rotary vane pump is a positive-displacement pump that consists of vanes mounted to a rotor that rotates inside a cavity.

18.83 ROTATIONAL ENERGY

Rotational energy is due to the rotation of the molecule about the three orthogonal axes in space. The sources of the rotational energy are (a) the rotational kinetic energy associated with the molecule's rotational velocity and (b) its moment of inertia. But, for the diatomic molecule, the moment of inertia about the internucleus axis (the z-axis) is very small, and therefore, the rotational kinetic energy about the z-axis is negligible in comparison to the rotation about the x- and y-axes. Thus, the diatomic molecule has only two geometric and two thermal degrees of freedom. The same is true for a linear polyatomic molecule such as CO_2. However, for a nonlinear polyatomic molecule, such as H_2O, the number of geometric as well as thermal degrees of freedom in rotation is three.

18.84 ROTATIONAL FLOW

A flow in which the net rotation of the fluid element is not equal to zero is known as the rotational flow.
　or
　The flow in which the fluid particles also rotate about their axes while flowing is called rotational flow. Distortion in this case is less than the irrotational flow.

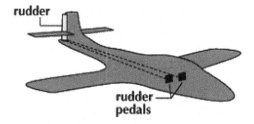

Figure 18.10 Aircraft rudder.

18.85 ROTATIONAL SPEED OF THE EARTH

The Earth rotates once every 23 h, 56 min, and 4.09053 s called the sidereal period, and its circumference is roughly 40,075 km. Thus, the surface of the Earth at the equator moves at a speed of 460 m/s – or roughly 1,000 mi/h.

18.86 ROTATING-VANE METER

Rotating-vane meter is a device for the measurement of volume flow rate. It is a positive-displacement meter.

18.87 RUDDER

This is a control surface meant for varying the side force on the vertical tail fin to control the yawing moment.

A rudder, shown in Figure 18.10, is a primary control surface used to steer a ship, boat, submarine, hovercraft, aircraft, or other conveyance that moves through a fluid medium (generally air or water).

On an aircraft, the rudder is used primarily to counter adverse yaw and p factor and is not the primary control used to turn the aeroplane. A rudder operates by redirecting the fluid past the hull (watercraft) or fuselage, thus, imparting a turning or yawing motion to the craft. In basic form, a rudder is a flat plane or sheet of material attached with hinges to the craft's stern, tail, or after end. Often, rudders are shaped so as to minimise the hydrodynamic or aerodynamic drag. On a simple watercraft, a tiller, essentially a stick or pole, acting as a lever arm may be attached to the top of the rudder to allow it to be turned by a helmsman. In larger vessels, cables, pushrods, or hydraulics may be used to link rudders to steering wheels. In a typical aircraft, the rudder is operated by pedals via mechanical linkages or hydraulics.

18.88 RUNWAY

According to the International Civil Aviation Organisation (ICAO), a runway is a 'defined rectangular area on a land aerodrome prepared for the landing and take-off of aircraft'. Runways may be a manufactured surfaces (often asphalt, concrete, or a mixture of both) or natural surfaces (grass, dirt, gravel, ice, sand, or salt). Runways as well as taxiways and ramps are sometimes referred to as 'tarmac', though very few runways are built using tarmac. Runways made of water for seaplanes are generally referred to as waterways. Runway lengths are now commonly given in metres worldwide, except in North America, where feet is used.

Chapter 19

Sake to System

19.1 SAKE

Sake is Japanese alcoholic beverage made from fermented rice. It is light in colour, noncarbonated, tastes sweet, and is up to 16% alcohol.

19.2 SALINE WATER

Saline water (more commonly known as salt water) contains a high concentration of dissolved salts (mainly sodium chloride). The salt concentration is usually expressed in parts per thousand (permille) or parts per million (ppm). Seawater has a salinity of roughly 35,000 ppm, equivalent to 35 g of salt per 1 L (or kg) of water.

19.3 SAND

Sand is a granular material composed of finely divided rock and mineral particles. It has various compositions but is defined by its grain size. Sand grains are smaller than gravel and coarser than silt. It is a non-renewable resource over human timescales. It is suitable for making concrete and is in high demand.

19.4 SAND FLOW

Sand flow is commonly used to diagnose a known sand production issue. It can also be used proactively to ensure downhole sand control measures are working optimally.

DOI: 10.1201/9781003348405-19

19.5 SANITATION

Sanitation refers to public health conditions related to clean drinking water and adequate treatment and disposal of human excreta and sewage. Preventing human contact with faeces is part of sanitation as is hand washing with soap. Sanitation systems aim to protect human health by providing a clean environment that will stop the transmission of disease, especially through the faecal-oral route. For example, diarrhoea, the main cause of malnutrition and stunted growth in children, can be reduced through adequate sanitation. There are many other diseases, which are easily transmitted in communities that have low levels of sanitation, such as ascariasis (a type of intestinal worm infection or helminthiasis), cholera, hepatitis, polio, schistosomiasis, and trachoma, to name just a few.

19.6 SATELLITE

A satellite is a moon, planet, or machine that orbits a planet or star.

19.7 SATURATED LIQUID

A liquid that is about to vaporise is called a saturated liquid (e.g., water at 1 atm and 100°C).

19.8 SATURATED SOLUTION

A saturated solution is a solution that contains the maximum amount of solute that is capable of being dissolved.

19.9 SATURATED VAPOUR

A vapour that is about to condense is called a saturated vapour.

19.10 SATURATION PRESSURE

At a given temperature, saturation pressure is the pressure at which a given liquid and its vapour or a given solid and its vapour can co-exist in equilibrium. For a pure substance, the saturation pressure is only a function of its temperature. The saturation pressure increases with temperature.

19.11 SATURATION STATE

A saturation state is the point where a phase change begins or ends. For example, the saturated liquid line represents the point where any further addition of energy will cause a small portion of the liquid to convert to vapour.

19.12 SATURATION TEMPERATURE

Saturation temperature means boiling point. The saturation temperature is the temperature for a corresponding saturation pressure at which a liquid boils into its vapour phase. The liquid can be said to be saturated with thermal energy.

19.13 SCALAR QUANTITIES

Scalar quantities only have magnitude and are without direction. In other words, scalar quantities require only a magnitude for a complete description. For example, temperature is a scalar quantity.

19.14 SCALE EFFECT

For dynamic similarity between the forces acting on an actual (or full-scale) machine and a scaled-down model used for testing (usually wind tunnel tests), the actual machine and the scale model must satisfy geometric and kinematic similarities.

19.15 SCHLICHTING JET

A Schlichting jet is a steady, laminar, round jet, emerging into a stationary fluid of the same kind with a very high Reynolds number.

19.16 SCHLIEREN TECHNIQUE

The Schlieren technique is a flow visualisation technique used to study high-speed flows in the transonic and supersonic Mach number ranges. This gives only a qualitative estimate of the density gradient of the field. This is used to visualise faint shock waves, expansion waves, etc. The Schlieren system gives the deflection angles of the incident rays. The optical patterns given by Schlieren are sensitive to the first derivative of flow density.

19.17 SCHMIDT NUMBER

Schmidt number is the ratio of momentum diffusivity to mass diffusivity. Schmidt number is similar to the Prandtl number in heat transfer.

19.18 SCRAMJET

The name SCRAMJET denotes the supersonic combustion ramjet, a concept that is being investigated for high flight Mach numbers. However, this concept is yet to become practical, and to date, ramjet combustion requires a subsonic airstream to provide stable combustion without excessive aerodynamic losses.

19.19 SEA

The salt water that covers large parts of the surface of the Earth is called the sea.

19.20 SEAWATER

Seawater is the water that makes up the oceans and seas, covering more than 70% of Earth's surface. Seawater is a complex mixture of 96.5% water, 2.5% salts, and smaller amounts of other substances, including dissolved inorganic and organic materials, particulates, and a few atmospheric gases.

19.21 SEA WAVES

Waves are created by energy passing through water, causing it to move in a circular motion, as seen in Figure 19.1. The ocean is never still. The friction between wind and surface water creates wind-driven waves or surface waves. As the wind blows across the surface of the ocean or a lake, the continual disturbance creates a wave crest.

19.22 SECOND LAW OF THERMODYNAMICS

The second law states that if the physical process is irreversible, the combined entropy of the system and the environment must increase. The final entropy must be greater than the initial entropy for an irreversible process.

Figure 19.1 Sea waves.

19.23 SECOND THROAT

The second throat is used to provide isentropic deceleration and highly efficient pressure recovery after the test section of a supersonic wind tunnel.

19.24 SEDIMENT TRANSPORT

Sediment transport is the movement of organic and inorganic particles by water. In general, the greater the flow, the more sediments will be conveyed.

19.25 SEDIMENTATION

Sedimentation is the process by which particles suspended in water settle out of the suspension under the effect of gravity. The particles that settle out of the suspension become sediments and form sludge during water treatment .

19.26 SEEBECK PRINCIPLE

The Seebeck principle states that 'heat flow in a metal is always accompanied by an emf'.

19.27 SEEPAGE

The slow escape of a liquid or gas through a porous material or small holes is called seepage.

19.28 SEISMOLOGY

Seismology is the scientific study of earthquakes and the propagation of elastic waves through the Earth or other planet-like bodies. The field also includes research on the environmental effects of earthquakes such as tsunamis as well as diverse seismic sources such as volcanic, tectonic, glacial, fluvial, oceanic, atmospheric, and artificial processes, such as explosions. A related field that uses geology to infer information regarding past earthquakes is paleoseismology. A recording of the Earth's motion as a function of time is called a seismogram. A seismologist is a scientist who does research in seismology.

19.29 SELF-SIMILAR FLOW

A flow that preserves its geometry in space or time or both is called a self-similar flow. In the simplest cases of flows, such motions are described by a single independent variable, referred to as the similarity variable.

19.30 SEMIFLUID

Semifluid is a substance with properties intermediate between those of a solid and a liquid.

19.31 SEMI-INFINITE VORTEX

A vortex is termed a semi-infinite vortex when one of its ends stretches to infinity.

19.32 SEMISOLID

A semisolid is a substance that is in between a solid and a liquid state.

19.33 SEPARATION BUBBLE

The forward flow separation region, shown in Figure 19.2, is usually referred to as a separation bubble.

19.34 SEPARATION POINT

The location where the flow leaves the body surface is termed the separation point. It is the position at which the boundary layer leaves the surface of

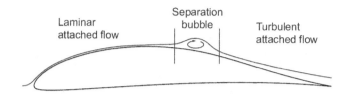

Figure 19.2 Separation bubble.

a solid body. If the separation takes place while the boundary layer is still laminar, the phenomenon is termed laminar separation. If it takes place for a turbulent boundary layer, it is called turbulent separation.

19.35 SERN

In rocketry, a SERN, which stands for single expansion ramp nozzle, is a type of physical linear expansion nozzle where the gas pressure transfers work only on one side. Traditional nozzles are axially symmetric, and therefore, surround the expanding gas. Linear nozzles are not axially symmetric, but consist of a two-dimensional configuration of two expansion ramps. A SERN could also be seen as a single-sided aero spike engine.

19.36 SEWAGE

It is the waste material produced from human bodies which is carried away from homes through large underground pipes (sewers).

19.37 SEWAGE TREATMENT

Sewage treatment (or domestic wastewater treatment, municipal wastewater treatment) is a type of wastewater treatment that aims to remove contaminants from sewage. Sewage contains wastewater from households and businesses and possibly pre-treated industrial wastewater. Physical, chemical, and biological processes are used to remove contaminants and produce treated wastewater (or treated effluent) that is safe enough for release into the environment. A by-product of sewage treatment is a semisolid waste or slurry, called sewage sludge. The sludge has to undergo further treatment before being disposed of or made suitable for land use.. The term 'sewage treatment plant' is often used interchangeably with the term 'wastewater treatment plant'.

19.38 SEWER

An underground pipe that carries human waste to a place where it can be treated.

19.39 SEWERAGE

Sewerage is the infrastructure that conveys sewage or surface runoff (storm-water, meltwater, rainwater) using sewers.

19.40 SHADOWGRAPH METHOD

The shadowgraph method is a flow visualisation technique meant for high-speed flows with transonic and supersonic Mach numbers. This is employed for fields with strong shock waves. A shadowgraph visualises the displacement experienced by an incident ray, which has crossed the high-speed flowing gas. The optical patterns given by the shadowgraph are sensitive to the second derivative of the flow density.

The shadowgraph is an optical method that reveals non-uniformities in transparent media like air, water, or glass. It is related to, but simpler than, the Schlieren and Schlieren photography methods that perform a similar function.

19.41 SHAFT POWER

The power input to the pump delivered to the pump shaft by the motor is called the shaft power.

19.42 SHAFT WORK

Open systems are capable of delivering work continuously, because in the system, the medium which transforms energy is continuously replaced. This useful work, which a machine continuously delivers, is called shaft work.

19.43 SHALLOW

Shallow is something that is not deep or someone who is concerned only about silly or inconsequential things.

19.44 SHALLOW FLOW

A shallow flow is generally defined as the situation in which the horizontal length scale of the flow is significantly larger than its depth.

19.45 SHALLOW WATER

It is water of such depth that surface waves are noticeably affected by bottom topography. Typically, this implies a water depth equivalent to less than half the wavelength.

19.46 SHALLOW WATER EFFECTS

Shallow water affects the 'draft & manoeuvrability' of ships. As the hull moves through shallow water, the area which it displaces is not so easily replaced by the surrounding water, therefore, leading to a state of partial vacuum as the propeller and rudder are still working.

19.47 SHAPE FACTOR

It is a dimensionless measure of the shape of the boundary layer velocity profile and is defined as the ratio of displacement thickness, δ^*, to momentum thickness, θ,

$$H = \frac{\delta^*}{\theta}$$

19.48 SHEAR FLOW

In fluid mechanics, the term shear flow (or shearing flow) refers to a type of fluid flow, which is caused by forces, rather than by the forces themselves. In a shear flow, the adjacent layers of fluid move parallel to each other at different speeds. Viscous fluids resist this shear motion.

19.49 SHEAR FOLDING

Shear folding, which is also referred to as slip folding, involves shear along planes that are oriented approximately parallel to the axial plane of the fold structure. These planes, which are typically axial-planar cleavage planes, facilitate high-angle reverse slip leading to fold limb rotation and amplification.

19.50 SHEAR FORCE

A shear force is a force applied perpendicular to a surface in opposition to an offset force acting in the opposite direction. This results in a shear strain. In simple terms, one part of the surface is pushed in one direction, while another part of the surface is pushed in the opposite direction.

19.51 SHERWOOD NUMBER

The Sherwood number (also called the mass-transfer Nusselt number) is a dimensionless number used in mass-transfer operation. It represents the ratio of the convective mass transfer to the rate of diffusive mass transport.

19.52 SHIP

A ship is a large watercraft that travels the world's oceans and other sufficiently deep waterways, carrying goods or passengers, or in support of specialised missions, such as defence, research, and fishing.

19.53 SHOCK

Shock may be described as a compression front in a supersonic flow field and flow process across which results in an abrupt change in fluid properties. The thickness of the shock is comparable to the mean free path of the gas molecules in the fluid.

The thin region where large gradients in temperature, pressure, and velocity occur, and where the transport phenomena of momentum (μ) and energy (K) are important, is called shock. Essentially, a shock is a compression front across which the flow properties jump.

For shocks in a calorically perfect gas flow or a chemically reacting equilibrium flow, the flow properties ahead of and behind the shock are uniform, and the gradients (that is, the jump) in flow properties take place almost discontinuously (that is, abruptly) within a thin region of not more than a few mean free path thickness ($\lambda \approx 6.6317 \times 10^{-8}$ m, for air at sea level).

However, in nonequilibrium flows, all chemical reactions and/or vibrational excitations take place at a finite rate. As the thickness of a shock wave is of the order of only a few mean free paths, the molecules in a fluid element can experience only a few collisions as the fluid element traverses the shock front. Consequently, the flow through the shock front is essentially frozen.

19.54 SHOCK ABSORBER

A shock absorber or damper is a mechanical or hydraulic device designed to absorb and damp shock impulses. It does this by converting the kinetic energy of the shock into another form of energy (typically heat), which is then dissipated.

19.55 SHOCK-ASSOCIATED NOISE

Broadband shock-associated noise and screech tones are generated only when a quasi-periodic shock-cell structure is present in the jet core. The quasi-periodicity of the shock-cells plays a crucial role in defining the characteristics of both the broadband and discrete frequency shock noises.

19.56 SHOCK-CELL

The expansion rays get reflected from the free boundary as compression waves, as illustrated in Figure 19.3, since the reflection of a wave from a free boundary is unlike (meaning opposite in nature; expansion wave to compression wave and vice-versa).

The supersonic flow in the zone (a) gets decelerated on passing through these reflected compression waves, and hence, the flow Mach number at (c) becomes less than that at (a). These compression waves coalesce to form shock waves that cross each other at the axis, proceed to the barrel shock, and reflect as expansion waves, as shown in the figure. The distance

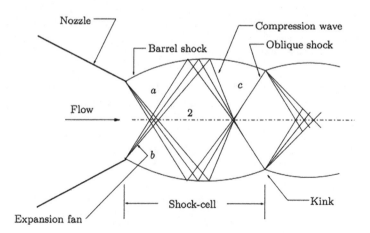

Figure 19.3 Waves in an underexpanded sonic jet.

from the nozzle exit to the first kink location on the barrel shock, where the shocks get reflected as expansion waves, is called a shock-cell in jet literature.

19.57 SHOCK DETACHMENT DISTANCE

An approximate expression for the shock detachment distance, δ, ahead of the nose of a blunt-nosed body with a spherical nose of radius R in terms of the density ratio across the detached shock is

$$\frac{\delta}{R} = \frac{\rho_1/\rho_2}{1+\sqrt{2(\rho_1/\rho_2)}}$$

where ρ_1 and ρ_2 are the density ahead of and behind the shock.

19.58 SHOCK DETACHMENT MACH NUMBER

At the Mach number corresponding to θ_{max}, the shock wave begins to detach. This is called the shock detachment Mach number. With a further decrease of M_1, the detached shock moves upstream of the nose.

19.59 SHOCK DIAMOND

Shock diamonds (also known as Mach diamonds or thrust diamonds) refer to formation of standing wave patterns that appear in the supersonic exhaust plume of an aerospace propulsion system, such as a supersonic jet engine, rocket, ramjet, or scramjet, when it is operated in an atmosphere. The 'diamonds' are actually a complex flow field made visible by abrupt changes in local density and pressure as the exhaust passes through a series of standing shock waves and expansion fans. Mach diamonds are named after Ernst Mach, the physicist who first described them.

19.60 SHOCK-EXPANSION THEORY

The shock and expansion waves discussed in this chapter are the basis for analysing a large number of two-dimensional, supersonic flow problems by simply 'patching' together appropriate combinations of two or more solutions. That is, the aerodynamic forces acting on a body present in a supersonic flow are governed by the shock and expansion waves formed at the surface of the body. In a shock-expansion theory, the shock is essentially a non-isentropic wave causing a finite increase of entropy. Thus, the total pressure of the flow decreases across the shock.

19.61 SHOCK POLAR

Shock polar is the graphical representation of the Rankine–Hugoniot equation in the hodograph plane. It is the locus of all possible states after an oblique shock.

19.62 SHOCK STRENGTH

The ratio

$$\frac{p_2 - p_1}{p_1} = \frac{\Delta p}{p_1}$$

where p_1 and p_2 are the static pressure ahead of and behind the shock, respectively, which is called the shock strength.

19.63 SHOCK SWALLOWING

When the second throat area is larger than the minimum required for any given condition, the shock wave can jump from the test section to the downstream side of the diffuser throat. This is termed shock swallowing.

19.64 SHOCK-TRAIN

The oblique shock system in the isolator is referred to as an oblique shock-train or simply as a shock-train.

19.65 SHOCK TUBES

The shock tube is a device to produce high-speed flow with high temperatures, by traversing normal shock waves, which are generated by the rupture of a diaphragm, which separates a high-pressure gas from a low-pressure gas. Shock tube is a very useful research tool for investigating not only the shock phenomena but also the behaviour of the materials and objects when subjected to very high pressures and temperatures.

19.66 SHOCK TUNNEL

Shock tunnels are wind tunnels that operate at Mach numbers of the order 25 or higher for time intervals up to a few milliseconds by using air heated and compressed in a shock tube. It consists of two major parts, the shock tube and the wind tunnel portion.

19.67 SHOCK WAVE

Shock wave is a strong pressure wave in any elastic medium such as air, water, or a solid substance, produced by supersonic aircraft, explosions, lightning, or other phenomena that create violent changes in pressure.

19.68 SHOULDER WING

It is the wing which is mounted above the fuselage middle.

19.69 SIDESLIPPING

The crab-wise motion, shown in Figure 19.4, of an aircraft in a straight path is termed sideslipping.

19.70 SILICA

It is a chemical compound of silicon found in sand and in rocks such as quartz, used in making glass and cement.

19.71 SILICA GEL

Silica gel is an amorphous and porous form of silicon dioxide (silica), consisting of an irregular three-dimensional framework of alternating silicon and oxygen atoms with nanometre-scale voids and pores. The voids may contain water or some other liquids, or may be filled by gas or vacuum.

19.72 SIMILARITY

Similarity or similitude in a general sense is the indication of a known relationship between two phenomena. In fluid dynamics, this is usually the relation between a full-scale flow and a flow with smaller but geometrically similar boundaries.

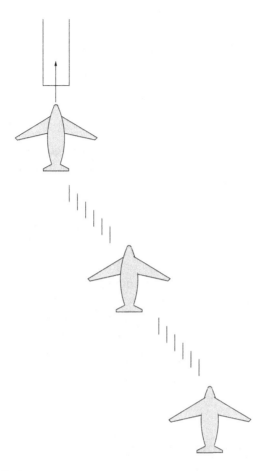

Figure 19.4 Sideslipping.

19.73 SIMILARITY RULE

An expression that relates the subsonic compressible flow past a certain profile to the incompressible flow past a second profile derived from the first principles through an affine transformation is called a similarity law.

19.74 SIMPLE-COMPRESSIBLE SUBSTANCE

A simple-compressible substance is one for which surface forces (e.g., surface tension) or body forces (gravity, magnetic, etc.) are unimportant. This term is usually used in conjunction with the term 'pure' (recall the earlier definition) to describe the substances familiar to engineers: air, water, etc.

19.75 SIMPLE-COMPRESSIBLE SYSTEM

A simple-compressible system is that system for which the electrical, gravitational, magnetic, motion, and surface tension effects are absent.

19.76 SIMPLE DISTILLATION

Simple distillation is a procedure by which two liquids with different boiling points can be separated. Simple distillation (the procedure outlined below) can be used effectively to separate liquids that have at least a 50° difference in their boiling points.

19.77 SIMPLE HARMONIC MOTION

In mechanics and physics, simple harmonic motion is a special type of periodic motion where the restoring force on the moving object is directly proportional to the magnitude of the object's displacement and acts towards the object's equilibrium position. Simple harmonic motion, in physics, is repetitive movement back and forth through an equilibrium, or central, position, so that the maximum displacement on one side of this position is equal to the maximum displacement on the other side. The time interval of each complete vibration is the same.

19.78 SIMPLE PENDULUM

A simple pendulum can be considered to be a point mass suspended from a string or rod of negligible mass. It is a resonant system with a single resonant frequency.

19.79 SIMPLE REGION

The simple region, in a supersonic flow field, is that in which both the left- and right-running characteristic lines are linear.

19.80 SIMPLE SYSTEM

A simple system is a single-state system with no internal boundaries and is not subjected to external force fields or inertial forces.

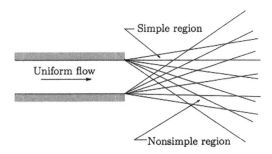

Figure 19.5 Simple and nonsimple regions in supersonic flow.

19.81 SIMPLE VORTEX

A simple or free vortex is a flow in which the fluid elements simply move along concentric circles without spinning about their axes. The fluid elements have only translatory motion in a free vortex. In addition to moving along concentric paths, if the fluid elements spin about their axes, the flow is termed a forced vortex.

19.82 SIMPLE AND NONSIMPLE REGIONS

A supersonic flow field with simple and nonsimple regions is shown in Figure 19.5. A supersonic expansion or compression zone with straight Mach lines is called a simple region.

The Mach lines which are straight in the simple region become curved in the nonsimple region after intersecting with other Mach lines. However, the wave segments between the adjacent cross-over points may be treated as linear, without introducing significant error to the calculated results.

19.83 SINK

Sink is a potential flow field in which flow gushes towards a point from all radial directions.

19.84 SIPHON

A siphon (also spelt syphon) is any of a wide variety of devices that involve the flow of liquids through tubes, as seen in Figure 19.6. In a narrower sense, the word refers particularly to a tube in an inverted 'U' shape, which causes a liquid to flow upwards, above the surface of a reservoir, with no

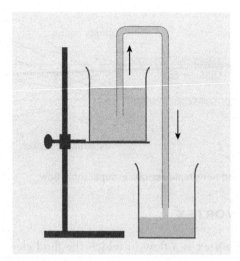

Figure 19.6 Siphon.

pump, but powered by the fall of the liquid as it flows down the tube under the pull of gravity, then discharging at a level lower than the surface of the reservoir from which it came.

19.85 SKIDDING

The aircraft may travel to the right or left along the lateral axis; such motion is called skidding.

19.86 SKIN FRICTION

Skin friction is a tangential traction or drag force acting in the direction of flow velocity. Skin friction is the drag at the surface of a body in a flow, caused by the viscosity of the flowing fluid.

It is important to note that the skin friction coefficient is referred to as the drag coefficient. This is because, for a streamlined body, such as a flat plat with flow over its surface, the skin friction drag is the major portion of the drag and the wake drag is negligible.

19.87 SKIN FRICTION DRAG

The friction between the surface of a body and the fluid causes viscous shear stress and this force is known as skin friction drag.

19.88 SKY

The sky is everything that lies above the surface of the Earth, including the atmosphere and outer space. In the field of astronomy, the sky is also called the celestial sphere. This is an abstract sphere, concentric to the Earth, on which the Sun, Moon, planets, and stars appear to be drifting. The celestial sphere is conventionally divided into designated areas called constellations.

19.89 SKYDIVING

Skydiving is a method of transiting from a high point in the atmosphere to the surface of Earth with the aid of gravity, involving the control of speed during the descent using a parachute.

19.90 SLANT HYDRAULIC JUMP

A hydraulic jump, which is along a line oblique to the flow direction, is called a slant hydraulic jump.

19.91 SLEET

Sleet is composed of rain and partially melted snow. Unlike hard ice pellets and freezing rain, which is fluid until striking an object, this precipitation is soft and translucent, but it contains some traces of ice crystals from partially fused snowflakes.

19.92 SLENDER BODY

Launch vehicles may be generally considered slender bodies with fineness ratios $l/s = O(10)$, where l is the length of the vehicle and s is the base of the vehicle.

19.93 SLENDER-BODY THEORY

The slender-body theory allows us to derive an approximate relationship between the velocity of the body at each point along its length and the force per unit length experienced by the body at that point.

19.94 SLENDER WING THEORY

Slender wing theory is used to calculate the effects of mounting the wing of a wing-body combination above or below the body axis, with and without wing-body angle on the lift and moment.

19.95 SLIP FLOW

The slip flow regime is the flow regime of slight rarefaction. The density of a gas is slightly lower than that of a complete continuum flow.

Slip flow $(0.01 < Kn < 0.1)$: Here again the continuum fluid dynamic analysis is applicable provided the slip boundary conditions are employed. That is, the no-slip boundary condition of continuum flows, dictating zero velocity at the surface of an object kept in the flow, is not valid. The fluid molecules move (slip) with a finite velocity, called the slip velocity, at the boundary.

19.96 SLIPSTREAM

The propeller produces thrust by forcing the air backwards, and the resultant stream of air that flows over the fuselage, tail units, and other parts of the aircraft is called the slipstream. The extent of slipstream may be taken roughly as being that of a cylinder of the same diameter as the propeller. Actually, there is a slight contraction of the diameter at a short distance behind the propeller.

Slipstream behind crossing shocks is a contact surface having two flow fields of different parameters $(T$ and $\rho)$ on either side of it. The contact surface may also be idealised as a surface of discontinuity. The contact surface can either be stationary or moving. Unlike the shock wave, there is no flow of matter across the contact surface. In the literature, we can find this contact surface being referred to by different names: material boundary, entropy discontinuity, slipstream or slip surface, vortex sheet, and tangential discontinuity.

It is essential to note that the contact surface is a fluid boundary across which there is no mass transport. Further, the surface can tolerate temperature and density gradients, but not pressure gradient. In other words, the temperature and density on either side of the slipstream can be different but the pressure on both sides must be equal.

The contact surface can tolerate thermal and concentration imbalance but not pressure imbalance.

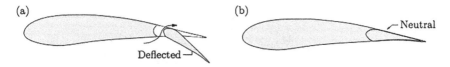

Figure 19.7 A slotted flap in (a) deflected and (b) neutral positions.

19.97 SLOTTED FLAP

In the case of a slotted flap, a gap or slot is opened up between the flap and the main wing when the flap is deflected, as illustrated in Figure 19.7.

The air beneath the wing is at a higher pressure than the air above it. Consequently, air blows through the slot when it is opened onto the upper surface of the flap. This flow from the bottom to the top re-energises the boundary layer and tends to prevent separation. In other words, this flap behaves in the same way as a plain flap, but the combination of variable geometry with that amounts to a measure of boundary layer control, resulting in further improvement of performance at all incidences. The increment in the maximum lift coefficient is larger than that provided by a plain or split flap. The drag increment, however, is much less, because of the prevention of separation. This also causes the moment effect to be relatively large.

19.98 SLOWEST FLYING BIRD

The American woodcock (*Scolopax minor*) and the Eurasian woodcock (*S. rusticola*) have both been timed flying at 5 mph without stalling during courtship displays.

19.99 SLUG FLOW

Slug flow is a typical two-phase flow where a wave is picked up periodically by the rapidly moving gas to form a frothy slug, which passes along the pipe at a greater velocity than the average liquid velocity.

19.100 SLUSH FLOW

A slush flow is a rapid mass movement of water and snow and is categorised as a type of debris flow. Slush flows are caused when water reaches a critical concentration in the snowpack due to more water inflow than outflow.

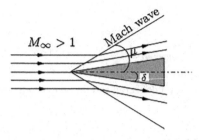

Figure 19.8 Mach cone.

19.101 SMALL-DENSITY-RATIO ASSUMPTION

An assumption common to hypersonic flow is that the ratio of the freestream density, ρ_∞, to the density just behind a shock, ρ_2, is extremely small, that is,

$$\varepsilon = \frac{\rho_\infty}{\rho_2} \ll 1$$

The equation is known as the small-density-ratio assumption.

19.102 SMALL DISTURBANCE

When the apex angle of wedge δ is vanishingly small, the disturbances will be small, and we can consider these to be identical to sound pulses. In such a case, the deviation of streamlines will be small and there will be an infinitesimally small increase in pressure across the Mach cone, as shown in Figure 19.8.

19. 103 SMALL PERTURBATION THEORY

The small perturbation theory postulates that the perturbation velocities are small compared to the main velocity components, that is, $u \ll V_\infty$, $v \ll V_\infty$, $w \ll V_\infty$.

19.104 SMOG

Smoke fog, or smog, is a type of intense air pollution. The word 'smog' was coined in the early twentieth century and is a contraction of the words smoke and fog to refer to smoky fog due to its opacity and odour.

Smog can irritate your eyes, nose, and throat. Or, it can worsen the existing heart and lung problems or perhaps cause lung cancer with regular long-term exposure. It also results in early death. Studies on ozone show that once it gets into your lungs, it can continue to cause damage even when you feel fine.

19.105 SMOKE

Smoke is a collection of airborne particulates and gases, emitted when a material undergoes combustion or pyrolysis, together with the quantity of air that is entrained or otherwise mixed into the mass.

19.106 SMOKE FLOW VISUALISATION

Smoke flow visualisation is one of the popular techniques used in low-speed flow fields with velocities up to about 30 m/s. Smoke visualisation is used to study problems like boundary layers, air pollution problems, design of exhaust systems of locomotives, cars, ships, topographical influence of disposal of stack gases, etc.

19.107 SMOKE TUNNEL

Smoke tunnel is basically an open circuit, low-speed tunnel. The flow through the tunnel is induced by the suction of atmospheric air through the test section using a simple axial fan. Schematic diagram of a typical smoke tunnel is shown in Figure 19.9.

Flow visualisation with smoke is generally done in a smoke tunnel. It is a low-speed wind tunnel carefully designed to produce a uniform steady flow in the test section with negligible turbulence. Smoke streaks are injected along the freestream or on the surface of the model for visualising flow patterns. White dense smoke is used for this purpose. When a beam of light is properly focused on the smoke filaments, the light gets scattered and reflected by the smoke particles making them distinguishably visible from the surroundings.

A smoke tunnel is generally used for demonstrating flow patterns such as flow around bodies of various shapes, flow separation, etc. Smoke pattern over a blunt body is shown in the Figure 19.10.

19.108 SNOW

Snow comprises individual ice crystals that grow while suspended in the atmosphere, and then fall, accumulating on the ground where they undergo further changes.

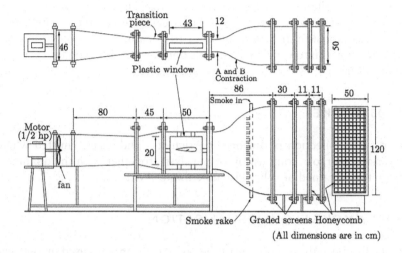

Figure 19.9 Smoke tunnel.

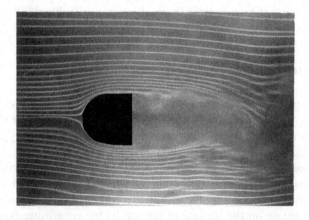

Figure 19.10 Smoke pattern over a blunt body.

19.109 SOAKING

Soaking means extremely wet. The main objective of soaking is to achieve quick and uniform water absorption. The lower the water temperature, the slower the soaking process.

19.110 SOAP BUBBLE

A soap bubble is an extremely thin film of soapy water enclosing air that forms a hollow sphere with an iridescent surface.

19.111 SOARING FLIGHT

Fight using thermal in the atmosphere is called soaring flight. Once a thermal is encountered, the pilot flies in circles to keep within the thermal, so gaining altitude before flying off to the next thermal and towards the destination. This is known as 'thermalling'. Climb rates depend on conditions, but rates of several metres per second are common. Thermals can also be formed in a line usually because of the wind or the terrain, creating cloud streets. These can allow flying straight while climbing in a continuous lift.

19.112 SODA WATER

Carbonated water (also known as soda water, sparkling water, fizzy water, water with gas or (especially in the US) as seltzer or seltzer water) is water containing dissolved carbon dioxide gas, either artificially injected under pressure or occurring due to natural geological processes.

19.113 SOFT WATER

Soft water is water that is free from dissolved salts of such metals as calcium, iron, or magnesium, which form insoluble deposits which appear as scale in boilers or soap curds in bathtubs and laundry equipment.

19.114 SOIL EROSION

Soil erosion is a gradual process that occurs when the impact of water or wind detaches and removes soil particles, causing the soil to deteriorate. The impact of soil erosion on water quality becomes significant, particularly as soil surface runoff. Sediment production and soil erosion are closely related.

19.115 SOIL PERCOLATION

Percolation in soil is simply the movement of the water through the soil and a soil percolation test is the means of measuring this movement.

19.116 SOLAR ENERGY

Solar energy is radiant light and heat from the Sun that is harnessed using a range of ever-evolving technologies such as solar heating, photovoltaics, solar thermal energy, solar architecture, molten salt power plants, and artificial photosynthesis.

19.117 SOLAR MASS

The solar mass is a standard unit of mass in astronomy, equal to $\sim 2 \times 10^{30}$ kg. It is often used to indicate the masses of other stars as well as stellar clusters, nebulae, galaxies, and black holes. It is approximately equal to the mass of the Sun.

19.118 SOLID

In a solid, the molecules are arranged in a three-dimensional pattern, which is repeated throughout its mass. The small distance between the molecules makes the attractive forces of molecules on each other very large and keeps the molecules at fixed position within the solid.

19.119 SOLID BLOCKING

The variation of static pressure along the test section produces a drag force known as horizontal buoyancy. It is usually small in closed test section and is negligible in open jets.

19.120 SOLID-PROPELLANT ROCKET

In a solid-propellant rocket, the fuel and the oxidiser are in solid form, and they are usually mixed to form the propellant. This propellant is carried within the combustion chamber.

19.121 SOLIDITY

Solidity is the ratio of the total blade area to the disc area. Solidity is a function of the ability to absorb power from the engine and the potential to provide rotor thrust.

19.122 SOLUBILITY

Solubility is the ability of a solid, liquid, or gaseous chemical substance (referred to as the solute) to dissolve in a solvent (usually a liquid) and form a solution.

19.123 SOLUTE

The liquid in which the substance dissolves is called the solvent, while the dissolved substance is called the solute.

19.124 SOLUTION

A solution is a homogeneous mixture of one or more solutes dissolved in a solvent. Solvent: the substance in which a solute dissolves to produce a homogeneous mixture. Solute: the substance that dissolves in a solvent to produce a homogeneous mixture.

19.125 SOLVENT

A solvent can be defined as a liquid that can dissolve, suspend, or extract other materials, without chemical change to the material or solvent.

19.126 SONIC BOOM

A sonic boom is a loud sound like an explosion. It is caused by shock waves created by any object that travels through the air faster than the speed of sound. Sonic booms create huge amounts of sound energy.

19.127 SONIC BOOM SPEED

A sonic boom is an impulsive noise similar to thunder. It is caused by an object moving faster than sound—about 750 miles per hour at sea level.

19.128 SOUND

Sound may be defined as any pressure variation (in air, water, or another medium) that the human ear can detect. An unpleasant or unwanted sound is termed noise.

19.129 SOUND BARRIER

The sound barrier or sonic barrier is the sudden increase in aerodynamic drag and other undesirable effects experienced by an aircraft or other object when it approaches the speed of sound. When an aircraft first approaches the speed of sound, these effects are seen as constituting a barrier, making faster speeds very difficult or impossible. The term sound barrier is still sometimes used today to refer to an aircraft reaching supersonic flight. Flying faster than sound produces a sonic boom.

19.130 SOUND FREQUENCY

The frequency of sound is defined as the number of pressure variations per second. The frequency is measured in hertz (Hz). The frequency of a sound produces its distinctive tone. For example, the rumble of distant thunder has a low frequency, while a whistle has a high frequency. The normal range of hearing for a healthy person extends from ~20 Hz up to 20,000 Hz. The range from the lowest to highest note of a piano is from 27.5 to 4,186 Hz.

An average human ear is not able to hear sound if the frequency is outside this range. Electronic detectors can detect waves of lower and higher frequencies as well. A dog can hear sound of the frequency of up to about 50 kHz and a bat up to about 100 kHz.

19.131 SOUND UNIT

Decibel is the commonly used sound unit. The decibel is a ratio between a measured quantity and an agreed reference level. Thus, a decibel is not an absolute unit of measurement. The reference level used is the hearing threshold of 20 μPa. The dB scale is logarithmic and the reference level 20 μPa is defined as 0 dB.

19.132 SOUND WAVE

A sound wave is a weak compression wave across which only infinitesimal changes in flow properties occur, that is, across these waves, there will be only infinitesimal pressure variations.

19.133 SOUNDING ROCKET

A sounding rocket is an instrument-carrying rocket designed to take measurements and perform scientific experiments during its sub-orbital flight.

19.134 SOUNDPROOFING

Soundproofing is any means of reducing the sound pressure concerning a specified sound source and receptor. There are several basic approaches to reducing sound: increasing the distance between source and receiver, using noise barriers to reflect or absorb the energy of the sound waves, using damping structures such as sound baffles, or using active antinoise sound generators.

19.135 SOURCE

A type of flow in which the fluid emanates from the origin and spreads radially outwards to infinity is called a source. It is a potential flow field in which flow emanating from a point spreads radially outwards.

19.136 SOURCE-SINK PAIR

This is a combination of a source and sink of equal strength, situated (located) at a distance apart. The stream function due to this combination is obtained simply by adding the stream functions of source and sink. When the distance between the source and sink is made negligibly small, in the limiting case, the combination results in a doublet.

19.137 SOY SAUCE

A thin dark brown sauce that is made from soya beans and has a salty taste, used in Chinese and Japanese cooking.

19.138 SPACE

Space is the boundless three-dimensional extent to which objects and events have relative positions and directions.

19.139 SPACE SCIENCE

Space science encompasses all of the scientific disciplines that involve space exploration and the study of natural phenomena and physical bodies occurring in outer space, such as space medicine and astrobiology.

19.140 SPACE-TIME

In physics, space-time is any mathematical model, which fuses the three dimensions of space and the one dimension of time into a single

four-dimensional manifold. Space-time diagrams can be used to visualise relativistic effects, such as why different observers perceive differently
where and when events occur.

19.141 SPACE VELOCITY

In chemical engineering and reactor engineering, space velocity refers to
the quotient of the entering volumetric flow rate of the reactants divided by
the reactor volume, which indicates how many reactor volumes of feed can
be treated in a unit time. It is commonly regarded as the reciprocal of the
reactor space-time.

19.142 SPACECRAFT

A spacecraft is a vehicle or machine designed to fly in outer space. A type
of artificial satellite, spacecraft is used for a variety of purposes, including
communications, Earth observation, meteorology, navigation, space colo-
nisation, planetary exploration, and transportation of humans and cargo.

19.143 SPAN

The distance between the wing tips is called the span.

19.144 SPARK-IGNITION ENGINE

If the combustion of the air-fuel mixture is initiated by a spark plug, the
reciprocating engine is called a spark-ignition (SI) engine.

19.145 SPECIAL PURPOSE TUNNELS

These are tunnels with layouts totally different from that of low-speed and
high-speed tunnels. Some of the popular special purpose tunnels are spinning tunnels, free-flight tunnels, stability tunnels, and low-density tunnels.

19.146 SPECIFIC GRAVITY

Specific gravity is the ratio of the density of a substance to the density of
some standard substance at a specified temperature.

19.147 SPECIFIC HEAT

Specific heat is defined as the amount of heat required to raise the temperature of a unit mass of medium by 1°. The value of the specific heat depends on the type of process involved in raising the temperature of the unit mass. Usually, constant volume process and constant pressure process are used for evaluating specific heat. The specific heats at constant volume and constant pressure processes, respectively, are designated by c_v and c_p.

19.148 SPECIFIC IMPULSE

A term that is widely used in defining the performance of rocket engines is the specific impulse, I_{sp}. This can either be defined as the thrust per unit mass flow rate of the propellant as

$$I_m = \frac{T}{\dot{m}} = \frac{T}{dm_f/dt}$$

or

it can be defined as the thrust per unit weight flow of the propellant as

$$I_w = \frac{T}{\dot{W}} = \frac{T}{\dot{m}g} = \frac{I_m}{g}$$

Specific impulse is a measure of how efficiently a reaction mass engine (a rocket using a propellant or a jet engine using fuel) creates thrust. For engines whose reaction mass is only the fuel they carry, specific impulse is exactly proportional to the exhaust gas velocity.

19.149 SPECIFIC SPEED

Specific speed is a number that defines the type of pump, such as radial-flow, axial-flow, or mixed-flow.

19.150 SPECIFIC THRUST

Thrust per unit mass flow rate is termed specific thrust.

19.151 SPECIFICS OF HELICOPTERS

The main difference between conventional aircraft and helicopters is that the helicopter is capable of vertical, horizontal, and translational (sideways) flights, whereas the fixed wing aircraft can fly only in the horizontal or forward direction. The forces that would act on a helicopter during its flight are the thrust, T, weight, W, drag, D, and lift, L. Another feature which is unique for a helicopter is that it can hover unlike an aircraft.

19.152 SPECULAR REFLECTION

Specular reflection is that in which the reflection is like a mirror image. That is, the wave angles of the incident and reflected waves will be the same. For this, the surface should be absolutely smooth. Specular reflection is essentially an assumption made to obtain simplified solutions for most practical problems.

19.153 SPEED OF LIGHT

The speed of light in a vacuum, commonly denoted by c is a universal physical constant important in many areas of physics. Its exact value is defined as 299,792,458 m/s.

19.154 SPEED OF LIGHT IN AIR

Light in air is 1.0003 times slower than light in a vacuum, which slows it all the way down from 299,792,458 to 299,702,547 m/s. That's a slowdown of 89,911 m/s, which looks like a lot but is only three ten-thousandths of the speed of light.

19.155 SPEED OF SOUND

The speed with which sound propagates in a medium is called the speed of sound and is denoted by a. In the limiting case of Δp and $\Delta \rho$, a becomes

$$a = \sqrt{\frac{dp}{d\rho}}$$

This is the Laplace equation and is valid for any fluid.

The sound wave is a weak compression wave; across which only infinitesimal change in fluid properties occurs. Further, the wave itself is extremely thin and changes in properties occur very rapidly. The rapidity of the process rules out the possibility of any heat transfer between the system of fluid particles and its surrounding.

For a perfect gas, the speed of sound takes the form

$$a = \sqrt{\gamma R T}$$

where γ is the specific heat ratio, R is the gas constant, and T is the temperature in kelvin.

19.156 SPILLWAYS

Spillways are structures constructed to provide the safe release of floodwaters from a dam to a downstream area.

19.157 SPIN

In flight dynamics, a spin is a special category of stall resulting in autorotation (uncommanded roll) about the aircraft's longitudinal axis and a shallow, rotating, downward path approximately centred on a vertical axis. Spins can be entered intentionally or unintentionally from any flight attitude if the aircraft has sufficient yaw while at the stall point. In a normal spin, the wing on the inside of the turn stalls while the outside wing remains flying. It is possible for both wings to stall, but the angle of attack of each wing, and consequently, its lift and drag, are different.

19.158 SPIRAL DIVERGENCE

Spiral divergence is characterised by an aeroplane that is very stable directionally but not very stable laterally; for example, a large-finned aeroplane with no dihedral. The bank angle increases and the aeroplane continues to turn into the sideslip in an ever-tightening spiral.

19.159 SPIRAL INSTABILITY

Spiral instability is instability about the longitudinal axis. For example, spiral instability means that if the right-wing tip moves down, it continues to move down rolling the plane to the right. It is simple to detect and address with left aileron.

19.160 SPLIT FLAP

The split flap forms a part of the bottom portion at the rear of the wing, as shown in Figure 19.11.

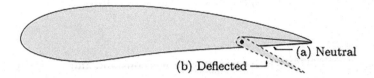

Figure 19.11 Profile with split flap in (a) neutral and (b) deflected positions.

Deflection of a split flap alters only the lower surface of the aerofoil. That is, in the case of a split flap, only the lower surface of the rear part of the aerofoil is movable, leading to the upper surface geometry unchanged when the flap is deflected. The split flap deflection also increases the effective camber, giving a reduction in zero-lift incidence. The overall effect is similar to that of a plain flap. However, because the upper surface is not so highly cambered, separation effects are less marked so that the performance at high incidence is improved. However, at flow incidence, the performance is adversely affected because of the large wake behind the deflected flap. This is relatively unimportant since the objective of the flap is to give the wing an improved performance at high incidence.

19.161 SPOILER CONTROL

Spoilers are long narrow solid strips normally fitted to the upper surface of the wing. In a normal mode of flight, they lie flush with the surface and do not affect the performance of the wing. However, the spoilers can be connected to the aileron controls in such a way that when an aileron is moved up beyond a certain angle, the spoiler is raised at a large angle to the airflow or comes up through the slit, causing turbulence, decreasing in lift, and increase in drag.

19.162 SPRAY

Spray is water or other liquid broken up into minute droplets and blown, ejected into, or falling through the air.

19.163 SPRAY POND

It is a pond where warm water is sprayed into the air and is cooled by the air as it falls into the pond.

19.164 SPRAYER

A sprayer is a device used to spray a liquid. Sprayers are commonly used for the projection of water, weed killers, crop performance materials, pest maintenance chemicals as well as manufacturing and production line ingredients. In agriculture, a sprayer is a piece of equipment that is used to apply herbicides, pesticides, and fertilisers to crops.

19.165 STABILITY

The ability of the aircraft to return to the same flying mode, when slightly disturbed from that condition, without any effort on the part of the pilot is called stability.

19.166 STAGNANT

Stagnant means not flowing in a current or stream.

19.167 STAGNANT WATER

Stagnant water is water that has remained in place for hours, , days, or even weeks. Although it's most common after a flood, it can happen any time when water cannot drain properly.

19.168 STAGNATION

The state of not flowing or moving is called stagnation.

19.169 STAGNATION ENTHALPY

The stagnation enthalpy represents the enthalpy of a fluid when it is brought to rest adiabatically.

In fluid dynamics, stagnation pressure (or pitot pressure) is the static pressure at a stagnation point in a fluid flow. At stagnation point, the fluid velocity is zero. In an incompressible flow, stagnation pressure is equal to the sum of the freestream static pressure and the freestream dynamic pressure.

19.170 STAGNATION PRESSURE

The stagnation pressure is the pressure that the fluid would obtain if brought to rest without loss of mechanical energy.

19.171 STAGNATION STATE

It is a state of zero flow velocity. That is, it is the state achieved by decelerating a flow to zero velocity. The properties of a fluid at the stagnation state are called stagnation properties.

19.172 STAGNATION TEMPERATURE

In thermodynamics and fluid mechanics, stagnation temperature is the temperature at a stagnation point in a fluid flow. At stagnation point, the speed of the fluid is zero and all of the kinetic energy has been converted to internal energy and is added to the local static enthalpy.

19.173 STALL

Stalling of an aircraft is characterised by loss of lift and increase of drag. This is due to the effects of separation. Thus, if stalling occurs in a flight, the aircraft will lose height, unless some action is taken to prevent it. The aircraft's behaviour and handling at and near the stall depends on the design of the wing.

In fluid dynamics, a stall is a reduction in the lift coefficient generated by a foil as the angle of attack increases. This occurs when the critical angle of attack of the foil is exceeded. The critical angle of attack is typically about 15°, but it may vary significantly depending on the fluid, foil, and Reynolds number.

19.174 STALL ANGLE OF ATTACK

A stall occurs when the angle of attack of an aerofoil exceeds the value, which creates maximum lift as a consequence of the airflow across it. This angle varies very little in response to the cross-section of the (clean) aerofoil and is typically around 15°.

19.175 STALLED STATE

In a stalled state, the airflow on the suction side of the aerofoil is turbulent. It is found that just before the stalled state sets in, the lift coefficient attains its maximum value, and the corresponding speed is called the stalling speed.

19.176 STALLING ANGLE

The angle of attack α around which the flow separates and the lift reaches a maximum and begins to decrease is called the stalling angle, and the corresponding value of lift coefficient is denoted by $C_{L_{\max}}$. A typical stalling angle would be about 15°, and a typical $C_{L_{\max}}$ for a wing without a high lift device is about 1.2–1.4.

19.177 STANDING WAVES

When two trains of a wave of equal amplitude, wavelength, and period, but travelling in opposite directions, are combined, the result is a set of standing or stationary waves.

In physics, a standing wave, also known as a stationary wave, is a wave which oscillates in time but whose peak amplitude profile does not move in space. The peak amplitude of the wave oscillations at any point in space is constant with time, and the oscillations at different points throughout the wave are in phase. The locations at which the absolute value of the amplitude is minimum are called nodes, and the locations where the absolute value of the amplitude is maximum are called antinodes.

19.178 STANTON NUMBER

The Stanton number is a dimensionless number that measures the ratio of heat transferred into a fluid to the thermal capacity of the fluid. Stanton number,

$$St = \frac{h_x}{\rho \ c_p \ V_\infty}$$

is a measure of the heat flux to the heat capacity of the fluid flow. The Stanton number is also known as the modified Nusselt number.

19.179 STARTING VORTEX

Soon after start-up, the separation point is dipped to the trailing edge, as per the Kutta hypothesis, and the slipstream rolls up. The vortex thus formed is pushed downstream and positioned at a location behind the aerofoil. This vortex is called the starting vortex. It is essentially a free vortex because it is formed by the kinematics of the flow and not by the viscous effect.

The starting vortex forms in the air adjacent to the trailing edge of an aerofoil as it is accelerated from rest in a fluid. It leaves the aerofoil (which now has an equal but opposite 'bound vortex' around it) and remains (nearly) stationary in the flow. It rapidly decays through the action of viscosity.

19.180 STATE EQUATION

State equation is an equation relating to the pressure, temperature and specific volume of a substance. Among the several equations of state, the simplest and the best-known equation for substances in the gas phase is the ideal-gas equation of state.

19.181 STATE POSTULATE

The state postulate says 'the state of a simple-compressible system is completely specified by two independent, intensive properties', but not from the first principle.

19.182 STATES OF MATTER

Four states of matter are observable in everyday life: solid, liquid, gas, and plasma. Many other states are known such as Bose–Einstein condensates and neutron-degenerate matter but these only occur in extreme situations such as in ultra-cold or ultra-dense matter.

19.183 STATIC PRESSURE

Static pressure of a flow is that pressure which is acting normal to the flow direction. It acts equally in all directions.

19.184 STATIC TEMPERATURE

In any form of matter the molecules are in motion relative to each other. By virtue of this motion, the molecules possess kinetic energy and this energy is sensed as the temperature of the solid, liquid, or gas. In the case of a gas in motion, it is called static temperature.

19.185 STATIC TEMPERATURE DETERMINATION

Direct measurement of static temperature is not possible. A direct measurement of adiabatic wall temperature can be used to determine T_{t_∞}, and then by measuring p_{t_∞} and p_∞, the Mach number M_∞ and the static temperature can be calculated.

19.186 STATIC THRUST

For a low-speed aircraft, the thrust developed by a fixed-pitch propeller is found to be the greatest when there is no forward speed, that is, when the aircraft is stationary on the ground. The thrust developed under this condition is called the static thrust. It is desirable to have a large static thrust since it serves to give the aircraft a good acceleration when starting from rest, and thus, reduces the take-off run distance required. However, for high-speed aircraft, a fixed-pitch propeller designed for maximum speed would have a large pitch, and therefore, a steep pitch angle. Some portion of such blades would strike the air at an angle as high as 70° or more when there is no forward speed, leading to poor static thrust. To overcome this difficulty, a variable-pitch propeller has to be employed for a high-speed aircraft.

19.187 STATICS

Statics deals with fluid elements at rest with respect to one another, and therefore, is free of shearing stresses. The static pressure distributions in a fluid and on bodies immersed in a fluid can be determined from a static analysis.

19.188 STATISTICAL THERMODYNAMICS

A macroscopic approach to the study of thermodynamics that does not require knowledge of the behaviour of the individual particles of the substance is called classical thermodynamics. An elaborate approach based on the behaviour of individual particles is called statistical thermodynamics.

19.189 STEADY

The term steady implies no change with time. The opposite of steady is unsteady or transient.

19.190 STEADY FLOW

If properties and flow characteristics at each position in space remain invariant with time, the flow is called steady flow.

19.191 STEADY-FLOW DEVICES

A large number of engineering devices such as nozzles, compressors, and turbines operate for long periods under the same conditions and they are classified as steady-flow devices.

19.192 STEADY-FLOW PROCESS

A steady-flow process may be defined as a process during which a fluid flows through a control volume steadily. That is, the fluid properties can change from point to point within a control volume, but at any fixed location, they remain the same during the entire process.

19.193 STEAM

Steam is water in the gas phase.

19.194 STEAM TURBINE

A steam turbine is a device that extracts thermal energy from pressurised steam and uses it to do mechanical work on a rotating output shaft.

19.195 STEEP BANK

As the aircraft banks steeper, the rudder gradually will take the place of elevators and vice versa. However, this has to be handled with caution, because in a vertical bank, for instance, the rudder will not be powerful in raising or lowering the nose as the elevators in horizontal flight. Therefore, a vertical bank without a sideslip is impossible, since in such a bank, the lift will be horizontal and will provide no contribution towards lifting the weight. If such a bank has to be executed, then a straight upward inclination of the fuselage together with the propeller thrust provides sufficient lift.

19.196 STELLAR MASS

Stellar mass is a phrase that is used by astronomers to describe the mass of a star. It is usually enumerated in terms of the Sun's mass as a proportion of a solar mass.

19.197 STIRLING AND ERICSSON CYCLES

These are cycles involving an isothermal heat addition process at high temperature T_H and an isothermal heat rejection process at low temperature T_L.

19.198 STOICHIOMETRIC AIR

The minimum amount of air needed for the complete combustion of a fuel is called stoichiometric or theoretical air.

19.199 STOKES' FLOW

This is a typical situation in flows where the fluid velocities are very slow, the viscosities are very large, or the length scales of the flow are very small. Creeping flow was first studied to understand lubrication. In nature, this type of flow occurs in the swimming of microorganisms and sperm and the flow of lava.

Stokes flow (named after George Gabriel Stokes), also named creeping flow or creeping motion, is a type of fluid flow where advective inertial forces are small compared with viscous forces. The Reynolds number is low, i.e., Re < 1.

19.200 STOKES' PARADOX

In the science of fluid flow, Stokes' paradox is the phenomenon where there can be no creeping flow of a fluid around a disc in two dimensions; or, equivalently, the fact there is no non-trivial steady-state solution for the Stokes equations around an infinitely long cylinder. This is opposed to the three-dimensional case, where Stokes' method provides a solution to the problem of flow around a sphere.

19.201 STOKES' THEOREM

Stoke's theorem relates the surface integral over an open surface to a line integral along the bounded curve. Let S be a simply connected surface, which is otherwise of arbitrary shape, whose boundary is c, and let u be any arbitrary vector. Also, we know that any arbitrary close curve on an arbitrary shape can be shrunk to a single point. The Stoke's integral theorem states that 'the line integral $\int u \, dx$ about the closed curve c is equal to the surface integral $\iint (\nabla \times u) \cdot n \, ds$ over any surface of arbitrary shape which has c as its boundary'.

19.202 STORM

Storm is a disturbance of the normal condition of the atmosphere, manifesting itself by winds of unusual force or direction, often accompanied by rain.

A storm is any disturbed state of an environment or in an astronomical body's atmosphere especially affecting its surface and strongly implying severe weather. It may be marked by significant disruptions to normal conditions such as strong wind, tornadoes, hail, thunder and lightning (a thunderstorm), heavy precipitation (snowstorm, rainstorm), heavy freezing rain (ice storm), strong winds (tropical cyclone, windstorm), or wind transporting some substance through the atmosphere as in a dust storm, blizzard, sandstorm, etc.

19.203 STOVL

Short take-off and vertical landing is STOVL. It is a hybrid of vertical take-off and landing (VTOL) and short take-off and landing (STOL).

19.204 STRAP-ON MOTOR

A strap-on motor is a booster that is attached to the side of a main stage of a rocket.

19.205 STRATIFIED FLOW

When the Reynolds number is very low due to high viscosity, the continuum concept is perfectly valid and such a flow is termed stratified flow. Flow of tar, honey. etc. are stratified flows.

19.206 STREAKLINE

Streakline may be defined as the instantaneous loci of all the fluid elements that have passed the point of injection at some earlier time. Consider a continuous tracer injection at a fixed-point Q in space. The connection of all elements passing through the point Q over a period is called the streakline.

19.207 STREAM

A stream is a body of water with surface water flowing within the bed and banks of a channel.

19.208 STREAM AVAILABILITY

The availability of a fluid stream is called stream availability.

19.209 STREAM FUNCTION

Based on the streamline concept, a function ψ called stream function can be defined. The velocity components of a flow field can be obtained by differentiating the stream function. In terms of stream function, ψ, the velocity components of a two-dimensional incompressible flow are given as

$$V_x = \frac{\partial \psi}{\partial y}, \quad V_y = -\frac{\partial \psi}{\partial x}$$

If the flow is compressible, the velocity components become

$$V_x = \frac{1}{\rho}\frac{\partial \psi}{\partial y}, \quad V_y = -\frac{1}{\rho}\frac{\partial \psi}{\partial x}$$

It is important to note that the stream function is defined only for two-dimensional flows and the definition does not exist for three-dimensional flows. Even though some books define ψ for axisymmetric flows, they again prove to be equivalent to two-dimensional flows. We must realise that the definition of ψ does not exist for three-dimensional flows. This is because such a definition demands a single tangent at any point on a streamline, which is possible only in two-dimensional flows.

19.210 STREAMFLOW

Streamflow, which is also known as channel runoff, refers to the flow of water in natural watercourses such as streams and rivers.

19.211 STREAMLINE FLOW

A streamline flow or laminar flow is defined as one in which there are no turbulent velocity fluctuations. The definition of a streamline is such that at one instant in time streamlines cannot cross; if one streamline forms a closed curve, this represents a boundary across which fluid particles cannot pass.

19.212 STREAMLINED BODY

In a streamlined body, the skin friction drag accounts for the major portion of the total drag and the wake drag is very small.

19.213 STREAMLINES

Streamlines are imaginary lines in a fluid flow drawn in such a manner that the flow velocity is always tangential to it. Flows are usually depicted graphically with the aid of streamlines. Streamlines proceeding through the periphery of an infinitesimal area at some instant of time t will form a tube called a streamtube, which is useful in the study of fluid flow. Flow cannot cross a streamline and the mass flow between two streamlines is confined. Based on the streamline concept, a function ψ called stream function can be defined. The velocity components of a flow field can be obtained by differentiating the stream function.

19.214 STREAMLINING

Streamlining is to give a vehicle, etc. a long smooth shape so that it will move easily through air or water. Streamlining, in aerodynamics, is the contouring of an object, such as an aircraft body, to reduce its drag or resistance to motion through a stream of air.

19.215 STREAMTUBE

Streamlines proceeding through the periphery of an infinitesimal area at some time 't' form a tube called a streamtube.

19.216 STREAMWISE VORTICES

Vortices with their rotational axis along the streamlines are streamwise vortices.

19.217 STRONG OBLIQUE SHOCK

If the downstream Mach number becomes less than unity, then the shock is called strong oblique shock.

19.218 STRONG SHOCK

In a strong shock, p_2/p_1, the pressure ahead of and behind the shock is very large.

19.219 STROUHAL NUMBER

The Strouhal number represents the ratio of inertial forces due to the local acceleration of the flow to the inertial forces due to the convective acceleration. The Strouhal number is defined as $(n\,d)/U$, where n is the frequency of vortex shedding, d is the diameter of the cylinder, and U is the velocity of flow upstream of the cylinder. The Strouhal number can be important when analysing unsteady, oscillating flow problems.

19.220 SUBCRITICAL CIRCULATION

The circulation that positions the stagnation points in proximity is called subcritical circulation.

19.221 SUBCRITICAL OPERATION

The operating condition in which a detached shock is positioned ahead of the inlet and there is flow spillage is called subcritical operation.

19.222 SUBCRITICAL REYNOLDS NUMBER

The Reynolds number below which the entire flow is laminar is called the subcritical Reynolds number.

19.223 SUBLIMATION

Sublimation is the transition of a substance directly from the solid to the gaseous state without passing through the liquid state.

19.224 SUBMARINE

A submarine is a watercraft capable of independent operation underwater. It differs from a submersible, which has very limited underwater capability.

19.225 SUBMERGE

Submerge means to place under or cover with water or the like; plunge into water, inundate, etc.

19.226 SUBMERGED JET

A jet propagating through a medium at rest is called a submerged jet.

19.227 SUBSONIC FLOW

Subsonic aerodynamics is the study of fluid motion that is slower than the speed of sound. There are several branches of subsonic flow, but one special case arises when the flow is inviscid, incompressible, and irrotational. This case is called potential flow. For this case, the differential equations used are a simplified version of the governing equations of fluid dynamics, thus, making a range of quick and easy solutions available to the aerodynamicist.

19.228 SUBSONIC INLETS

An air-breathing engine installed in an aircraft or a missile must be provided with an air intake and a ducting system. We know that for turbojet engines, the airflow entering the compressor or fan must have a Mach number in the range of 0.4–0.7. Usually, the upper part of this range is suitable only for transonic compressors. For engines designed for a subsonic cruise, say Mach 0.8, the inlet must act as a diffuser with a gradual diffusion from Mach 0.8 to 0.6. Usually, part of this deceleration occurs upstream of the inlet entrance.

19.229 SUBSONIC JETS

Subsonic jets are those with Mach numbers between 0.3 and 1.0, are always correctly expanded, and develop with an included angle of about 10°.

19.230 SUBSTANTIVE RATE OF CHANGE

The rate of change of a property experienced by a material particle is termed the material or the substantive rate of change.

19.231 SUCROSE

Sucrose is common sugar. It is a disaccharide, a molecule composed of two monosaccharides: glucose and fructose. Sucrose is produced naturally in plants from which table sugar is refined. It has the molecular formula $C_{12} H_{22} O_{11}$.

19.232 SUCTION

The negative pressure is referred to as suction. It is essential to note that what is referred to as suction pressure is relative to some reference, such as the freestream pressure in the present case, and there is no question of suction in any absolute sense, that is, there can be no negative pressure. Also, suction is the colloquial term to describe the air pressure differential between areas. Removing air from a space results in a pressure differential. Suction pressure is therefore limited by external air pressure. Even a perfect vacuum cannot suck with more pressure than is available in the surrounding environment. Suctions can form on the sea, for example, when a ship founders.

19.233 SUCTION FORCE

Suction is often defined as a force that causes a fluid or a mixture to be drawn into an interior space. Some dictionaries more accurately define suction as a force or condition produced by a difference in pressures.

19.234 SUDDENLY EXPANDED FLOW

The sudden expansion of flow in both subsonic and supersonic regimes is an important problem with a wide range of applications. The use of a jet and a shroud configuration in the form of a supersonic parallel diffuser is an excellent application for sudden expansion problems.

19.235 SUGAR

Sugar is the generic name for sweet-tasting, soluble carbohydrates, many of which are used in food. Table sugar, granulated sugar, or regular sugar refers to sucrose, a disaccharide composed of glucose and fructose.

19.236 SUGARCANE

Sugarcane is a perennial grass of the family Poaceae, primarily cultivated for its juice from which sugar is processed.

19.237 SUGARCANE JUICE

Sugarcane juice is the liquid extracted from pressed sugarcane. It is consumed as a beverage in many places, especially where sugarcane is commercially grown, such as Southeast Asia, the Indian Subcontinent, North Africa, and Latin America.

19.238 SULPHURIC ACID

Sulphuric acid, also known as oil of vitriol, is a mineral acid composed of the elements sulphur, oxygen, and hydrogen with molecular formula H_2SO_4. It is a colourless, odourless, and viscous liquid that is miscible with water at all concentrations.

19.239 SUN

The Sun is the star at the centre of the Solar System. It is a nearly perfect sphere of hot plasma. The temperature of the Sun at its core is 27 million degrees but is only about 10 million degrees on its surface. The core of the Sun is considered to extend from the centre to about 0.2–0.25 of the solar radius. It is the hottest part of the Sun and the Solar System. It has a density of 150 g/cm^3 at the centre and a temperature of 15 million kelvin.

19.240 SUNLIGHT

Sunlight is a portion of the electromagnetic radiation given off by the Sun, in particular infrared, visible, and ultraviolet light. On Earth, sunlight is scattered and filtered through Earth's atmosphere and is obvious as daylight when the Sun is above the horizon.

19.241 SUNFLOWER OIL

Sunflower oil is the non-volatile oil pressed from the seeds of a sunflower.

19.242 SUPERCHARGER

A supercharger is an air compressor that increases the pressure or density of air supplied to an internal combustion engine. This gives each intake cycle of the engine more oxygen, letting it burn more fuel and do more work, thus increasing the power output.

19.243 SUPERCOOLED LIQUID

A liquid at a state where it is not about to vaporise is called a supercooled liquid. For example, water, say, at 1 atm and 20°C, is a supercooled liquid.

19.244 SUPERCRITICAL AEROFOIL

A supercritical aerofoil is designed primarily to delay the onset of wave drag in the transonic speed range.

19.245 SUPERCRITICAL CIRCULATION

The circulation that makes the stagnation points coincide and take a position outside the surface of the cylinder is called supercritical circulation.

19.246 SUPERCRITICAL FLUID

A supercritical fluid is any substance at a temperature and pressure above its critical point, where distinct liquid and gas phases do not exist, but below the pressure required to compress it into a solid.

19.247 SUPERCRITICAL OPERATION

The operation where the shock is swallowed is termed a supercritical operation.

19.248 SUPERCRITICAL REYNOLDS NUMBER

The limiting Reynolds number above which the entire flow is turbulent is called the supercritical Reynolds number.

19.249 SUPERHEATED VAPOUR

A vapour, which is not about to condense, is called a superheated vapour.

19.250 SUPERSONIC AIRCRAFT

A supersonic aircraft is an aircraft capable of supersonic flight, which can fly faster than the speed of sound (Mach number, $M > 1$).

19.251 SUPERSONIC COMBUSTION

Supersonic combustion demands that the fuel be injected and mixed with the oxidiser (usually air) flowing at a supersonic speed. The air stream enters the combustor at supersonic speed. Hydrogen fuel in a gaseous state is injected at an angle to the airflow direction. The combined perturbation of fuel injection, fuel-air mixing, and combustion results in a complex shock pattern. In hypersonic missile programmes, the aim is to establish supersonic combustion with kerosene as the fuel. In this conceptual engine, the hypersonic air stream decelerated in the intake enters the combustor at a supersonic Mach number of about 2. The kerosene is injected into this supersonic stream as tiny droplets. These droplets will position a bow shock ahead of them. Also, these droplets would vaporise rapidly. Therefore, the detached shock ahead of a droplet will undergo a rapid change in its shape and strength. There are many droplets in the combustor. We know that a shock causes a significant increase in entropy, and thus, the shocks in the combustor cause a large increase in entropy of the fuel-air mixture. This increase in entropy will induce a thorough mixing of the fuel and air, leading to high combustion efficiency.

19.252 SUPERSONIC DIFFUSERS

We know that in the supersonic flow regime a decrease in the area of a duct will result in deceleration of flow. This phenomenon is exploited in the form of a second throat in supersonic wind tunnels, that is, attaching a convergent-divergent duct at the end of the wind tunnel test section will decelerate the supersonic flow by forming a normal shock in the duct. To minimise the pressure loss, it is usual to position the shock just downstream of the second throat where the Mach number is slightly greater than one. The concept of using a convergent-divergent duct, which is usually referred to as a nozzle in supersonic flow studies, for decelerating supersonic flow to subsonic speed is termed a reverse nozzle diffuser in inlet studies.

19.253 SUPERSONIC EXPANSION

In a supersonic expansion process, the Mach lines are divergent. Consequently, there is a tendency to decrease the pressure, density, and temperature of the flow passing through them. In other words, an expansion is isentropic throughout. It is essential to note that the statement 'expansion is isentropic throughout' is not true always. Let us examine the centred and continuous expansion processes illustrated in Figures 19.12a and b.

We know that the expansion rays in an expansion fan are isentropic waves across which the change of pressure, temperature, density, and Mach

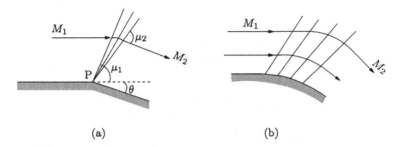

Figure 19.12 Cantered and continuous expansion process. (a) Cantered expansion and (b) continuous (simple) expansion.

number are small but finite. But when such small changes coalesce, they can give rise to a large change. One such point where such a large change of flow properties occurs due to the amalgamation of the effect due to a large number of isentropic expansion waves is point P, which is the vertex of the centred expansion fan in Figure 19.12a. As illustrated in Figure 19.12a, the pressure at the wall suddenly drops from p_1 to p_2 at the vertex of the expansion fan. Similarly, the temperature and density also drop suddenly at point P. The Mach number at P suddenly decreases from M_1 to M_2. The entropy change across the vertex of the expansion fan is

$$s_2 - s_1 = c_p \ln \frac{T_2}{T_1} - R \ln \frac{p_2}{p_1}$$

It is seen that the entropy change associated with the expansion process at point P is finite. Thus, the expansion process at point P is nonisentropic. Therefore, it is essential to realise that a centred expansion process is isentropic everywhere except at the vertex of the expansion fan, where it is nonisentropic.

But for the continuous expansion illustrated in Figure 19.12b, there is no sudden change in flow properties. Even at the wall surface, the properties change gradually as shown in Figure 19.12, due to the absence of any point such as P in Figure 19.12a, where all the expansion rays are concentrated. Therefore, continuous expansion is isentropic everywhere.

19.254 SUPERSONIC FLOW

Supersonic flow is the flow in which the flow velocity and the speed of sound are of comparable magnitude but $V > a$. The changes in Mach number M take place through substantial variation in both V and a.

A flow with a Mach number greater than unity is termed supersonic flow. In a supersonic flow, $V > a$ and the flow upstream of a given point remains unaffected by changes in conditions at that point.

A flow in the Mach number range $1.2 < M < 5$ is called supersonic flow.

Supersonic aerodynamic problems are those involving flow speeds greater than the speed of sound. Calculating the lift on the Concorde during cruises can be an example of a supersonic aerodynamic problem.

19.255 SUPERSONIC INLETS

For gas turbine engines, the flow leaving the engine inlet system should be subsonic, even when they fly at supersonic speeds. This is because it is difficult to pass a fully supersonic stream through the compressor without excessive shock losses.

19.256 SUPERSONIC JETS

Jets with a Mach number of more than 1.0 are called supersonic jets.

19.257 SUPERSONIC SHEAR LAYERS

A shear layer having a supersonic convective Mach number is termed supersonic shear layer.

Mixing in supersonic shear layers is critically dependent upon the compressibility effects in addition to the velocity and density ratios across the shear layer. The compressibility level is best described by a parameter called the convective Mach number.

19.258 SUPERSONIC STATIC PROBE

Locating the static pressure orifice at the proper location for measuring static pressure with reasonable accuracy poses serious problems in the probe design. In general, static probes with measuring orifice located sufficiently far downstream are suitable for the measurement of static pressure in supersonic flows. A typical probe, shown in Figure 19.13, measures static pressure within ±0.5% at Mach 1.6, if the pressure tapings are located more than ten tube diameters downstream of the shoulder. This distance increases with increase of Mach number in the supersonic regime.

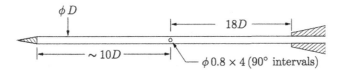

Figure 19.13 Supersonic static probe.

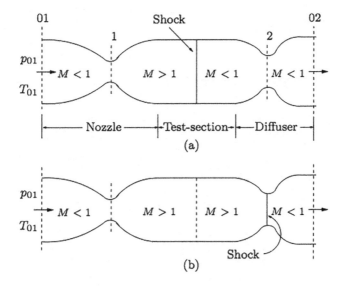

Figure 19.14 Schematic of a part of a supersonic tunnel circuit: (a) running with a normal shockinthetestsectionand(b)runningwithanormalshockatthediffuserthroat.

19.259 SUPERSONIC WIND TUNNEL DIFFUSERS

Supersonic wind tunnel diffusers are essentially convergent-divergent ducts attached to the end of the test section as shown in Figure 19.14.

19.260 SURFACE FORCES

All forces exerted on a boundary by its surroundings through direct contact are termed surface forces, e.g., pressure.

19.261 SURFACE TENSION

Liquids behave as if their free surfaces were perfectly flexible membranes having a constant tension σ per unit width; this tension is called the surface tension. It is important to note that this is neither a force nor a stress but a force per unit length. The value of surface tension depends on the nature of the fluid, the nature of the substance with which it is in contact at the surface, and the temperature and pressure.

19.262 SURGING

One of the most troublesome problems associated with wind tunnels is tunnel surging. It is a low-frequency vibration in velocity that may run as high as 5% of the dynamic pressure.

19.263 SURROUNDINGS WORK

The work done by or against the surroundings during a process is called surroundings work.

19.264 SUTHERLAND'S RELATION

Sutherland's theory of viscosity expresses the viscosity coefficient, at temperature T, as

$$\frac{\mu}{\mu_0} = \left(\frac{T}{T_0}\right)^{3/2} \frac{T_0 + S}{T + S}$$

where μ_0 is the viscosity at the reference temperature T_0, and S is a constant, which assumes the value 110 K for air. For air, Sutherland's relation can also be expressed as

$$\mu = 1.46 \times 10^{-6} \left(\frac{T^{3/2}}{T + 111}\right) (\text{Ns}) / \text{m}^2$$

where T is in kelvin. This equation is valid for the static pressure range of 0.01–100 atm, which is commonly encountered in atmospheric flight. The temperature range in which this equation is valid is from 0 to 3000 K.

19.265 SUSPENSION

A suspension is a heterogeneous mixture of a finely distributed solid in a liquid. The solid is not dissolved in the liquid as is the case with a mixture of salt and water.

19.266 SWEEP ANGLE

Sweep angle is usually measured as the angle between the line of 25% chord and a perpendicular to the root chord. The sweep of a wing affects the changes in the maximum lift, the stall characteristics, and the effects of compressibility.

19.267 SWEEP FORWARD

Sweep forward is a feature in which the tip of the wing is forward than the root. It is to be noted that only sweepback wings are commonly used and sweep forward is rare.

19.268 SWEEPBACK

Sweepback is a feature in which the lines of reference such as the leading and trailing edges of the wing are not normal to the flow direction, and the tip is aft of the root. A wing with a sweepback is also referred to as a swept wing.

19.269 SWEPT WING

A swept wing is a wing that angles either backwards or occasionally forward from its root rather than in a straight sideways direction. Swept wings have been flown since the pioneer days of aviation.

Wing sweep at high speeds was first investigated in Germany as early as 1935 by Albert Betz and Adolph Busemann, finding application just before the end of the Second World War. It has the effect of delaying the shock waves and accompanying aerodynamic drag rise caused by fluid compressibility near the speed of sound, improving performance. Swept wings are, therefore, almost always used on jet aircraft designed to fly at these speeds. Swept wings are also sometimes used for other reasons, such as low drag, low observability, structural convenience, or pilot visibility.

19.270 SWING

An oscillation is a particular kind of motion in which an object repeats the same movement over and over.

19.271 SWING ON TAKE-OFF

Swing is a tendency to turn to the nose side during take-off owing to some asymmetry of the aircraft.

19.272 SWIRL

To make or cause something to make fast circular movements.

19.273 SWIRL AROUND

To move around in a twisting, winding, gyrating motion. To encircle and move around someone or something in a twisting, winding, gyrating motion.

19.274 SYMMETRIC FLIGHT

When the aircraft velocity V is in the plane of symmetry, the flight is termed symmetric. There are three types of symmetrical flight: gliding, horizontal, and climbing. Among these, gliding is the only flight possible without the use of an engine.

19.275 SYMMETRICAL AEROFOIL

An aerofoil that has the same shape on both sides of its centreline (the centreline is thus straight) is called a symmetrical aerofoil. For example, the NACA 2412 aerofoil has a maximum camber of 2% located 40% (0.4 chords) from the leading edge with a maximum thickness of 12% of the chord. The NACA 0015 aerofoil is symmetrical—the 00 indicating that it has no camber.

19.276 SYPHON BAROMETER

The syphon barometer is essentially a U-tube of glass with one limb very much shorter than the other, as shown in Figure 19.15. The short limb is open to the atmosphere and the long limb end is closed and a Torricelli vacuum is created in this limb. A common scale is used for both limbs. If h_1 is the height of the mercury column in the closed limb and h_2 is that at the open limb, then the barometric height $h = h_1 - h_2$. Thus, the atmospheric pressure is measured easily with the syphon barometer.

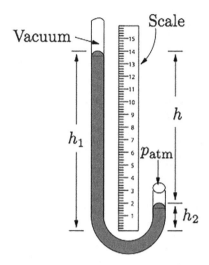

Figure 19.15 Syphon barometer.

19.277 SYRUP

A thick sweet liquid, often made by boiling sugar with water or fruit juice.

19.278 SYSTEM

A system is an identified quantity of matter or an identified region in space chosen for a study.

Chapter 20

Tail-First Aircraft to Typhoon

20.1 TAIL-FIRST AIRCRAFT

There are some designs in which the tail of the aircraft is located ahead of the wing. This type of configuration is termed tail-first or canard configuration. The tail in the front can hardly be called a tail, and this surface is commonly known as the fore plane.

20.2 TAILERONS

Tailerons are aircraft control surfaces that combine the functions of the elevator (used for pitch control) and the aileron (used for roll control). They are frequently used on tailless aircraft such as flying wings.

20.3 TAILLESS AIRCRAFT

Tailless aircraft are essentially flying machines with a large degree of sweep-back or even delta-shaped wings, as in Figure 20.1.

20.4 TAILPLANE

A tailplane, also known as a horizontal stabiliser, shown in Figure 20.2, is a small lifting surface located on the tail (empennage) behind the main lifting surfaces of a fixed-wing aircraft as well as other non-fixed-wing aircraft such as helicopters and gyroplanes. Not all fixed-wing aircraft have tailplanes. Canards, tailless, and flying wing aircraft have no separate tailplane, while in V-tail aircraft, the vertical stabiliser, rudder, and the tailplane and elevator are combined to form two diagonal surfaces in a 'V' layout.

DOI: 10.1201/9781003348405-20

Figure 20.1 Tailless aircraft.

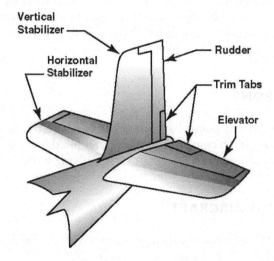

Figure 20.2 Tailplane.

20.5 TAILWIND

A tailwind is a wind that blows in the direction of travel of an object while a headwind blows against the direction of travel. A tailwind increases the object's speed and reduces the time required to reach its destination, while a headwind has the opposite effect.

20.6 TAKE-OFF

Take-off is the phase of flight in which an aerospace vehicle leaves the ground and becomes airborne. For an aircraft travelling vertically, this is known as a lift-off. For an aircraft that takes off horizontally, this usually involves starting with a transition from moving along the ground on a runway. For balloons, helicopters, and some specialised fixed-wing aircraft (VTOL aircraft such as the Harrier), no runway is needed.

20.7 TAPER RATIO

Taper ratio, λ, is the ratio of the tip chord, c_t, to the root chord, c_r, for the wing planforms with straight leading and trailing edges:

$$\lambda = \frac{c_t}{c_r}$$

The taper ratio affects the lift distribution of the wing. A rectangular wing has a taper ratio of 1.0, while the pointed tip delta wing has a taper ratio of 0.0. A tapered wing has a tip chord less than the root chord.

20.8 TARTARIC ACID

Tartaric acid is a white, crystalline organic acid that occurs naturally in many fruits, most notably in grapes. It is also found in bananas, tamarinds, and citrus fruits. Its salt, potassium bitartrate, commonly known as cream of tartar, develops naturally in the process of fermentation.

20.9 TAYLOR–COUETTE FLOW

The Taylor–Couette flow is the flow of a viscous fluid sheared in the gap between two rotating coaxial cylinders.

20.10 TAYLOR–PROUDMAN THEOREM

The relative velocity field does not vary in the direction of the rotation axis and the flow tends to be two-dimensional in planes perpendicular to the rotational axis. This is known as the Taylor–Proudman theorem.

20.11 TEA

Tea is an aromatic beverage prepared by pouring hot or boiling water over cured or fresh leaves of *Camellia sinensis*, an evergreen shrub native to China and East Asia.

20.12 TEMPERATURE

Temperature is a measure of the intensity of 'hotness' or 'coldness'; it is not easy to give an exact definition for it. By virtue of this motion, the molecules possess kinetic energy and this energy is sensed as the temperature of the solid, liquid, or gas. In the case of a gas in motion, it is called static temperature. Temperature's unit is kelvin (K) or degree Celsius (°C) in SI units. For all calculations in this book, the temperature will be expressed in kelvin, i.e., from absolute zero. At standard sea-level conditions, the atmospheric temperature is 288.15 K.

20.13 TEMPERATURE GAUGES USING FLUIDS

Heating of a fluid in a confined space results in its pressure rise. The fluid can fill the space completely or partially. The change of pressure because of the change of temperature of the fluid is used for temperature measurement. A typical temperature gauge using fluid is illustrated in Figure 20.3.

This gauge has a temperature pick-up unit, A, which is fitted into a suitable pocket at the point where the temperature is to be measured. The pick-up unit is filled with a working fluid, such as mercury. The unit is coupled using a long metal capillary tube, B, to a Bourdon pressure gauge, C, that is calibrated to measure temperature, instead of pressure, which it usually measures. During operation, as the temperature changes, the pressure in the fluid system also changes, so the Bourdon gauge pointer keeps moving on the scale.

20.14 TEMPERATURE MEASUREMENT IN FLUID FLOWS

In fluid flow analysis, we are interested in measuring the static and total temperatures of the flow at any specified location. In principle, for direct

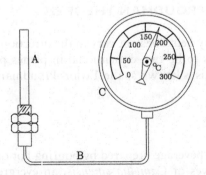

Figure 20.3 Temperature gauge using fluid.

measurement of the static temperature, the measuring device should travel at the velocity of the flow without disturbing the flow. But it is impossible to make such an infinitely thin device.

Direct measurement of adiabatic wall temperature can be used to determine the static temperature.

20.15 TEMPERATURE-MEASURING PROBLEMS IN FLUID

In the measurement of temperature in flowing fluids, certain types of problems are encountered irrespective of the nature of the device being used. These problems are mainly due to the heat transfer between the probe and its environment. Conduction errors and radiation errors are some of them.

20.16 TEMPERATURE SCALES

Temperature scales aim at using a common basis for temperature measurements. The temperature scales in SI and English systems are the Celsius scale (formerly called the centigrade scale) and the Fahrenheit scale.

20.17 TENDER COCONUT WATER

Tender coconut water is rich in potassium, and hence, helps keep the kidneys healthy.

20.18 TENSORS

Tensors require the specification of nine or more scalar components for a complete description. For example, stress, strain, and mass moment of inertia are tensor quantities.

20.19 TERMINAL VELOCITY

Terminal velocity is the maximum velocity (speed) attainable by an object as it falls through a fluid (air is the most common example). It occurs when the sum of the drag force and the buoyancy is equal to the downward force of gravity acting on the object. In a vertical dive, an aircraft will eventually reach a steady velocity called the terminal velocity.

20.20 TERMINAL VELOCITY DIVE

The extreme altitude at which the aircraft is diving vertically is called terminal velocity dive. In this case, the lift vanishes, the incidence is that of zero lift, and if the dive is undertaken from a sufficiently great height, the weight just balances the drag, the speed being the terminal speed, which may be five or six times the stalling speed. The attitude of the aircraft will then be about $-90°$.

20.21 TEST SECTION

The portion of the tunnel with constant flow characteristics across its entire section is termed the test or working section. The model to be tested is placed in the airstream, leaving the downstream end of the effuser, and the required measurements and observations are made. If rigid walls bind the test section, the tunnel is called a closed-throat tunnel. If it is binded by air at different velocities (usually at rest), the tunnel is called an open-jet tunnel. The test section is also referred to as working section.

20.22 TEST-SECTION NOISE

The test-section noise is defined as pressure fluctuations. Noise may result from unsteady settling chamber pressure fluctuations due to upstream flow conditions. It may also be due to weak unsteady shocks originating in a turbulent boundary layer on the tunnel wall.

20.23 THEOREMS OF COMPLEX NUMBERS

Theorem 1: The real part of the difference between two conjugate complex numbers is zero.
Theorem 2: The imaginary part of the sum of two conjugate complex numbers is zero.

20.24 THEOREM OF EQUIPARTITION OF ENERGY

The theorem of equipartition of energy of kinetic theory of gases states that 'each thermal degree of freedom of the molecule contributes $\frac{1}{2}kT$ to the energy of each molecule, or $\frac{1}{2}RT$ to the energy per unit mass of gas'.

20.25 THERMAL

Thermals are columns of rising air that are formed on the ground through the warming of the surface by sunlight. If the air contains enough moisture, the water will condense from the rising air and form cumulus clouds. Birds, such as raptors, vultures and storks, often use a thermal lift.

20.26 THERMAL BOUNDARY LAYER THICKNESS

The thermal boundary layer thickness is the distance across a boundary layer from the wall to a point where the flow temperature has essentially reached the freestream temperature.

20.27 THERMAL CONDUCTION

Thermal conduction is the transfer of internal energy by microscopic collisions of particles and the movement of electrons within a body. The colliding particles, which include molecules, atoms and electrons, transfer disorganised microscopic kinetic and potential energy, jointly known as internal energy. Conduction takes place in all phases: solid, liquid, and gas.

20.28 THERMAL CONDUCTIVITY

Thermal conductivity is a measure of heat flow in a given material.

20.29 THERMAL DIFFUSIVITY

In heat transfer analysis, thermal diffusivity is the thermal conductivity divided by density and specific heat capacity at constant pressure. It measures the rate of transfer of heat of a material from the hot end to the cold end. It has the SI-derived unit of m^2/s.

20.30 THERMAL ENERGY

Thermal energy is defined as the fraction of heat input that is converted to the net work output. It is a measure of the performance of a heat engine.

20.31 THERMAL ENERGY RESERVOIR

A thermal energy reservoir is a hypothetical body with a relatively large thermal energy capacity (mass×specific heat) that can supply or absorb finite amounts of heat without undergoing any change in its temperature. The atmospheric air and large quantities of water such as oceans, lakes, and rivers can be modelled as thermal energy reservoirs since they have large thermal energy storage capabilities or thermal masses. For example, megajoules of waste energy dumped in the ocean or in large rivers by power plants do not cause any significant change in the water temperature.

A thermal reservoir that supplies energy in the form of heat is called a source and the one that absorbs energy in the form of heat is called a sink. Thermal energy reservoirs are also referred to as heat reservoirs.

20.32 THERMAL EQUATION OF STATE

The thermal equation of state is $p = \rho R\,T$.

20.33 THERMAL EQUILIBRIUM

A system is in thermal equilibrium if the temperature is the same throughout the system.

20.34 THERMAL GRADIENT

Thermal gradient is defined as the ratio of the temperature difference and the distance between two points (equivalently, it's the change in temperature over a given length).

20.35 THERMAL INERTIA OF HOT-WIRE

It is the inability of the wire to get heated up or cooled down fully in time with the velocity fluctuations.

20.36 THERMAL PAINTS

These are special paints, which are applied on surfaces whose temperature distribution is required. When the temperature of the surface increases, establishing a distribution over the area, the colour of the paint, being a property dictated by the temperature level, changes. Thus, temperature distribution is determined by colour calibration against known surface temperatures.

20.37 THERMAL SHUNTING

An act of altering the measurement temperature by inserting a measurement transducer is termed thermal shunting. The problem of thermal shunting is comparatively more for RTDs than thermocouples since the physical size of an RTD is larger than the thermocouple.

20.38 THERMALLY PERFECT GAS

A gas is said to be thermally perfect when its internal energy and enthalpy are functions of temperature alone.

For a thermally perfect gas, $c_p = c_p(T)$ and $c_v = c_v(T)$; that is, both c_p and c_v are functions of temperature. But even though the specific heats c_p and c_v vary with temperature, their ratio, γ, becomes a constant and independent of temperature, that is, $\gamma = \text{constant} \neq \gamma(T)$.

20.39 THERMISTORS

A thermistor is a semiconductor device that has a negative temperature coefficient of resistance, in contrast to the positive coefficient displayed by most metals. Like the resistance temperature detector (RTD), the thermistor is also a temperature-sensitive resistor. While the thermocouple is the most versatile temperature transducer and RTD is the most stable temperature transducer, the best description of the thermistor is that it is the most sensitive temperature transducer. Of the above three temperature sensors, the thermistor exhibits the largest parameter change with temperature.

Thermistors are generally composed of semiconductor materials. Most thermistors have a negative temperature coefficient; that is, their resistance decreases with increasing temperature. The negative temperature coefficient can be as large as several percent per degree Celsius, allowing the thermistor to detect minute changes in temperature that could not be observed with a thermocouple or RTD.

20.40 THERMOCOUPLES

Thermocouples are widely used for temperature measurements in fluid streams. These are devices that operate on the principle that 'a flow of current in a metal accompanies a flow of heat'. This principle is popularly known as the Seebeck effect. In some metals, such as copper, platinum, iron, and chromal, the flow of current is in the direction of heat flow. In some other metals, such as constantan, alumel, and rhodium, the flow of current is in the direction opposite to that of the heat flow. These two groups are called dissimilar metals. Thermocouples consist of two dissimilar metals joined together at two points, one point being the place where the temperature is to be measured and the other point being a place where the temperature is known, termed the reference junction.

20.41 THERMODYNAMICS

Thermodynamics may be defined as the study of energy, its forms and transformations, and the interaction of energy with matter. Thermodynamics deals with the conservation of energy from one form to another. It is a branch of physics that deals with heat, work, and temperature, and their relation to energy, radiation, and physical properties of matter. The behaviour of these quantities is governed by the four laws of thermodynamics, which convey a quantitative description using measurable macroscopic physical quantities, but may be explained in terms of microscopic constituents by statistical mechanics. Thermodynamics applies to a wide variety of topics in science and engineering, especially physical chemistry, biochemistry, chemical engineering, and mechanical engineering, but also in other complex fields such as meteorology.

20.42 THERMODYNAMIC EQUILIBRIUM

When a system satisfies the conditions for all modes of equilibrium, it is said to be in thermodynamic equilibrium.

20.43 THERMODYNAMIC PROBABILITY

The number of processes by which the state of a physical system can be realised is thermodynamic probability. It is a system characterised by specific values of density, pressure, temperature, and other measurable quantities. Each given particle distribution is called a microstate of the system.

20.44 THERMODYNAMIC SYSTEM

A thermodynamic system is a quantity of matter of fixed identity, around which we can draw a boundary.

20.45 THERMODYNAMIC TEMPERATURE SCALE

A temperature scale that is independent of the properties of the substances that are used to measure temperature is called a thermodynamic temperature scale.

20.46 THERMOELECTRICITY

Thermoelectricity, also called the Peltier–Seebeck effect, is the direct conversion of heat into electricity or electricity into heat through two related mechanisms, the Seebeck effect and the Peltier effect.

20.47 THERMOMETER

Mercury-in-glass thermometer is the most common type of thermometer. A typical form of mercury-in-glass thermometer is illustrated in Figure 20.4.

It consists of a fine-bore glass tube, called the capillary tube, on the bottom of which is fused a thin-wall glass tube that is generally cylindrical but can sometimes be spherical. The tube and bulb are then completely filled with mercury by repeated heating and cooling, and finally, in the filled condition, it is brought to a temperature higher than its intended operational range. It is then sealed. Upon cooling, the mercury will contract and the mercury level in the capillary tube will fall to some level depending upon the prevailing ambient temperature. The scale is then etched onto, or fitted to, the glass tube.

Mercury is well suited for a thermometer since it does not wet the glass and it has a reasonable expansion. It has a freezing point of −38.86°C and a boiling point of 356°C. These temperatures, therefore, limit the range of the mercury-in-glass thermometer.

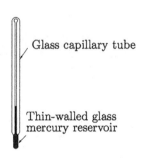

Glass capillary tube

Thin-walled glass mercury reservoir

Figure 20.4 Mercury-in-glass thermometer.

20.48 THERMOSPHERE

The thermosphere is the layer in the Earth's atmosphere directly above the mesosphere and below the exosphere. Within this layer of the atmosphere, ultraviolet radiation causes photoionisation/photo dissociation of molecules, creating ions; the thermosphere, thus, constitutes the larger part of the ionosphere.

20.49 THE RESISTANCE TEMPERATURE DETECTOR

Temperature can also be measured using resistance temperature detector (RTD) devices. They work on the principle that metals have marked temperature dependence.

20.50 THIN AEROFOIL THEORY

This theory is based on the assumption that the aerofoil is thin so that its shape is effectively that of its camber line and the camber line shape deviates only slightly from the chord line. In other words, the theory should be restricted to low angles of incidence.

This theory is applicable as long as the shocks are attached. This theory may be further simplified by approximating it by using the approximate relations for the weak shocks and expansion when the aerofoil is thin and is kept at a small angle of attack, that is, if the flow inclinations are small. This approximation will result in simple analytical expressions for lift and drag.

The difference between the shock-expansion theory and thin aerofoil theory is the following:

> In the shock-expansion theory, the shock is essentially a non-isentropic wave causing a finite increase of entropy. Thus, the total pressure of the flow decreases across the shock. But in the thin aerofoil theory, even the shock is regarded as an isentropic compression wave. Therefore, the flow across this compression wave is assumed to be isentropic. Thus, the pressure loss across the compression wave is assumed to be negligibly small.

In the thin aerofoil theory, the drag is split into drag due to lift, drag due to camber, and drag due to thickness. But the lift coefficient depends only on the mean angle of attack.

20.51 THIRD LAW OF THERMODYNAMICS

The entropy change of a pure crystalline substance at absolute zero temperature is zero is the third law of thermodynamics.

20.52 THOMSON EFFECT

The Thomson effect is the generation of reversible heat when an electrical current is sent through and subjected to a temperature gradient.

20.53 THOMSON'S VORTEX THEOREM

This theorem states that 'in a flow of inviscid and barotropic fluid, with conservative body forces, the circulation around a closed curve (material line) moving with the fluid remains constant with time', if the motion is observed from a non-rotating frame.

The vortex theorem can be interpreted as follows.

'The position of a curve c in a flow field, at any instant of time can be located by following the motion of all the fluid elements on the curve'. That is, Kelvin's circulation theorem states that, the circulation around the curve c at the two locations is the same.

20.54 THREE-DIMENSIONAL WING

For a three-dimensional wing, the span is finite, and the flow at the wing tips can easily establish a cross-flow, moving from a higher pressure to a lower pressure. For a wing experiencing positive lift, the pressure over the lower surface is higher than the pressure over the upper surface. This would cause a flow communication from the bottom to the top, at the wing tips. This tip communication would establish span-wise variations in the flow.

20.55 THREE-HOLE YAW PROBES

Three-hole yaw probes are used for measuring the flow direction in two-dimensional flows.

20.56 THROAT

The minimum area that divides the convergent and divergent sections of the duct is called the throat.

20.57 THROTTLING VALVE

Any kind of flow-restricting device that causes a significant pressure drop in the fluid is called a throttling valve. For example, the adjustable valve, capillary tube, and porous plug are throttling devices.

20.58 THRUST

Thrust is the force that opposes drag and enables the aeroplane to go forward. In a steady level flight, the thrust must be equal to the drag. To accelerate the aircraft, the thrust must be greater than the drag. In a climbing flight also, the thrust must be greater than the drag. The performance of the aircraft largely depends on the amount of thrust provided by its engines.

All propulsive systems deliver thrust as a result of giving momentum to air or other gases. Thus, the amount of thrust generated will be equal to the rate at which the momentum is imparted to the air.

If $\dot{m}$ is the mass flow rate of air and V is the velocity imparted to the air by the propulsion device, the thrust generated is

$$T = \dot{m} \, V$$

From this expression for thrust, it is evident that the thrust generated can be controlled by the adjustments of either mass flow rate or velocity.

Thrust is a reaction force described quantitatively by Newton's third law. When a system expels or accelerates mass in one direction, the accelerated mass will cause a force of equal magnitude but opposite direction to be applied to that system. The force applied on a surface in a direction perpendicular or normal to the surface is also called thrust. Force, and thus, thrust, is measured using the International System of Units (SI) in newton (symbol: N) and represents the amount needed to accelerate 1 kg of mass at the rate of 1 m/s. In mechanical engineering, force orthogonal to the main load (such as in parallel helical gears) is referred to as static thrust.

20.59 THRUST GENERATION

Thrust is the forward force that pushes the engine, and therefore, the aeroplane forward. Sir Isaac Newton discovered that for 'every action there is an equal and opposite reaction'. An engine uses this principle. The engine takes in a large volume of air. The air is decelerated and compressed and slowed down. The air is forced through many spinning blades. Mixing this air with jet fuel and burning the fuel–air mixture can increase the temperature of the burnt gas to as high as 3,000 K. The power of this gas is used to turn the turbine. Finally, when the air leaves, it is pushed backwards out of the engine. This causes the aeroplane to move forward.

20.60 THRUST AND MOMENTUM

All propulsive systems deliver thrust as a result of giving momentum to air or other gases. Thus, the amount of thrust generated will be equal to the rate at which momentum is imparted to the air.

20.61 THRUST-TO-WEIGHT RATIO

Thrust-to-weight ratio is a dimensionless ratio of thrust to weight of a rocket, jet engine, propeller engine, or a vehicle propelled by such an engine that is an indicator of the performance of the engine or vehicle.

20.62 THUNDER

Thunder is the sound that follows a flash of lightning and is caused by the sudden expansion of the air in the path of the electrical discharge.

20.63 THUNDERSTORM

A thunderstorm is a violent short-lived weather disturbance that is almost always associated with lightning, thunder, dense clouds, heavy rain or hail, and strong gusty winds.

20.64 TIDAL ENERGY

Tidal energy is power produced by the surge of ocean waters during the rise and fall of tides. It is a renewable source of energy.

20.65 TIMELINES

In modern fluid flow analysis, yet another graphical representation, namely a timeline, is used. When a pulse input is periodically imposed on a line of tracer source placed normal to a flow, a change in the flow profile can be observed. The tracer image is generally termed timeline. Timelines are often generated in the flow field to aid the understanding of flow behaviour such as the velocity and velocity gradient.

20.66 TIP-LOSS FACTOR

The tip-loss factor B is defined as

$$B = 1 - \frac{2C_T}{b}$$

where C_T is the thrust coefficient and b is the number of blades in the helicopter rotor.

20.67 TITANIUM TETRACHLORIDE

It is a liquid which when exposed to moist air produces copious fumes of titanium dioxide along with hydrochloric acid. This smoke though very dense is somewhat toxic and forms deposits, and hence, is not recommended for flow visualisation, except for very short periods of use. Stannic chloride also has the same property as titanium tetrachloride.

20.68 TITRATION

Titration is a technique where a solution of known concentration is used to determine the concentration of an unknown solution. Typically, the titrant (the know solution) is added from a buret to a known quantity of the analyse (the unknown solution) until the reaction is complete.

20.69 TOLLMIEN–SCHLICHTING WAVE

A Tollmien–Schlichting wave is a streamwise unstable wave which arises in a bounded shear flow. It is one of the more common methods by which a laminar bounded shear flow transitions to turbulence.

20.70 TOLUENE

Toluene, also known as toluol, is an aromatic hydrocarbon. It is a colourless, water-insoluble liquid with the smell associated with paint thinners. It is a mono-substituted benzene derivative, consisting of a methyl group (CH_3) attached to a phenyl group.

20.71 TONIC WATER

Tonic water is a carbonated soft drink in which quinine is dissolved.

Figure 20.5 Tornado.

20.72 TORNADO

A tornado is a violently rotating column of air that is in contact with both the surface of the Earth and a cumulonimbus cloud, or, in rare cases, the base of a cumulus cloud (Figure 20.5).

20.73 TORPEDO

A torpedo, shown in Figure 20.6, is a bomb, shaped like a long narrow tube that is fired from a submarine and explodes when it hits another ship.

20.74 TORR

A unit of pressure equivalent to 1 mm of mercury in a barometer and equal to 133.32 Pa.

20.75 TOTAL DRAG OF A WING

The total drag of a wing is the sum of profile drag and induced drag. If C_D is the total drag coefficient, then

$$C_D = C_{D_0} + C_{D_v}$$

Figure 20.6 Torpedo.

where C_{D_0} is the profile drag coefficient and C_{D_v} is the induced drag coefficient.

20.76 TOTAL ENERGY

The total energy of a system is the sum of chemical, electrical, kinetic, potential, mechanical, thermal, and nuclear energy.

20.77 TOTAL HEAD

The total mechanical energy divided by the weight of liquid is termed the total head.

20.78 TOTAL PRESSURE

The pressure that a fluid flow will experience if it is brought to rest isentropically is termed total pressure. The total pressure is also called impact pressure.

20.79 TOTAL TEMPERATURE MEASUREMENT

The measurement of total or stagnation temperature is simple in principle. The temperature inside a pitot tube, where the flow is brought to rest, should be the stagnation temperature, in both subsonic and supersonic flows, and can be measured by a thermometer placed inside the tube.

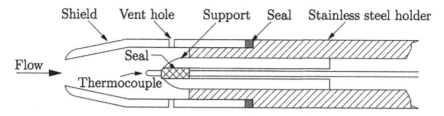

Figure 20.7 Total temperature probe.

In the absence of a wall, the stagnation temperature probe such as that shown in Figure 20.7 can be used to obtain the freestream total temperature T_∞.

This total temperature probe uses a thermocouple as the sensing element. The shield and support have to be designed to keep the rate of heat loss by conduction and radiation to a minimum. To replenish some of the lost energy, a small amount of flow through the probe is permitted by the vents provided.

When the flow is supersonic, there will be a detached shock standing in front of the probe. However, the measurement of T_∞ is unaffected by the presence of the shock, since the flow across a shock is adiabatic.

20.80 TOTALLY REVERSIBLE PROCESS

A process for which no irreversibilities occur either within the system or its surroundings is called a totally reversible process or simply reversible process. This process involves no heat transfer through a finite temperature difference, no friction or other dissipative effects, and no non-quasi-equilibrium changes.

20.81 TOWING TANK

A towing tank is essentially a channel in which water is made to flow in a controlled manner with any desired depth. A schematic of the towing tank used for the present study is shown in Figure 20.8. Two rails are provided for the carriage along and parallel to the length of the tank. The carriage is driven by four synchronous electric motors attached to each of the four wheels. The carriage speed can be varied over a wide range from a few mm/s to a few m/s with these meters.

A chronograph is provided to record the actual speed of the carriage. From the time versus displacement plot, the velocity and acceleration at any instant can be determined.

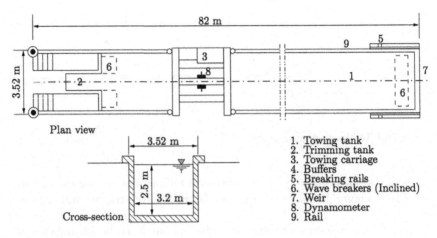

Figure 20.8 Towing tank.

20.82 TOXIC GASES

Toxic gases (or noxious gases) are gases that are harmful for living things. They can easily build up in confined working spaces when the production process uses noxious gases. They may also result in the biological chemical breakdown of a substance that is being stored in a tank.

20.83 TRACTOR

If a propeller is in front of the engine, it will cause tension in the shaft, and so will pull the aircraft – such an airscrew is called a tractor (refer Figure 20.9).

20.84 TRAILING EDGE FLAP

Trailing edge flap is a small auxiliary aerofoil located near the rear of the wing which can be deflected about a given line where it is hinged. The flap deflection modifies the geometry of the aerofoil, resulting in increased camber, leading to a higher lift.

20.85 TRAILING VORTEX DRAG

A finite wing spins the airflow near the tips into what eventually become two trailing vortices of considerable core size. The generation of these vortices requires a certain quantity of kinetic energy. The constant expenditure

Figure 20.9 Tractor.

of energy appears to the aerofoil as the trailing vortex drag also known as the induced drag.

20.86 TRANSITION FLOW

Transition flow ($0.10 < Kn < 5$): In this regime of flow, the fluid cannot be treated as continuum. At the same time, it cannot be treated as a free molecular flow since such a flow demands the intermolecular force of attraction to be negligible. Hence, it is a flow regime between continuum and free molecular flow. The kinetic theory of gases must be employed to adequately describe this flow.

20.87 TRANSITION POINT

Transition point may be defined as the end of the region at which the flow in the boundary layer on the surface ceases to be laminar and begins to become turbulent. It is essential to note that the transition from laminar to turbulent nature takes place over a length and not at a single point. Thus, the transition point marks the beginning of the transition process from laminar to turbulent nature.

20.88 TRANSITION REGION

This is the region where the centreline velocity begins to decay. This characteristic decay zone extends from about $5D_e$ to $10D_e$ downstream over which the turbulence changes from its annular to a somewhat pseudo-cylindrical distribution. As a result, the velocity difference between the ambient fluid and the high-speed core region of the jet decreases and attenuates the shear that supports the vortical rings in the jet, and thus, the velocity profiles become smoother with jet propagation.

The transition region is characterised by a growth of three-dimensional flow due to wave instability of the cores of the vortex rings. The merging of these distorted vortices produces large eddies which can remain coherent around the potential core region of the jet.

20.89 TRANSLATIONAL ENERGY

Translational energy – the translational kinetic energy of the centre of mass of the molecule is the source of this energy. A molecule has three geometric degrees of freedom in translation. As motion along the x -, y -, and z-coordinate directions constitutes the total kinetic energy, the molecule is also said to have three thermal degrees of freedom.

20.90 TRANSLATIONAL MOTION

Translational motion is the motion by which a body shifts from one point in space to another.

20.91 TRANSONIC FLOW

Transonic flow is the flow in which the difference between the flow velocity and the speed of sound is small compared to either V or a. The changes in V and a are of comparable magnitude.

The term transonic refers to a range of velocities just below and above the local speed of sound (generally taken as Mach 0.8–1.2). It is defined as the range of speeds between the critical Mach number, when some parts of the airflow over an aircraft become supersonic, and a higher speed, typically near Mach 1.2, when all of the airflow is supersonic. Between these speeds, some of the airflow is supersonic and some is not.

The exact range of speeds depends on the object's critical Mach number, but the transonic flow is seen at flight speeds close to the speed of sound (343 m/s at sea level), typically between Mach 0.8 and 1.2.

20.92 TRANSPORT PHENOMENA

The motion of molecules causes the transport of mass, momentum, and energy, popularly termed *transport phenomena*.

Transport phenomenon, in physics, is any of the phenomena involving the movement of various entities, such as mass, momentum, or energy, through a medium, fluid, or solid, by virtue of non-uniform conditions existing within the medium.

20.93 TRANSPORT PROPERTIES

The diffusion coefficient, viscosity coefficient, and thermal conductivity are known as transport properties. The properties characterising the transport of mass, momentum, and energy, respectively are diffusion coefficient, D; viscosity coefficient, μ; and the thermal conduction coefficient, K.

20.94 TRANSVERSE WAVE

In physics, it is a wave whose oscillations are perpendicular to the direction of the wave's advance.

20.95 TRIM DRAG

The tail surface producing L_t also produces a drag force, which is known as the trim drag. The trim drag may vary from 0.5% to 5% of the total drag of the aircraft.

20.96 TRIM TAB

Trim tab is a small-hinged surface in addition to the main tab. In practice, for a small adjustment to the trim, it is a common practice to provide a very small-hinged surface or trim tab in addition to the main elevator.

20.97 TRIMMED AIRCRAFT

An aircraft is said to be trimmed when the sum of the moments about the cg is zero.

20.98 TRIPLANE

A triplane, shown in Figure 20.10, is a fixed-wing aircraft equipped with three vertically stacked wing planes.

Figure 20.10 Triplane.

20.99 TRIPLE POINT

The triple point is the temperature and pressure at which solid, liquid, and vapour phases of a particular substance coexist in equilibrium. It is a specific case of thermodynamic phase equilibrium.

20.100 TROPICAL CYCLONE

A tropical cyclone is a rapidly rotating storm system characterised by a low-pressure centre, a closed low-level atmospheric circulation, strong winds, and a spiral arrangement of thunderstorms that produce heavy rain and/or squalls.

Tropical cyclones form only over warm ocean waters near the equator. To form a cyclone, warm, moist air over the ocean rises upwards from near the surface. As this air moves up and away from the ocean surface, it leaves less air near the surface.

20.101 TROPICAL WEATHER

Tropical climates are characterised by monthly average temperatures of 18°C or higher all the year round and feature hot temperatures. There are normally only two seasons in tropical climates, a wet season and a dry season. The annual temperature range in tropical climates is normally very small. The sunlight is intense.

20.102 TROPOPAUSE

The tropopause is the boundary in the Earth's atmosphere between the troposphere and the stratosphere. It is a thermodynamic gradient stratification layer, marking the end of the troposphere. It lies, on average, at 17 km above equatorial regions and about 9 km over the Polar Regions.

20.103 TROPOSPHERE

The layer of air above the Earth's surface up to 11 km altitude is called the troposphere.

The temperature-altitude variation in the troposphere is linear and may be expressed as

$$T = T_0 - \lambda\, z$$

where T_0 = 288 K is the standard sea-level temperature and λ is known as the lapse rate. In the international standard atmosphere, the lapse rate is assigned a value of 6.5 K/km.

20.104 TRUE AIRSPEED

The true airspeed (TAS; also KTAS, for knots true airspeed) of an aircraft is the speed of the aircraft relative to the air mass through which it is flying. It is important information for accurate navigation of an aircraft.

20.105 TSUNAMI

Tsunami (see Figure 20.11) (Japanese: 'harbour wave'), also called seismic sea wave or tidal wave, catastrophic ocean wave, is usually caused by a submarine earthquake, an underwater or coastal landslide, or a volcanic eruption.

20.106 TUFTS

Tufts, shown in Figure 20.12, are used to visualise flow fields in the speed range from 40 to 150 m/s. This visualisation method is usually employed to study boundary layer flow, wake flow, flow separation, stall spread, and so on.

Figure 20.11 Tsunami.

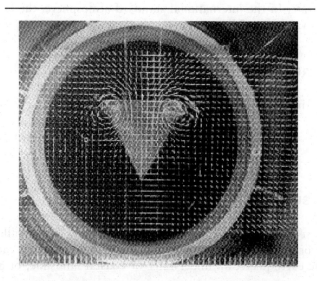

Figure 20.12 Vortices visualised with tufts.

20.107 TURBINES

The devices meant for converting fluid energy to mechanical energy are termed turbines.

The high-energy airflow coming out of the combustor goes into the turbine, causing the turbine blades to rotate. The turbines are linked by a shaft to turn the blades in the compressor and spin the intake fan at the front. This rotation takes some energy from the high-energy flow that is used to drive the fan and the compressor. The gases produced in the combustion chamber move through the turbine and spin its blades. The turbines of the jet engine spin around thousands of times per second. (For example, large jet engines operate around 10,000–25,000 rpm, while micro turbines spin as fast as 500,000 rpm.) They are fixed on shafts that have several sets of ball bearings in between them.

A turbine is a rotary mechanical device that extracts energy from a fluid flow and converts it into useful work. The work produced by a turbine can be used for generating electrical power when combined with a generator. A turbine is a turbomachine with at least one moving part called a rotor assembly, which is a shaft or drum with blades attached. Moving fluid acts on the blades so that they move and impart rotational energy to the rotor. Early turbine examples are windmills and waterwheels.

20.108 TURBINE PUMPS

Turbine pumps are special types of centrifugal pumps, which use turbine-like impellers with radially oriented teeth to move fluid.

20.109 TURBOCHARGER

A turbocharger, colloquially known as turbo, is a turbine-driven, forced induction device that increases an internal combustion engine's power output by forcing extra compressed air into the combustion chamber.

20.110 TURBOFAN ENGINES

Turbofan engines are the most widely used ones in aircraft propulsion. In turbofan engines, a large fan driven by the turbine forces a large amount of air through a duct surrounding the engine.

The turbofan or fanjet is a type of air-breathing jet engine that is widely used in aircraft propulsion. The word 'turbofan' is a portmanteau of 'turbine' and 'fan': the turbo portion refers to a gas turbine engine which achieves mechanical energy from combustion, and the fan is a ducted fan that uses the mechanical energy from the gas turbine to accelerate air rearwards. Thus, whereas all the air taken in by a turbojet passes through the combustion chamber and turbines, in a turbofan, some of that air bypasses these components. A turbofan, thus, can be thought of as a turbojet being used to drive a ducted fan, with both of these contributing to the thrust.

20.111 TURBOJETS

The turbojet engine is a reaction engine. In a reaction engine, expanding gases push hard against the front of the engine. The turbojet sucks in air and compresses or squeezes it. The gases flow through the turbine and make it spin. These gases bounce back and shoot out of the rear of the exhaust, pushing the plane forward.

The turbojet has no reciprocating parts, such as the piston-cylinder in the piston engine. Because of the absence of reciprocating parts, the wear and vibrations associated with the turbojet are insignificant. Another important advantage is that the turbojet produces significantly larger thrusts for a given weight at high speed compared with a piston engine. Furthermore, it will work efficiently close to and beyond the speed of sound, where propellers cannot be used.

20.112 TURBOPROPS

A turboprop engine is a jet engine attached to a propeller. The turbine at the back is turned by the hot gases, and this turns a shaft that drives the propeller. Turboprops power some small airliners and transport aircraft.

Like the turbojet, the turboprop engine consists of a compressor, combustion chamber, and turbine. The air and gas pressure is used to run the turbine, which then creates power to drive the compressor. Compared with a turbojet engine, the turboprop has better propulsion efficiency at flight speeds below about 800 km/h. Modern turboprop engines are equipped with propellers that have a smaller diameter but a larger number of blades for efficient operation at much higher flight speeds. To accommodate the higher flight speeds, the blades are scimitar shaped (shaped like a scimitar sword with increasing sweep along the leading edge) with swept-back leading edges at the blade tips. Engines featuring such propellers are called prop-fans. An artistic view of the turboprop engine is shown in Figure 20.13.

20.113 TURBOSHAFTS

This is another form of gas turbine engine that operates much like a turboprop system. It does not drive a propeller. Instead, it provides power for a helicopter rotor. The turboshaft engine is designed so that the speed of the helicopter rotor is independent of the rotating speed of the gas generator. This permits the rotor speed to be kept constant even when the speed of the generator is varied to modulate the amount of power produced.

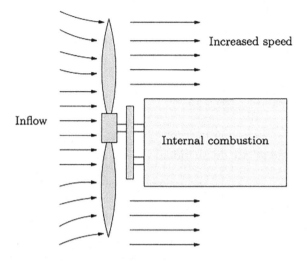

Figure 20.13 A view of turboprop engine.

20.114 TURBULENCE

Turbulent flow is usually described as a flow with irregular fluctuations. In nature, most of the flows are turbulent. Turbulent flows have characteristics which are appreciably different from those of laminar flows. We have to explain all the characteristics of turbulent flow to completely describe it. Incorporating all the important characteristics, turbulence may be described as 'a three-dimensional, random phenomenon, exhibiting multiplicity of scales, possessing vorticity, and showing very high dissipation'. Turbulence is described as a three-dimensional phenomenon. This means that even in a one-dimensional flow field the turbulent fluctuations are always three-dimensional. In other words, the mean flow may be one- or two- or three-dimensional, but turbulence is always three-dimensional. From the above, it is evident that turbulence can only be described and cannot be defined.

20.115 TURBULENCE FACTOR

The turbulence factor, TF, is the ratio of the effective Reynolds number, Re_e to the critical Reynolds number, Re_c

$$TF = \frac{Re_e}{Re_c}$$

For wind tunnels, the turbulence factor varies from 1.0 to about 3.0.

20.116 TURBULENCE NUMBER

The turbulence level for any given flow with a mean velocity $\bar{U}$ is expressed as a turbulence number n, defined as

$$n = 100 \; \frac{\sqrt{\overline{u'^2} + \overline{v'^2} + \overline{w'^2}}}{3\bar{U}}$$

where u', v' and w' are fluctuational velocities.

20.117 TURBULENT MIXING NOISE

The turbulent mixing noise is from both large-scale turbulence structures and fine-scale turbulence of the jet flow. The large-scale turbulence structures generate the dominant part of the turbulent mixing noise. The fine-scale turbulence is responsible for the background noise.

20.118 TURBULENT SPOT

The turbulent spot can be thought of as the 'building block' for a turbulent boundary layer flow. Spots first appear in the laminar boundary layer at the start of the laminar to the turbulent transition process. Their numbers increase in the flow direction, and they grow in size until individual spots merge with their neighbours to produce extensive regions of turbulent flow. The flow becomes fully turbulent once it is saturated with spots.

20.119 TURBULENT STRESSES

The terms, $\rho\overline{u'^2}$, $\rho\overline{v'^2}$, $\rho\overline{w'^2}$, $\rho\overline{u'v'}$, $-\rho\overline{v'w'}$, and $-\rho\overline{w'u'}$ are known as turbulent stresses.

20.120 TWO-PHASE FLOW

Two-phase flow is a flow in which two different aggregate states of a substance or two different substances are simultaneously present. The possible combinations include gaseous/liquid (see Gas content of fluid handled), gaseous/solid, and liquid/solid (see Solids transport).

20.121 TYNDALL EFFECT

The Tyndall effect is a phenomenon in which the particles in a colloid scatter the beams of light that are directed at them. This scattering makes the path of the light beam visible.

20.122 TYPES OF CHEMICAL ROCKETS

There are two types of chemical rockets, liquid-propellant rockets and solid-propellant rockets. In a liquid-propellant rocket, the fuel and the oxidiser are stored in the rocket in liquid form and pumped into the combustion chamber.

20.123 TYPES OF FLUID FLOW

There are six different types of fluid flow: (a) Steady and Unsteady, (b) Uniform and Non-Uniform, (c) Laminar and Turbulent, (d) Compressible and Incompressible, (e) Rotational and Irrotational, and (f) One-, Two-, and Three-Dimensional.

20.124 TYPES OF FLUIDS

Fluids are separated into five basic types: ideal fluid, real fluid, Newtonian fluid, non-Newtonian fluid, and ideal plastic fluid.

20.125 TYPES OF INTERNAL BALANCES

Two general groups of internal balances exist. The first group consists of the so-called box balance. These can be constructed from one solid piece of metal or can be assembled from several parts. Their main characteristic is that their outer shape most often appears cubic, such that the loads are transferred from the top to the bottom of the balance. The second type of internal balance is termed string balance. These balances have a cylindrical shape such that the loads are transferred from one end of the cylinder to the other in the longitudinal direction.

20.126 TYPES OF JET ENGINES

Jet engines can be broadly classed as turbojets, turboprops, turbofans, turboshafts, and ramjets.

20.127 TYPES OF JET FLOW

Based on the initial discharge conditions, jets can be categorised into three major groups: Free, wall, and surface jets, which discharge into an unbounded ambient fluid, near or at a solid surface, and near or at a free surface, respectively.

20.128 TYPES OF LIQUID PROPELLANTS

Most liquid chemical rockets use two separate propellants: a fuel and an oxidiser. Typical fuels include kerosene, alcohol, hydrazine and its derivatives, and liquid hydrogen. Many others have been tested and used. Oxidisers include nitric acid, nitrogen tetroxide, liquid oxygen, and liquid fluorine.

20.129 TYPES OF PLASMA ENGINES

Helicon double-layer thruster, Magnetoplasmadynamic thrusters, Hall effect thrusters, Electrode-less plasma thrusters, Variable Specific Impulse Magnetoplasma Rocket or VASIMR are all plasma engines.

20.130 TYPES OF ROCKETS

Solid-fuel rocket, liquid-fuel rocket, ion rocket, and plasma rocket are the major classification of rockets.

20.131 TYPES OF TRANSLATIONAL MOTION

Translational motion can be of two types: rectilinear and curvilinear.

20.132 TYPES OF WAVES

Different types of waves have different sets of characteristics. Based on the orientation of particle motion and direction of energy, waves are categorised as mechanical waves and electromagnetic waves.

20.133 TYPHOON

Typhoon is a mature tropical cyclone that develops between 180° and 100° E in the Northern Hemisphere.

This region is referred to as the Northwestern Pacific Basin and is the most active tropical cyclone basin on Earth, accounting for almost one-third of the world's annual tropical cyclones.

Chapter 21

Ultrapure Water to Upper Critical Reynolds Number

21.1 ULTRAPURE WATER

Ultrapure water is a term commonly used in the semiconductor industry to emphasise the fact that water is treated to the highest levels of purity for all contaminant types.

21.2 ULTRASONIC FLOW METERS

The fact that small-magnitude pressure disturbances travel through a fluid medium at a definite velocity, the speed of sound, relative to the fluid, is made use of in ultrasonic flow meters. The word ultrasonic refers to the fact that the pressure disturbances generally are short bursts of sine waves whose frequency is above the range audible to human hearing (about 20 kHz). A typical frequency may be about 10 MHz.

21.3 ULTRASONIC WAVES

Sound waves with a frequency above the audible range are called ultrasonic. Ultrasonic wave is defined as 'inaudible sound with high frequency for human', the frequency of which generally exceeds 20 kHz. These days, a sound wave, which is not intended to be heard, is also called an ultrasonic wave.

21.4 UNCERTAINTY

An uncertainty is a possible value that the error might take on in a given measurement. Since the uncertainty can take on various values over a range, it is basically a statistical variable. Uncertainty can be considered as a histogram of values.

DOI: 10.1201/9781003348405-21

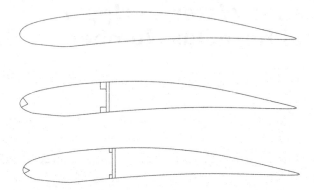

Figure 21.1 Undercambered aerofoil.

21.5 UNDERCAMBERED AEROFOIL

The undercambered aerofoil, shown in Figure 21.1, is what is often called a 'one-speed aerofoil', and a flat-bottom or semi-symmetrical foil is more adept at cutting through the air with less drag.

21.6 UNDEREXPANDED JETS

A jet is said to be underexpanded when the nozzle exit pressure (p_e) is higher than the backpressure (p_b). Since the nozzle exit pressure is higher than the backpressure, wedge-shaped expansion waves occur at the edge of the nozzle. These waves cross one another and are reflected from the boundaries of the jet flow field as compression waves. The compression waves again cross one another and are reflected on the boundaries of the jet as expansion waves. A sketch of an underexpanded jet is shown in Figure 21.2.

For an underexpanded jet, in addition to the expansion fan caused by the level of underexpansion, there will be another expansion effect due to the relaxation experienced by the jet on exiting the nozzle into a large space. Thus, the combined effect of these two causes establishes a stronger expansion fan than what the underexpansion level alone can establish. The essential difference between the moderately underexpanded and highly underexpanded jet is that at the end of the first shock cell, there is no Mach disc in the former but there is a Mach disc in the latter.

21.7 UNDEREXPANDED NOZZLE

When the exit pressure, p_e, is higher than the backpressure, p_b, the nozzle is said to be underexpanded.

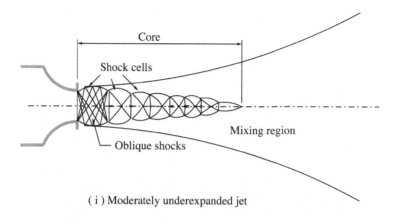

(i) Moderately underexpanded jet

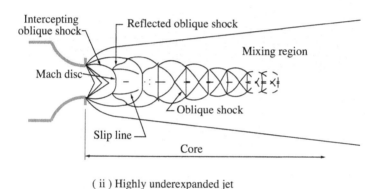

(ii) Highly underexpanded jet

Figure 21.2 Schematic of the waves prevailing in an underexpanded jet.

21.8 UNIFORM

The term uniform implies no change with location over a specified region.

21.9 UNIFORM FLOW

A flow in which both families' characteristics are straight, as in Figure 21.3, is called uniform flow. Thus, a uniform flow has parallel streamlines, as in the test section of a Hele-Shaw apparatus.

21.10 UNIFORM OPEN CHANNEL FLOW

A channel flow is termed uniform when the magnitude and direction of the velocity of the liquid remains invariant throughout the channel. This

Figure 21.3   Streamlines in the test-section.

condition is achieved only when the cross-section of the flow does not change along the length of the channel, and thus, the depth of liquid must be unchanged. Consequently, uniform flow is characterised by the liquid surface being parallel to the base of the channel.

A flow in which the liquid surface is not parallel to the base of the channel is said to be non-uniform.

21.11 UNIVERSAL GAS CONSTANT

The universal gas constant is a constant, 8.314 J/K equal to the product of the pressure and the volume of a one-gram molecule of an ideal gas divided by the absolute temperature.

21.12 UNLIKE REFLECTION

The reflection of an incident wave from a free boundary is called unlike reflection (opposite manner), that is, a shock wave reflects as an expansion wave and an expansion wave reflects as a shock (a compression wave).

21.13 UNSATURATED SOLUTIONS

Unsaturated solutions are solutions in which the amount of dissolved solute is less than the saturation point of the solvent (at that specific temperature gradient).

21.14 UNSTABLE

An aircraft that tends to move further away from the original position, when disturbed, is said to be unstable.

21.15 UNSTABLE AIR

Unstable air means that the weather might change quickly with very little warning. Unstable air leads to sudden thunderstorms.

21.16 UNSTEADY FLOW

A time-dependent flow is referred to as unsteady flow.

21.17 UNTWISTED AEROFOIL

When the axes of zero lift of all the profiles of the aerofoil are parallel, each profile meets the freestream wind at the same absolute incidence and the incidence is the same at every point on the span of the aerofoil, the aerofoil is said to be aerodynamically untwisted.

21.18 UPPER ATMOSPHERE

The upper atmosphere, the mesosphere, lies between the altitudes of about 50 and 80 km on Earth.

21.19 UPPER CRITICAL REYNOLDS NUMBER

The upper critical Reynolds number is that Reynolds number above which the entire flow is turbulent.

Chapter 22

Vacuum Cleaner to Vorticity

22.1 VACUUM CLEANER

A vacuum cleaner, also known simply as a vacuum or a hoover, is a device that causes suction to remove debris from floors, upholstery, draperies, and other surfaces. It is generally electrically driven.

22.2 VACUUM GAUGE

A vacuum gauge is a low-pressure gauge.

22.3 VACUUM SPACE

A vacuum is a space in which there is no matter or in which the pressure is so low that any particles in space do not affect any processes being carried on there. It is a condition well below normal atmospheric pressure and is measured in units of pressure (the pascal).

22.4 VALLEY GLACIERS

Valley glaciers, shown in Figure 22.1, are streams of flowing ice that are confined within steep-walled valleys, often following the course of an ancient river valley.

22.5 VANE ANEMOMETER

The vane anemometer is a classic wind speed meter, which is now very commonly used, not only outdoors, but also indoors.

DOI: 10.1201/9781003348405-22

Figure 22.1 Valley glacier.

22.6 VAPOUR-COMPRESSION CYCLE

The vapour-compression cycle involves four main components, namely a compressor, a condenser, an expansion valve, and an evaporator.

The refrigeration cycle process is as follows: the refrigerant enters the compressor as a vapour and is compressed to the condenser pressure. It leaves the compressor at a high temperature and cools down and condenses as it flows through the coils of the condenser by rejecting heat from the surrounding medium. Then it enters a capillary tube where its pressure and temperature drop drastically due to the throttling effect. The low-temperature refrigerant then enters the evaporator, where it evaporates by absorbing heat from the refrigerated space. The cycle is completed as the refrigerant leaves the evaporator and re-enters the compressor.

22.7 VARIABLE-HEAD METERS

All meters that measure the pressure drop across a restriction in a pipeline are variable-head meters. Several types of flow meters fall under the category of variable-head meters. Such devices are also called flow-obstruction meters or differential-pressure meters.

22.8 VECTOR QUANTITIES

Vector quantities require, in addition to magnitude, a complete directional specification for their description. Usually, three values associated with orthogonal directions are used to specify a vector. These quantities are called scalar components of a vector. For example, velocity is a vector quantity.

22.9 VEGETABLE OILS

Vegetable oils are triglycerides extracted from plants. Some of these oils have been part of human culture for millennia. Edible vegetable oils are used in food, both in cooking and as supplements. Many oils, edible and otherwise, are burned as fuel, such as in oil lamps and as substitutes for petroleum-based fuels.

22.10 VELOCITY

The speed at which something moves in a particular direction.

22.11 VELOCITY DISTRIBUTION FUNCTION

The velocity distribution function $f(r, C)$ is defined as the number of molecules per unit volume of physical space at r, with velocities per unit volume of velocity space at C.

22.12 VELOCITY GRADIENT

The difference in velocity between adjacent layers of the fluid is known as a velocity gradient.

22.13 VELOCITY MEASUREMENT

Some of the popularly used techniques are velocity measurement through pressure measurements, through optical properties like Doppler shift using laser Doppler anemometer, through measurement of the vortex-shedding frequency, and through the heat transfer principle using a hot-wire anemometer.

22.14 VELOCITY POTENTIAL

For irrotational flows (the fluid elements in the field are free of angular motion), there exists a function ϕ called velocity potential or potential function. For two-dimensional flows, ϕ must be a function of x, y, and t. The velocity components are given by

$$V_x = \frac{\partial \phi}{\partial x}, \; V_y = \frac{\partial \phi}{\partial y}$$

These relations between the stream function and potential function are given by the equations

$$\frac{\partial \psi}{\partial y} = \frac{\partial \phi}{\partial x}, \; \frac{\partial \psi}{\partial x} = -\frac{\partial \phi}{\partial y}$$

which are the famous Cauchy–Riemann equations of the complex-variable theory. It can be shown that the lines of constant ψ or potential lines form a family of curves which intersect the streamlines in such a manner as to have the tangents of the respective curves always at right angles at the point of intersection. Hence, the two sets of curves given by ψ = constant and ϕ = constant form an orthogonal grid system or flow net.

Unlike the stream function, the potential function exists for three-dimensional flows also. This is because there is no condition such as the local flow velocity must be tangential to the potential lines imposed in the definition of ψ. The only requirement for the existence of ϕ is that the flow must be potential.

22.15 VELOCITY OF SOUND

A sound wave is a weak compression wave across which only infinitesimal changes in flow properties occur, that is, across these waves, there will be only infinitesimal pressure variations.

The velocity of sound is used as the limiting value for differentiating the subsonic flow from the supersonic flow. Flows with velocity more than the speed of sound are called supersonic flows and those with velocities less than the speed of sound are called subsonic flows. Flows with velocities close to the speed of sound are classified under a special category called transonic flows.

22.16 VENA CONTRACTA

Vena contracta is the point in a fluid stream where the diameter of the stream is the least, and fluid velocity is at its maximum, such as in the case of a stream issuing out of a nozzle (orifice). The effect is also observed in flow from a tank into a pipe or a sudden contraction in pipe diameter.

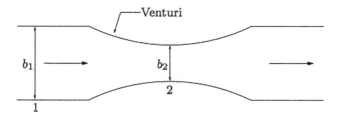

Figure 22.2 Flow through a Venturi in a channel.

22.17 VENOM

This is a poisonous liquid that some snakes, spiders, etc. produce when they bite or sting a person.

22.18 VENTURI

A Venturi is a constriction as shown in Figure 22.2. The mid-section of the Venturi with the minimum cross-sectional area is known as the throat and the discharge per unit area of the throat is the maximum.

22.19 VENTURI EFFECT

The Venturi effect is the reduction in fluid pressure that results when a fluid flows through a constricted section (or choke) of a pipe. The Venturi effect is named after its discoverer, Giovanni Battista Venturi.

22.20 VENTURI METER

A Venturi meter constricts the flow in a smooth throat and pressure sensors measure the differential pressure before and within the constriction. The reduction in pressure due to the increased velocity in the constriction allows the calculation of the velocity and flow rate from Bernoulli's equation.

22.21 VERTICAL STABILISER

A vertical stabiliser, vertical stabiliser, or fin, is a structure designed to reduce the aerodynamic sideslip and provide directional stability. They are most commonly found on vehicles such as aircraft or cars. It is analogous to a skeg on boats and ships. Other objects such as missiles or bombs utilise them too. They are typically found on the aft end of the fuselage or body.

22.22 VERTICAL WIND TUNNEL

A vertical wind tunnel (VWT) is a wind tunnel which moves air up in a vertical column.

22.23 VIBRATION

Vibration is a mechanical phenomenon whereby oscillations occur about an equilibrium point. The word comes from the Latin word *vibrationem* ('shaking, brandishing'). The oscillations may be periodic, such as the motion of a pendulum, or random, such as the movement of a tyre on a gravel road.

22.24 VIBRATIONAL ENERGY

In vibrational energy, the molecules and atoms are vibrating with respect to an equilibrium location within the molecule. For a diatomic molecule, this vibration may be modelled by a spring connecting the two atoms. Thus, the molecule has vibrational energy. The sources of this vibrational energy are (a) the kinetic energy of the linear motion of the atoms as they vibrate back and forth and (b) the potential energy associated with the intramolecular force. Therefore, although a diatomic molecule vibrates along one direction, namely the internucleus axis only, and has only one geometric degree of freedom, it has two thermal degrees of freedom because of the contribution of kinetic and potential energies. For polyatomic molecules, the vibrational motion is more complex and numerous fundamental vibrational modes can exist with a consequent large number of degrees of freedom.

22.25 VIBRATIONAL RATE EQUATION

Vibrational rate equation for a moving fluid element is

$$\frac{De_{vib}}{Dt} = \frac{1}{\tau}\left(e_{vib}^{eq} - e_{vib}\right)$$

where e_{vib}^{eq} is the equilibrium value of vibrational energy per unit mass of gas and e_{vib} is the local nonequilibrium value of vibrational energy per unit mass of gas, and τ is the vibrational relaxation time.

22.26 VINEGAR

Vinegar is an aqueous solution of acetic acid and trace compounds that may include flavourings. Vinegar typically contains 5%–8% acetic acid by volume. Usually, acetic acid is produced by double fermentation converting simple sugars to ethanol using yeast and ethanol to acetic acid by the acetic acid bacteria.

22.27 VISCOMETER

A viscometer is an instrument used to measure the viscosity of a fluid. For liquids with viscosities, which vary with flow conditions, an instrument called a rheometer is used. Thus, a rheometer can be considered a special type of viscometer. Viscometers only measure less than one flow condition.

22.28 VISCOSITY

The property, which characterises the resistance that a fluid offers to applied shear force, is termed viscosity. This resistance, unlike for solids, does not depend upon the deformation itself but on the rate of deformation. Viscosity is often regarded as the stickiness of a fluid, and its tendency is to resist sliding between layers.

The viscosity of a fluid is a measure of the reluctance of the fluid to flow when subjected to a shearing force. Basically, it is the property of a fluid which relates applied stress to the resulting strain rate.

The viscosity of gases increases with an increase in temperature because there is greater molecular activity and momentum transfer at higher temperatures.

In the case of liquids, the molecular cohesion between molecules plays a major role in affecting viscosity. At higher temperatures, molecular cohesion diminishes, and although the momentum transfer between layers increases, the net result is a decrease in liquid viscosity with increasing temperature.

22.29 VISCOSITY COEFFICIENT

Viscosity coefficient is the tangential force per unit area.

22.30 VISCOUS INCOMPRESSIBLE FLOWS

The theory of viscous incompressible fluids assumes fluid density to be constant. It finds widespread application in the flow of liquids and the flow of air at low velocity. The phenomena involving viscous forces flow separation and eddy flows are studied with the help of this theory.

22.31 VISUALISATION TECHNIQUES

The general principle for flow visualisation is to render the 'fluid elements' visible either by observing the motion of suitable selected foreign materials added to the flowing fluid or by using an optical pattern resulting from the variation of the optical properties of the fluid, such as refractive index, due to the variation of the properties of the flowing fluid itself.

22.32 VISUALISING COMPRESSIBLE FLOWS

For visualising compressible flows, optical flow visualisation techniques are commonly used. Interferometer, Schlieren, and shadowgraph are the three popularly employed optical flow visualisation techniques for visualising shocks and expansion waves in supersonic flows.

22.33 VODKA

Vodka is a clear distilled alcoholic beverage from Europe. It has different varieties originating in Poland, Russia, and Sweden.

22.34 VOLATILE

Volatile means evaporating rapidly or passing off readily in the form of vapour. Acetone is a volatile solvent.

22.35 VOLATILITY

In chemistry, volatility is a material quality, which describes how readily a substance vaporises. At a given temperature and pressure, a substance with high volatility is more likely to exist as a vapour, while a substance with low volatility is more likely to be a liquid or solid.

22.36 VOLCANO

A volcano, seen in Figure 22.3, is a rupture in the crust of a planetary-mass object, such as Earth, that allows hot lava, volcanic ash, and gases to escape from a magma chamber below the surface. On Earth, volcanoes are most often found where tectonic plates are diverging or converging and most are found underwater.

Volcanoes are categorised into three main categories: active, dormant, and extinct.

Figure 22.3 Volcano.

An active volcano is the one which has recently erupted, and there is a possibility that it may erupt soon.

A dormant volcano is the one that has not erupted in a long time but there is a possibility that it can erupt in the future.

An extinct volcano is the one which has erupted thousands of years ago and there's no possibility of an eruption.

22.37 VOLUME

Volume is the quantity of three-dimensional space occupied by a liquid, solid, or gas.

22.38 VOLUME MODULUS OF ELASTICITY

Volume modulus of elasticity, E, is a quantitative measure of compressibility. It is defined as

$$\Delta p = -E \frac{\Delta V}{V_i}$$

where Δp is the change in static pressure, ΔV is the change in volume, and V_i is the initial volume.

22.39 VOLUMETRIC FLOW RATE

The volumetric flow rate (also known as volume flow rate, rate of fluid flow, or volume velocity) is the volume of fluid that passes per unit time.

22.40 VOLUMETRIC FLUX

In fluid dynamics, the volumetric flux is the rate of volume flow across a unit area.

22.41 VORTEX

Vortex is a fluid flow in which the streamlines are concentric circles. The vortex motions encountered in practice may in general be classified as free vortex or potential vortex and forced vortex or flywheel vortex.

Vortex is a flow field in which the fluid elements translate along circular orbits.

A vortex in which the movement of fluid elements is pure translation is termed free or potential vortex for which the circulation is finite whereas the vorticity is zero.

If the elements, in addition to moving along circular orbits, spin about their axes, the vortex is termed a forced vortex, for which both circulation and vorticity are finite.

The specific difference between the pure and forced vortices is that the vorticity for free vortex is zero, but for forced vortex, the vorticity is finite.

A vortex is a flow system in which a finite area in a plane normal to the axis of a vortex contains vorticity.

22.42 VORTEX DYNAMICS

Vortex dynamics is a natural paradigm for the field of chaotic motion and modern dynamical system theory.

22.43 VORTEX FILAMENT

A very thin vortex tube is referred to as a vortex filament.

22.44 VORTEX GENERATORS

Vortex generators are small aerofoils of large camber so placed as to introduce swirling motions in the boundary layer. This energising of the boundary layer tends to prevent separation.

22.45 VORTEX-INDUCED VIBRATION

In fluid dynamics, vortex-induced vibrations (VIV) are motions induced on bodies interacting with an external fluid flow, produced by, or the motion producing, periodic irregularities on this flow. A classic example is the VIV of an underwater cylinder. How this happens can be seen by putting a cylinder into the water (a swimming pool or even a bucket) and moving it through the water in a direction perpendicular to its axis. Since real fluids always present some viscosity, the flow around the cylinder will be slow while in contact with its surface forming a so-called boundary layer. At some point, however, that layer can separate from the body because of its excessive curvature. A vortex is then formed changing the pressure distribution along the surface.

When the vortex does not form symmetrically around the body (with respect to its midplane), different lift forces develop on each side of the body, thus leading to motion transverse to the flow. This motion changes the nature of the vortex formation in such a way as to lead to a limited motion amplitude (differently, than, from what would be expected in a typical case of resonance). This process then repeats until the flow rate changes substantially.

22.46 VORTEX MOTION

Vortex is a fluid flow in which the streamlines are concentric circles. The vortex motions encountered in practice may in general be classified as free vortex or potential vortex and forced vortex or flywheel vortex. The streamlined pattern for a vortex may be represented as concentric circles.

22.47 VORTEX NAME

A vortex is named after the axis about which it is rotating. A vortex rotating about the streamline (with streamline as the axis of rotation) is called a stream-wise vortex.

22.48 VORTEX PAIR

A vortex pair is two vortices of equal strength but opposite nature (one rotating clockwise and the other rotating counterclockwise).

22.49 VORTEX RING

A vortex ring, also called a toroidal vortex, is a torus-shaped vortex in a fluid or gas; that is, a region where the fluid mostly spins around an imaginary axis line that forms a closed loop. The dominant flow in a vortex ring is said to be toroidal, more precisely poloidal.

Figure 22.4 Vortex shedding.

22.50 VORTEX SHEDDING

Vortex shedding is an oscillating flow, as illustrated in Figure 22.4, that takes place when a fluid such as air or water flows past a bluff (as opposed to streamlined) body at certain velocities depending on the size and shape of the body.

22.51 VORTEX SIZE

The size of a vortex formed is proportional to the radius of curvature of the edge from which it is shed. For example, the size of the vortex shed from the extremities of the minor-axis of an elliptical opening is proportional to the semi-major axis.

22.52 VORTEX SPEED

The vortex speed is adjustable and usually ranges from 100 to 3,200 rpm.

22.53 VORTEX-SHEDDING FLOW METERS

The shedding frequency of the vortex behind a blunt body kept in a steady flow is made use of for the measurement of flow in vortex-shedding flow meters.

Some typical vortex-shedding shapes used in flow meters and the flow meter principle are shown in Figure 22.5.

The vortices shed cause alternative forces or local pressures on the shedder. Strain gauge or piezoelectric methods can be employed to detect these forces. Hot film thermal anemometer sensors embedded in the shoulder of the body can detect periodic flow-velocity fluctuations. The interruption of ultrasonic beams by the passing vortices may be used for counting the vortices. The differential pressures induced by the vortices may be detected by making use of oscillations created by the pressure in a small-caged bellow hose motion with a magnetic proximity pickup.

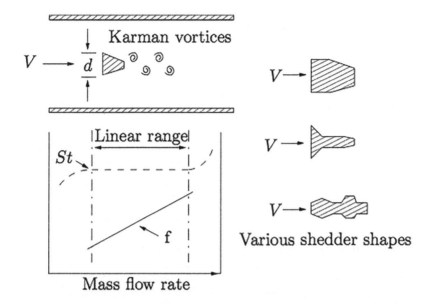

Figure 22.5 Vortex-shedding body and flow meter principle.

22.54 VORTEX-SHEDDING TECHNIQUE

Measurement of low velocities, below 2 m/s, is difficult to do with instruments like a pitot-static probe in conjunction with a manometer. For measuring such low velocities, vortex-shedding techniques could be used.

22.55 VORTEX TUBE

The axis of a vortex, in general, is a curve in space, and area S is a finite size. It is convenient to consider that the area S is made up of several elemental areas. In other words, a vortex consists of a bundle of elemental vortex lines or filaments. Such a bundle is termed a vortex tube and is bounded by vortex filaments.

Vortex-tube cabinet cooling systems have been solving problems associated with overheated electronic control systems for decades. Cabinet coolers provide advantages for everyone involved in the purchasing cycle – engineers, maintenance and plant personnel and purchasing agents. Paired with a source of clean compressed air and a water/dirt filter separator, vortex-tube coolers provide a long-lasting cooling alternative.

Vortex tubes, which are the foundation of these coolers, use compressed air to provide cold air into an enclosure. A typical vortex tube is shown in Figure 22.6.

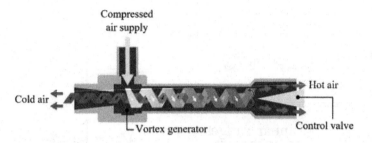

Figure 22.6 Vortex tube.

22.56 VORTICITY

Vorticity is the circulation per unit area. It is twice the angular velocity.

In continuum mechanics, vorticity is a pseudovector field that describes the local spinning motion of a continuum near some point as would be seen by an observer located at that point and travelling along with the flow.

It is an important quantity in the dynamical theory of fluids and provides a convenient framework for understanding a variety of complex flow phenomena, such as the formation and motion of vortex rings.

Chapter 23

Wake to Working Fluid

23.1 WAKE

The separated flow behind an object, shown in Figure 23.1, is referred to as wake. It is the separated region behind an object (usually a bluff body) where the pressure loss is severe. It is essential to note that what is meant by pressure loss is total pressure loss and there is nothing like static pressure loss. Depending on the Reynolds number level, the wake may be laminar or turbulent.

23.2 WAKE BLOCKING

A lateral constraint to flow pattern about the wake is known as wake blocking. This effect increases with increase of wake size. For closed test section, wake blocking increases the drag of the model. Wake blocking is usually negligible for open test sections.

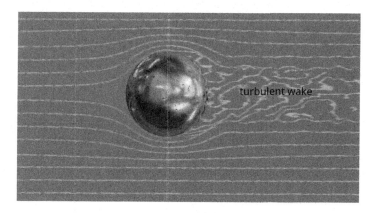

Figure 23.1 Wake behind a cylinder.

DOI: 10.1201/9781003348405-23

479

23.3 WAKE SURVEY

Wake surveys are a common method for measuring profile drag.

23.4 WAKE TURBULENCE

Wake turbulence is a disturbance in the atmosphere that forms behind an aircraft as it passes through the air.

It includes various components, the most important of which are wingtip vortices and jet wash. Jet wash refers to the rapidly moving gases expelled from a jet engine; it is extremely turbulent but of short duration. Wingtip vortices, however, are much more stable and can remain in the air for up to 3 min after the passage of an aircraft. It is, therefore, not true turbulence in the aerodynamic sense, as true turbulence would be chaotic. Instead, it refers to the similarity to atmospheric turbulence as experienced by an aircraft flying through this region of disturbed air.

23.5 WALL JET

A wall jet is a flow created when fluid is blown tangentially along a wall.

23.6 WALL-FRICTION VELOCITY

It is defined as

$$U^* = \sqrt{\frac{\tau_w}{\rho}}$$

where τ_w is the wall shear stress and ρ is the flow density.

23.7 WASHOUT

A washout or decrease of angle of incidence towards the wing tips. This means that when the centre portions of the wings are at their stalling angle, the outer portions are well below the angle, and therefore, the aileron will function in the normal way. The disadvantage of this arrangement is that the washout must be considerable to have any appreciable effect on the control and the result will be a corresponding loss of lift from the outer portion of the wing in normal flight.

23.8 WATER

Water is a substance composed of the chemical elements hydrogen and oxygen existing in gaseous, liquid, and solid states. It is one of the most plentiful and essential compounds. A tasteless and odourless liquid at room temperature, it has an important ability to dissolve many other substances.

23.9 WATER CYCLE

The water cycle shows the continuous movement of water within the Earth and atmosphere. It is a complex system that includes many different processes. Liquid water evaporates into water vapour, condenses to form clouds, and precipitates back to the Earth in the form of rain and snow.

23.10 WATER FLOW CHANNEL

Water flow channel is a simple duct through which water is made to flow with uniform velocity over a length of the channel. The uniform flow portion can be used to visualise the flow past objects.

Schematic diagram of a water flow channel of a rectangular cross-section is shown in Figure 23.2. Water from the chamber spills over the inclined plate and is conditioned using an array of wire meshes before reaching the test section. It is essential to ensure that the flow quality is fairly uniform in the test section. For this, a colour dye may be injected at specified locations across the test-section width. If the dye streaks in the test section are parallel and smooth, the quality of flow can be taken as good for any visualisation study. The velocity of the flow may be measured using the floating-particle method technique over the length of the test section.

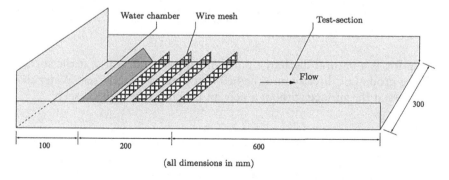

(all dimensions in mm)

Figure 23.2 Schematic diagram with the dimensions of the water flow channel.

23.11 WATER CONTENT IN HUMAN BODY

Up to 60% of the human adult body is water. According to H.H. Mitchell, *Journal of Biological Chemistry* 158, the brain and heart are composed of 73% water and the lungs have about 83% water. The skin contains 64% water, muscles and kidneys contain 79%, while the bones contain 31% water.

23.12 WATER IN HUMAN BLOOD

Blood is composed of 90% water.

23.13 WATER JET

Cutting with a water jet is a method of engineering for cutting objects using the energy from high-speed, high-density, and ultra-high-pressure water.

23.14 WATER POLLUTION

Water pollution is the contamination of water bodies, usually as a result of human activities. Water pollution occurs when harmful substances – often chemicals or microorganisms – contaminate a stream, river, lake, ocean, aquifer, or other body of water, degrading water quality and rendering it toxic to humans or the environment.

23.15 WATER RETENTION IN SOIL

Water retention in the soil can be understood as the water retained by the soil after it runs through the soil pores to join water bodies such as ground-water or surface streams.

23.16 WATER SPRING

A spring is a natural discharge point of subterranean water at the surface of the ground or directly into the bed of a stream, lake, or sea. Water that emerges at the surface without a perceptible current is called a seep.

23.17 WATER STREAM

A stream is a body of water with surface water flowing within the bed and banks of a channel. The flow of a stream is controlled by three inputs:

surface water, subsurface water and groundwater. The surface and subsurface water are highly variable between periods of rainfall.

23.18 WATER TABLE

The water table is an underground boundary between the soil surface and the area where groundwater saturates spaces between sediments and cracks in the rock. The soil surface above the water table is called the unsaturated zone, where both oxygen and water fill the spaces between sediments.

23.19 WATER VAPOUR

Water vapour, or aqueous vapour, is the gaseous phase of water. It is one state of water within the hydrosphere. Water vapour can be produced from the evaporation or boiling of liquid water or from the sublimation of ice. Water vapour is transparent, like most constituents of the atmosphere.

23.20 WATER VEIL

A very thin water veil like a thin transparent sheet of vinyl has been produced by shaping a double-disc nozzle in such a way to accelerate the water flow inside the nozzle.

23.21 WATER WHEEL

A water wheel, shown in Figure 23.3, is a machine for converting the energy of flowing or falling water into useful forms of power, often in a watermill. Some water wheels are fed by water from a millpond, which is formed when a flowing stream is dammed. A channel for the water flowing to or from a water wheel is called a millrace.

23.22 WATERFALL

A waterfall is an area where water flows over a vertical drop or a series of steep drops in the course of a stream or river.

23.23 WATERLOGGING

Waterlogging occurs when there is too much water in a plant's root zone, which decreases the oxygen available to roots. Waterlogging can be a major

Figure 23.3 Water wheel.

constraint to plant growth and production and, under certain conditions, will cause plant death.

23.24 WAVE

Wave is the propagation of disturbances from place to place in a regular and organised way. The most familiar waves are surface waves that travel on water, but sound, light, and the motion of subatomic particles all exhibit wavelike properties.

23.25 WAVE DRAG

For supersonic flow, drag exists even in the idealised, nonviscous fluid. This new component of drag encountered when the flow is supersonic is called wave drag, and is fundamentally different from the skin-friction drag and separation drag which are associated with the boundary layer in a viscous fluid. The wave drag is related to the loss of total pressure and increase of entropy across the oblique shock waves generated by the aerofoil.

Wave drag is a sudden rise in drag on the aircraft caused by air building up in front of it. At lower speeds, this air has time to 'get out of the way', guided by the air in front of it that is in contact with the aircraft. However, at the speed of sound, this can no longer happen, and the air that was previously following the streamline around the aircraft now hits it directly. The amount of power needed to overcome this effect is considerable. The critical Mach number is the speed at which some of the air passing over the aircraft becomes supersonic.

In aeronautics, wave drag is a component of the aerodynamic drag on aircraft wings and fuselage, propeller blade tips and projectiles moving at transonic and supersonic speeds, due to the presence of shock waves. Wave drag is independent of viscous effects and tends to present itself as a sudden and dramatic increase in drag as the vehicle increases speed to the critical Mach number. It is the sudden and dramatic rise of wave drag that leads to the concept of a sound barrier.

23.26 WAVE ENERGY

Wave energy is the sum of the contribution from gravitational energy, kinetic energy and free-surface energy.

23.27 WAVE FUNCTION

A wave function in quantum physics is a mathematical description of the quantum state of an isolated quantum system. The wave function is complex-valued probability amplitude, and the probabilities for the possible results of measurements made on the system can be derived from it.

23.28 WAVELENGTH

The wavelength, defined as the distance from one wave top or pressure peak to the next, as shown in Figure 23.4, can be calculated if the speed and the frequency of sound are known.

$$\text{Wavelength}(\lambda) = \frac{\text{Speed of sound}}{\text{Frequency}}$$

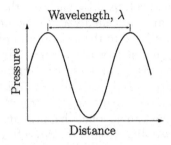

Figure 23.4 Sound pressure propagation.

In physics, the wavelength is the spatial period of a periodic wave and the distance over which the wave's shape repeats. It is the distance between consecutive corresponding points of the same phase on the wave, such as two adjacent crests, troughs, or zero crossings, and is a characteristic of both travelling waves and standing waves as well as other spatial wave patterns. The inverse of the wavelength is called the spatial frequency. Wavelength is commonly designated by the Greek letter lambda (λ). The term wavelength is also sometimes applied to modulated waves, and the sinusoidal envelopes of modulated waves or waves formed by interference of several sinusoids.

23.29 WAVENUMBER

In physics, the wavenumber is also known as the propagation number or angular wavenumber. It is defined as the number of wavelengths per unit distance of the spatial wave frequency and is known as spatial frequency.

23.30 WAVES PREVAILING IN AN OVEREXPANDED JET

A jet is said to be overexpanded when the nozzle exit pressure p_e is lower than the ambient pressure p_a to which it is discharging. To increase p_e to the level of p_a, oblique shock waves are formed at the edge of the nozzle exit. These shocks of the opposite families cross each other at the jet axis and travel to the jet boundary and are reflected as expansion waves. Figure 23.5 schematically shows the shock and expansion waves prevailing in an overexpanded jet. Owing to these waves, a periodic shock-cell structure is generated in the jet flow field and the cell length of these periodic structures is found to increase with an increase in the jet Mach number. It is essential to note that overexpansion is possible only for supersonic jets; subsonic jets are always correctly expanded and a sonic jet can be either correctly expanded or underexpanded but can never be overexpanded.

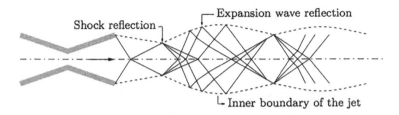

Figure 23.5 Schematic of the waves prevailing in an overexpanded jet.

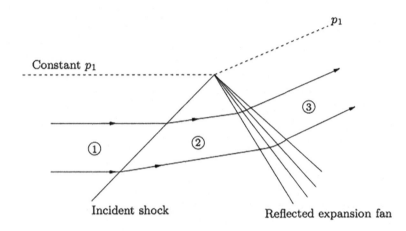

Figure 23.6 Shock wave reflection from a free boundary.

23.31 WAVE REFLECTION FROM A FREE BOUNDARY

When the boundary is a free boundary (such as a fluid boundary), the reflection will not be a like reflection.

Examine the reflection of an oblique shock from a free boundary, as shown in Figure 23.6. The pressure p_1 in region 1 is equal to the surrounding atmosphere. The pressure in the region downstream of the incident shock is p_2, which is higher than p_1 ($p_2 > p_1$). At the edge of the jet boundary, shown by the dashed line in Figure 23.6, the pressure must always be p_1. Therefore, when the incident shock impinges on the boundary, it must be reflected in such a manner as to result in p_1, in region 3 behind the reflected wave. Hence, we have $p_3 = p_1$ and $p_1 < p_2$.

This situation demands that the reflected wave must be an expansion wave, as shown in the figure. The flow in turn is deflected upwards by both the incident shock and the reflected expansion fan, resulting in the upwards deflection of the free boundary, as shown in the figure.

23.32 WEAK MIXTURE

Weak mixture is approximately the correct mixture to burn the fuel efficiently.

23.33 WEAK OBLIQUE SHOCK

A weak oblique shock is a compression wave which causes a small flow defection and the Mach number downstream of the shock $M_2 > 1$. It compresses the flow with an entropy increase almost close to zero.

It is important to note that an oblique shock is termed as weak when the downstream Mach number M_2 is supersonic (even though less than the upstream Mach number M_1). When the flow traversed by an oblique shock becomes subsonic (that is, $M_2 < 1$), the shock is termed strong. But when the flow turning θ caused by a weak oblique shock is very small, then the weak shock assumes a special significance. This kind of weak shock with both decrease of flow Mach number $(M_1 - M_2)$ and flow turning angle θ, which is small, can be regarded as isentropic compression waves.

23.34 WEAK SHOCK

A weak shock is that for which the normalised pressure jump is very small, i.e.,

$$\frac{\Delta p}{p_1} = \frac{p_2 - p_1}{p_1} \leq 1$$

where p_1 and p_2 are the static pressure ahead of and behind the shock.

23.35 WEATHER

Weather is the state of the atmosphere, describing the degree to which it is hot or cold, wet or dry, calm or stormy, clear or cloudy.

23.36 WEATHER RADAR

Weather radar, also called weather surveillance radar (WSR) and Doppler weather radar, is a type of radar used to locate precipitation, calculate its motion, and estimate its type (rain, snow, hail, etc.).

Figure 23.7 Weathercock.

23.37 WEATHERCOCK

Weathercock, shown in Figure 23.7, is an instrument used for showing the direction of the wind. Weathercock is also known as a weather vane or wind vane.

23.38 WEBER NUMBER

Weber number is the ratio of inertia force to the surface tension. The Weber number (We) is a dimensionless number in fluid mechanics that is often useful in analysing fluid flows where there is an interface between two different fluids, especially for multiphase flows with strongly curved surfaces. It is named after Moritz Weber (1871–1951). It can be thought of as a measure of the relative importance of the fluid's inertia compared to its surface tension. The quantity is useful in analysing thin film flows and the formation of droplets and bubbles.

23.39 WEIR

Weir is an obstruction, which extends across the full width of the stream.

23.40 WET

Wet is the condition of containing liquid or being covered or saturated in liquid.

23.41 WET BULB TEMPERATURE

A simple and practical approach for measuring the saturation temperature is to use a thermometer whose bulb is covered with a cotton wick saturated with water and to blow air over the wick. The temperature measured in this way is called the wet bulb temperature.

23.42 WET COOLING TOWERS

These are devices that recirculate cooling water to serve as a transport medium for heat between the source and the sink. It is essentially a semi-enclosed evaporative cooler.

23.43 WETNESS

Wetness is the ability of a liquid to adhere to the surface of a solid, so when we say that something is wet, we mean that the liquid is sticking to the surface of a material. Cohesive forces are attractive forces within the liquid that cause the molecules in the liquid to prefer to stick together.

23.44 WETTED AREA

This is appropriate for surface ships and barges.

23.45 WHEY

Whey is the thin liquid that is left from sour milk after the solid parts (curds) have been removed.

23.46 WHISKY

Whisky is a type of distilled alcoholic beverage made from fermented grain mash. Various grains are used in different varieties, including barley, corn, rye, and wheat. Whisky is typically aged in wooden casks, which are often old sherry casks or may also be made of charred white oak.

23.47 WHISPER

Whisper is to speak very quietly into somebody's ear so that other people cannot hear what you are saying.

23.48 WHISPERING DECIBELS

Sound is measured in decibels (dB). A whisper is about 30 dB, a normal conversation is about 60 dB, and a motorcycle engine running is about 95 dB.

23.49 WHITE FUMES

Dense white fumes are formed when a glass rod dipped in ammonium hydroxide is brought near the mouth of a bottle full of HCL gas. This happens because of the presence of the hydrochloric gas in ammonium hydroxide. It comes out as white fumes which are vapours and gases containing hydroxide.

23.50 WHITE HOLE

In general relativity, a white hole is a hypothetical region of space-time and singularity that cannot be entered from the outside, although energy-matter, light and information can escape from it.

23.51 WHITE NOISE

White noise is a noise having frequencies evenly distributed throughout the audible range. White noise will sound like rushing water.

23.52 WIND

Wind is the natural movement of air or other gases relative to a planet's surface.

23.53 WIND ENERGY

Wind energy is a form of solar energy. Wind energy (or wind power) describes the process by which wind is used to generate electricity. Wind turbines convert the kinetic energy in the wind into mechanical power. A generator can convert mechanical power into electricity.

23.54 WIND TUNNELS

Wind tunnels are devices which provide air streams flowing under controlled conditions so that models of interest can be tested using them. From an operational point of view, wind tunnels are generally classified as low-speed, high-speed, and special purpose tunnels.

23.55 WIND TUNNEL BALANCE

Wind tunnel balance is a device to measure the actual forces and moments acting on a model placed in the test-section stream.

23.56 WIND TURBINE

Wind turbines convert the kinetic energy in the wind into mechanical power.

23.57 WIND IN A VERTICAL TUNNEL

A recreational wind tunnel enables human beings to experience the sensation of flight without planes or parachutes, through the force of wind being generated vertically. Air moves upwards at ~195 km/h (120 mph or 55 m/s).

23.58 WINDMILL

A windmill, shown in Figure 23.8, is a type of working engine. It converts the wind's energy into rotational energy.

23.59 WINDMILLING

Windmilling is the aerodynamic effect that causes an idle fan or a pinwheel to rotate when air flows across its blades.

23.60 WINE

Wine is an alcoholic drink that is made from grapes or sometimes, other fruits.

23.61 WING

A wing is a type of fin that produces lift while moving through air or some other fluid. Accordingly, wings have streamlined cross-sections that are

Figure 23.8 Windmill.

subject to aerodynamic forces and act as aerofoils. A wing's aerodynamic efficiency is expressed as its lift-to-drag ratio. The lift a wing generates at a given speed and angle of attack can be one to two orders of magnitude greater than the total drag on the wing. A high lift-to-drag ratio requires a significantly smaller thrust to propel the wings through the air at a sufficient lift.

23.62 WING AREA

This is the planform area of the wing, viewed in the direction normal to the wing-span, from the top of the wing. Thus, the representative area of the wing may be regarded as the product of the span and the average chord. Although a portion of the area may be covered by fuselage, the pressure distribution over the fuselage surface is accounted for in the representative wing area.

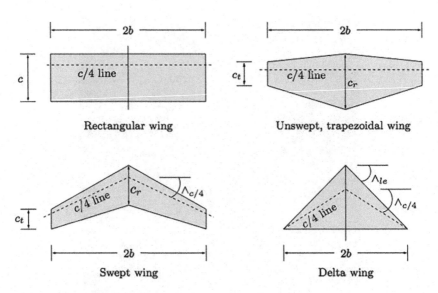

Figure 23.9 Geometric parameters of some wing planforms: (a) rectangular wing, (b) unswept, trapezoidal wing, (c) swept wing, and (d) delta wing.

23.63 WING GEOMETRICAL PARAMETERS

Aircraft wings are made up of aerofoil sections, placed along the span. In an aircraft, the geometry of the horizontal and vertical tails, high-lifting devices such as flaps on the wings and tails, and control surfaces such as ailerons are also made by placing the aerofoil sections in span-wise combinations.

The relevant parameters used to define the aerodynamic characteristics of a wing of rectangular, unswept trapezoidal, swept, and delta configurations are illustrated in Figure 23.9.

23.64 WING LOADING

Wing loading, w, is the average load per unit area of the wing plan, $w = W/S$, where W is the total weight of the aircraft and S is the wing planform area.

In aerodynamics, wing loading is the total mass of an aircraft or flying animal divided by the area of its wing. The stalling speed of an aircraft in a straight, level flight is partly determined by its wing loading. An aircraft or animal with a low wing loading has a larger wing area relative to its mass, as compared to the one with a high wing loading.

23.65 WING TWIST

Wing twist is an aerodynamic feature added to aircraft wings to adjust lift distribution along the wing. Often, the purpose of lift redistribution is to ensure that the wing tip is the last part of the wing surface to stall, for example, when executing a roll or steep climb; it involves twisting the wing-tip slightly downwards in relation to the rest of the wing. This ensures that the effective angle of attack is always lower at the wingtip than at the root, meaning the root will stall before the tip. This is desirable because the air-craft's flight control surfaces are often located at the wingtip, and the vari-able stall characteristics of a twisted wing alert the pilot to the advancing stall while still allowing the control surfaces to remain effective, meaning the pilot can usually prevent the aircraft from stalling fully before control is completely lost.

23.66 WINGLET

Winglet is a vertical projection on the tip of an aircraft wing, as in Figure 23.10, for reducing drag. Winglets prevent the formation of vortices wing tips due to the higher pressure at the bottom surface and the lower pressure at the top surface of the aeroplane's wing.

23.67 WINGSPAN

This is the distance between the tips of the port and starboard wings.

Figure 23.10 Winglet.

The wingspan (or just span) of a bird or an aeroplane is the distance from one wingtip to the other wingtip. For example, the Boeing 777–200 has a wingspan of 60.93 m (199 ft 11 in) while a wandering albatross (*Diomedea exulans*) caught in 1965 had a wingspan of 3.63 m (11 ft 11 in) – the official record for a living bird.

The term wingspan, more technically extent, is also used for other winged animals such as the pterosaurs, bats, insects, etc. and other aircraft such as ornithopters. In humans, the term wingspan also refers to the arm span, which is the distance between the length from one end of an individual's arms (measured at the fingertips) to the other when raised parallel to the ground at shoulder height at a 90° angle. Former professional basketball player Manute Bol stands at 2.31 m and owns one of the largest wingspans at 2.59 m.

23.68 WINGTIP VORTICES

Wingtip vortices, shown in Figure 23.11, are circular patterns of rotating air left behind a wing as it generates lift.

One wingtip vortex trails from the tip of each wing. Wingtip vortices are sometimes named trailing or lift-induced vortices because they also occur at points other than at the wing tips. Indeed, vorticity is trailed at any point on the wing where the lift varies span-wise (a fact described and quantified by the lifting-line theory); it eventually rolls up into large vortices near the wingtip, at the edge of flap devices, or other abrupt changes in wing planform.

Wingtip vortices are circular patterns of rotating air left behind a wing as it generates lift.

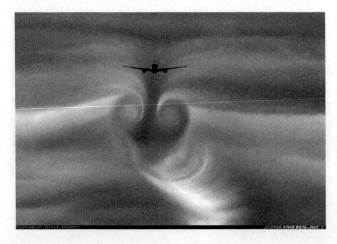

Figure 23.11 Wingtip vortices.

23.69 WOOD SMOKE

Smoke can be produced by partial combustion of damp wood shavings, as shown in Figure 23.12. However, this smoke is rich in tar and water vapour. It has to be initially filtered thoroughly by passing through a filter and a water condenser. White and dense smoke can be produced by this method. But handling the apparatus is somewhat cumbersome since the burnt wood shavings have to be removed often. By choosing the proper quality of wood, smoke having an agreeable aroma can be obtained. If one can afford it, sandalwood shavings are recommended for a good aroma. Addition of a small quantity of liquid paraffin will appreciably increase the optical density of the smoke.

Figure 23.12 Wood smoke.

Figure 23.13 Woolaston wire.

23.70 WOOLASTON WIRE

Woolaston wire is a fine platinum wire (or platinum-rhodium alloy wire) coated with a thick film of silver, as shown in Figure 23.13.

23.71 WORK

Work is an energy interaction between a system and its surroundings. Energy can cross the boundary of a closed system in the form of heat or work. In fact, heat and work are the only two mechanisms by which the energy of a closed system can be changed. Therefore, if the energy crossing the boundary is not heat, it must be work.

23.72 WORKING FLUID

All cyclic devices including heat engines usually involve a fluid to and from which heat is transferred while undergoing a cycle. This fluid is called the working fluid.

Chapter 24

Xenon

24.1 XENON

Xenon is a chemical element with the symbol Xe and atomic number 54. It is a colourless, dense, and odourless noble gas found in the Earth's atmosphere in trace amounts. Xenon is used in flash lamps and arc lamps and as a general anaesthetic.

DOI: 10.1201/9781003348405-24

Chapter 25

Yacht to Yoghurt

25.1 YACHT

A yacht, shown in Figure 25.1, is a large boat with sails or a motor, used for racing or pleasure trips.

25.2 YAW PROBES

The instruments used for determining the flow direction are termed yaw probes. These yaw probes are generally used for determining the flow direction as well as its magnitude.

Figure 25.1 Yacht.

DOI: 10.1201/9781003348405-25

25.3 YAW RATE

The yaw rate or yaw velocity of a car, aircraft, projectile, or other rigid body is the angular velocity of this rotation or the rate of change of the heading angle when the aircraft is horizontal. It is commonly measured in degrees per second or radians per second.

25.4 YAW SENSITIVITY OF PITOT PRESSURE PROBE

A pitot tube is insensitive to yaw up to about 20°. For instance, at 20° yaw, the pressure measured is only about 1% less than that at zero yaw.

25.5 YAW SENSITIVITY OF STATIC PRESSURE PROBE

The yaw sensitivity of this probe with string support is about a 1% reduction in the measured pressure for 3°–5° yaw.

25.6 YAW SPHERE

Yaw sphere is a probe used to determine the flow direction. It is like the one shown in Figure 25.2 and can be used to determine the flow direction.

For a two-dimensional flow, just two holes are sufficient to measure the flow angularity. But, for a three-dimensional flow, four holes are necessary since both the yaw and pitch of the flow have to be measured simultaneously to determine the flow direction.

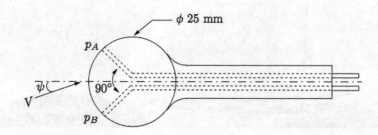

Figure 25.2 Yaw sphere.

25.7 YAWING

Rotary motion of the aircraft about the normal axis is called yawing.

25.8 YAWING MOMENT

Yawing moment is the moment acting about the vertical, z-axis, of the aircraft. Yawing moment tends to rotate the aircraft's nose to the right and is regarded as positive.

25.9 YEAST

Yeast is a single-cell organism, called *Saccharomyces cerevisiae*, which needs food, warmth, and moisture to thrive. It converts its food – sugar and starch – through fermentation into carbon dioxide and alcohol. It's the carbon dioxide that makes baked goods rise.

25.10 YOGHURT

Yoghurt, shown in Figure 25.3, is a food produced by the bacterial fermentation of milk. The bacteria used to make yoghurt are known as yoghurt cultures.

Figure 25.3 Yoghurt.

Chapter 26

Zap Flap to Zone of Silence

26.1 ZAP FLAP

This is a combination of the split-flap concept and the Fowler flap concept. A typical zap flap is illustrated in Figure 26.1.

26.2 ZERO GRAVITY

Zero gravity or zero-G can simply be defined as the state or condition of weightlessness.

26.3 ZEROTH LAW OF THERMODYNAMICS

The Zeroth law of thermodynamics states that if two bodies are in thermal equilibrium with a third body, they are also in thermal equilibrium with each other. It serves as a basis for the validity of temperature measurement. By replacing the third body with a thermometer, the Zeroth law can be restated as 'two bodies are in thermal equilibrium if both have the same temperature reading even if they are not in contact'.

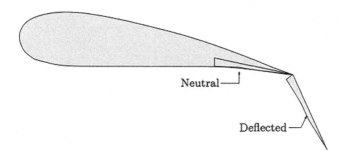

Figure 26.1 A zap flap in neutral and deflected positions.

DOI: 10.1201/9781003348405-26

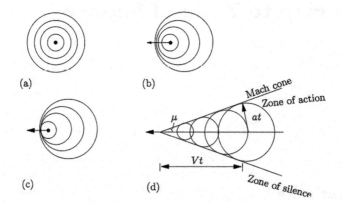

Figure 26.2 Zone of action. (a) $V=0$, (b) $V=a/2$, (c) $V=a$, and (d) $V>a$.

26.4 ZONE OF ACTION

For supersonic motion of an object, there is a well-defined conical zone in the flow field with the object located at the nose of the cone, and the disturbance created by the moving object is confined only to the field included inside the cone, as illustrated in Figure 26.2. The flow field zone outside the cone does not even feel the disturbance. For this reason, von Karman termed the region inside the cone as the zone of action.

26.5 ZONE OF SILENCE

The region outside the cone in a supersonic motion of an object is called the zone of silence.

Printed in the United States
by Baker & Taylor Publisher Services